# Montego Owners Workshop Manual

## John S Mead

**Models covered**
Austin/MG (Rover) Montego 2.0 litre models
(inc. Vanden Plas) with 1994 cc petrol engines; Saloon
and Estate, including Turbo and special/limited
editions

*Does not cover Diesel or 1.3/1.6 litre petrol engine models*

(1067-9U7)

ABCDE
FGHIJ
K

2

**Haynes Publishing Group**
Sparkford Nr Yeovil
Somerset BA22 7JJ England

**Haynes Publications, Inc**
861 Lawrence Drive
Newbury Park
California 91320 USA

## Acknowledgements

Thanks are due to Champion Spark Plug who supplied the illustrations showing spark plug conditions, to Holt Lloyd Limited who supplied the illustrations showing bodywork repair, and to Duckhams Oils, who provided lubrication data. Thanks are also due to BL Cars Limited for the supply of technical information and to Unipart for their assistance. Sykes-Pickavant Limited provided some of the workshop tools. Special thanks are due to all those people at Sparkford who helped in the production of this manual.

© **Haynes Publishing Group 1992**

A book in the **Haynes Owners Workshop Manual Series**

Printed by J. H. Haynes & Co. Ltd., Sparkford, Nr Yeovil, Somerset BA22 7JJ, England

ISBN 1 85010 872 2

**British Library Cataloguing in Publication Data**
A catalogue record for this book is available from the British Library

We take great pride in the accuracy of information given in this manual, but vehicle manufacturers make alterations and design changes during the production run of a particular vehicle of which they do not inform us. No liability can be accepted by the authors or publishers for loss, damage or injury caused by any errors in, or omissions from, the information given.

# Restoring and Preserving our Motoring Heritage

Few people can have had the luck to realise their dreams to quite the same extent and in such a remarkable fashion as John Haynes, Founder and Chairman of the Haynes Publishing Group.

Since 1965 his unique approach to workshop manual publishing has proved so successful that millions of Haynes Manuals are now sold every year throughout the world, covering literally thousands of different makes and models of cars, vans and motorcycles.

A continuing passion for cars and motoring led to the founding in 1985 of a Charitable Trust dedicated to the restoration and preservation of our motoring heritage. To inaugurate the new Museum, John Haynes donated virtually his entire private collection of 52 cars.

Now with an unrivalled international collection of over 210 veteran, vintage and classic cars and motorcycles, the Haynes Motor Museum in Somerset is well on the way to becoming one of the most interesting Motor Museums in the world.

A 70 seat video cinema, a cafe and an extensive motoring bookshop, together with a specially constructed one kilometre motor circuit, make a visit to the Haynes Motor Museum a truly unforgettable experience.

Every vehicle in the museum is preserved in as near as possible mint condition and each car is run every six months on the motor circuit.

Enjoy the picnic area set amongst the rolling Somerset hills. Peer through the William Morris workshop windows at cars being restored, and browse through the extensive displays of fascinating motoring memorabilia.

From the 1903 Oldsmobile through such classics as an MG Midget to the mighty 'E' Type Jaguar, Lamborghini, Ferrari Berlinetta Boxer, and Graham Hill's Lola Cosworth, there is something for everyone, young and old alike, at this Somerset Museum.

## Haynes Motor Museum

*Situated mid-way between London and Penzance, the Haynes Motor Museum is located just off the A303 at Sparkford, Somerset (home of the Haynes Manual) and is open to the public 7 days a week all year round, except Christmas Day and Boxing Day.*

# Contents

*Spark plug condition and bodywork repair colour pages between pages 32 and 33*

Austin Montego 2.0 HLS

# About this manual

## Its aim

The aim of this manual is to help you get the best value from your vehicle. It can do so in several ways. It can help you decide what work must be done (even should you choose to get it done by a garage), provide information on routine maintenance and servicing, and give a logical course of action and diagnosis when random faults occur. However, it is hoped that you will use the manual by tackling the work yourself. On simpler jobs it may even be quicker than booking the car into a garage and going there twice, to leave and collect it. Perhaps most important, a lot of money can be saved by avoiding the costs a garage must charge to cover its labour and overheads.

The manual has drawings and descriptions to show the function of the various components so that their layout can be understood. Then the tasks are described and photographed in a step-by-step sequence so that even a novice can do the work.

## Its arrangement

The manual is divided into twelve Chapters, each covering a logical sub-division of the vehicle. The Chapters are each divided into Sections, numbered with single figures, eg 5; and the Sections into paragraphs (or sub-sections), with decimal numbers following on from the Section they are in, eg 5.1, 5.2, 5.3 etc.

It is freely illustrated, especially in those parts where there is a detailed sequence of operations to be carried out. There are two forms of illustration: figures and photographs. The figures are numbered in sequence with decimal numbers, according to their position in the Chapter – eg Fig. 6.4 is the fourth drawing/illustration in Chapter 6. Photographs carry the same number (either individually or in related groups) as the Section or sub-section to which they relate.

There is an alphabetical index at the back of the manual as well as a contents list at the front. Each Chapter is also preceded by its own individual contents list.

References to the 'left' or 'right' of the vehicle are in the sense of a person in the driver's seat facing forwards.

Unless otherwise stated, nuts and bolts are removed by turning anti-clockwise, and tightened by turning clockwise.

Vehicle manufacturers continually make changes to specifications and recommendations, and these, when notified, are incorporated into our manuals at the earliest opportunity.

**We take great pride in the accuracy of information given in this manual, but vehicle manufacturers make alterations and design changes during the production run of a particular vehicle of which they do not inform us. No liability can be accepted by the authors or publishers for loss, damage or injury caused by any errors in, or omissions from, the information given.**

# Introduction to the Austin Montego

The top of the range Montego models covered by this Manual are available in four-door notchback Saloon or five-door Estate versions, and are powered by a 2.0 litre overhead camshaft engine. This power unit is a refined version of the Austin Rover O-series engine and is available with carburettor induction and electronic engine management, or with electronic fuel injection or turbocharger according to model.

The cars feature independent front suspension by MacPherson struts with front wheel drive through a Honda five-speed manual gearbox. Rear suspension is semi-independent by a trailing twist axle. Conventional front disc/rear drum brakes are used with the front discs being of ventilated pattern.

A high standard of interior trim and comfort, particularly on the Vanden Plas version, is offered with an ultra high technology driver information system available on MG derivatives.

All models in the range have been designed with the emphasis on economical motoring with a high standard of handling, performance and comfort.

BL and BL Cars Limited are now known as the Rover Group (previously Austin Rover Group). All instances in which we recommend owners to seek the advice of a BL dealer should now, of course, be taken to mean seek the advice of a Rover Group dealer.

# General dimensions, weights and capacities

*For modifications, and information applicable to later models, see Supplement at end of manual*

## Dimensions

| | |
|---|---|
| Turning circle (between kerbs) | 34.5 ft (10.5 m) |
| Wheelbase | 101.1 in (2565 mm) |
| Overall length | 176.1 in (4468 mm) |
| Overall width (excluding mirrors) | 67.4 in (1710 mm) |
| Overall height | 55.9 in (1418 mm) |
| Ground clearance | 6.3 in (160 mm) |
| Track: | |
|     Front | 58.4 in (1481 mm) |
|     Rear | 57.3 in (1455 mm) |

## Weights

| | |
|---|---|
| Kerb weight: | |
|     2.0 HL | 2244 lb (1020 kg) |
|     2.0 HLS and MG EFi | 2266 lb (1030 kg) |
|     Vanden Plas | 2310 lb (1050 kg) |
| Maximum roof rack weight (distributed) | 154 lb (70 kg) |
| Towing hitch downward load | 75 to 100 lb (35 to 45 kg) |

## Capacities

| | |
|---|---|
| Engine oil (refill with filter change) | 7.0 pt (4.0 litre) |
| Gearbox | 3.5 pt (2.0 litre) |
| Cooling system | 15 pt (8.5 litre) |
| Power-assisted steering pump reservoir | 0.9 pt (0.5 litre) |
| Fuel tank | 11.25 gal (50 litre) |
| Fuel octane rating (all models) | 97 RON (4 star) |

# Buying spare parts and vehicle identification numbers

*Buying spare parts*

Spare parts are available from many sources, for example: BL garages, other garages and accessory shops, and motor factors. Our advice regarding spare parts sources is as follows:

*Officially appointed BL garages* – This is the best source for parts which are peculiar to your car and are not generally available (eg complete cylinder heads, internal gearbox components, badges, interior trim etc). It is also the only place at which you should buy parts if your vehicle is still under warranty – non-BL components may invalidate the warranty. To be sure of obtaining the correct parts it will always be necessary to give the storeman your car's vehicle identification number, and if possible, to take the 'old' part along for positive identification. Many parts are available under a factory exchange scheme – any parts returned should always be clean. It obviously makes good sense to go straight to the specialists on your car for this type of part as they are best equipped to supply you.

*Other garages and accessory shops* – These are often very good places to buy materials and components needed for the maintenance of your car (eg oil filters, spark plugs, bulbs, drivebelts, oils and greases, touch-up paint, filler paste, etc). They also sell general accessories, usually have convenient opening hours, charge lower prices and can often be found not far from home.

*Motor factors* – Good factors will stock all of the more important components which wear out relatively quickly (eg clutch components, pistons, valves, exhaust systems, brake pipes/seals and pads etc).

Motor factors will often provide new or reconditioned components on a part exchange basis – this can save a considerable amount of money.

*Vehicle identification numbers*

Modifications are a continuing and unpublicised process in vehicle manufacture quite apart from major model changes. Spare parts manuals and lists are compiled upon a numerical basis, the individual vehicle numbers being essential to correct identification of the component required.

When ordering spare parts, always give as much information as possible. Quote the car model, year of manufacture, body and engine numbers as appropriate.

The vehicle identification number is stamped on a plate attached to the front body panel, or to the left-hand front door pillar. The number is repeated on the right-hand front strut turret.

The engine number is stamped on a plate attached to the cylinder block below the spark plugs.

The gearbox number is stamped on the top left-hand side of the bellhousing.

The automatic transmission number is stamped on a plate attached to its bottom face.

The body number is located on the right-hand side of the spare wheel well.

# Tools and working facilities

## Introduction

A selection of good tools is a fundamental requirement for anyone contemplating the maintenance and repair of a motor vehicle. For the owner who does not possess any, their purchase will prove a considerable expense, offsetting some of the savings made by doing-it-yourself. However, provided that the tools purchased meet the relevant national safety standards and are of good quality, they will last for many years and prove an extremely worthwhile investment.

To help the average owner to decide which tools are needed to carry out the various tasks detailed in this manual, we have compiled three lists of tools under the following headings: *Maintenance and minor repair*, *Repair and overhaul*, and *Special*. The newcomer to practical mechanics should start off with the *Maintenance and minor repair* tool kit and confine himself to the simpler jobs around the vehicle. Then, as his confidence and experience grow, he can undertake more difficult tasks, buying extra tools as, and when, they are needed. In this way, a *Maintenance and minor repair* tool kit can be built-up into a *Repair and overhaul* tool kit over a considerable period of time without any major cash outlays. The experienced do-it-yourselfer will have a tool kit good enough for most repair and overhaul procedures and will add tools from the *Special* category when he feels the expense is justified by the amount of use to which these tools will be put.

It is obviously not possible to cover the subject of tools fully here. For those who wish to learn more about tools and their use there is a book entitled *How to Choose and Use Car Tools* available from the publishers of this manual.

Both UNF and metric threads to ISO standards are used on the Montego range.

## Maintenance and minor repair tool kit

The tools given in this list should be considered as a minimum requirement if routine maintenance, servicing and minor repair operations are to be undertaken. We recommend the purchase of combination spanners (ring one end, open-ended the other); although more expensive than open-ended ones, they do give the advantages of both types of spanner.

*Combination spanners - $\frac{7}{16}$, $\frac{1}{2}$, $\frac{9}{16}$, $\frac{5}{8}$, $\frac{3}{4}$, $\frac{3}{16}$, $\frac{7}{8}$ and $\frac{15}{16}$ in AF*
*Combination spanners - 10, 11, 12, 13, 14 and 17 mm*
*Adjustable spanner - 9 inch*
*Gearbox drain plug key*
*Spark plug spanner (with rubber insert)*
*Spark plug gap adjustment tool*
*Set of feeler gauges*
*Brake bleed nipple spanner*
*Screwdriver - 4 in long x $\frac{1}{4}$ in dia (flat blade)*
*Screwdriver - 4 in long x $\frac{1}{4}$ in dia (cross blade)*
*Combination pliers - 6 inch*
*Hacksaw (junior)*
*Tyre pump*
*Tyre pressure gauge*
*Oil can*
*Fine emery cloth (1 sheet)*
*Wire brush (small)*
*Funnel (medium size)*

## Repair and overhaul tool kit

These tools are virtually essential for anyone undertaking any major repairs to a motor vehicle, and are additional to those given in the *Maintenance and minor repair* list. Included in this list is a comprehensive set of sockets. Although these are expensive they will be found invaluable as they are so versatile - particularly if various drives are included in the set. We recommend the $\frac{1}{2}$ in square-drive type, as this can be used with most proprietary torque wrenches. If you cannot afford a socket set, even bought piecemeal, then inexpensive tubular box spanners are a useful alternative.

The tools in this list will occasionally need to be supplemented by tools from the *Special* list.

*Sockets (or box spanners) to cover range in previous list*
*Reversible ratchet drive (for use with sockets)*
*Extension piece, 10 inch (for use with sockets)*
*Universal joint (for use with sockets)*
*Torque wrench (for use with sockets)*
*'Mole' wrench - 8 inch*
*Ball pein hammer*
*Soft-faced hammer, plastic or rubber*
*Screwdriver - 6 in long x $\frac{5}{16}$ in dia (flat blade)*
*Screwdriver - 2 in long x $\frac{5}{16}$ in square (flat blade)*
*Screwdriver - 1$\frac{1}{2}$ in long x $\frac{1}{4}$ in dia (cross blade)*
*Screwdriver - 3 in long x $\frac{1}{8}$ in dia (electricians)*
*Pliers - electricians side cutters*
*Pliers - needle nosed*
*Pliers - circlip (internal and external)*
*Cold chisel - $\frac{1}{2}$ inch*
*Scriber*
*Scraper*
*Centre punch*
*Pin punch*
*Hacksaw*
*Valve grinding tool*
*Steel rule/straight-edge*
*Allen keys*
*Selection of files*
*Wire brush (large)*
*Axle-stands*
*Jack (strong scissor or hydraulic type)*

## Special tools

The tools in this list are those which are not used regularly, are expensive to buy, or which need to be used in accordance with their manufacturers' instructions. Unless relatively difficult mechanical jobs are undertaken frequently, it will not be economic to buy many of these tools. Where this is the case, you could consider clubbing together with friends (or joining a motorists' club) to make a joint purchase, or borrowing the tools against a deposit from a local garage or tool hire specialist.

The following list contains only those tools and instruments freely available to the public, and not those special tools produced by the vehicle manufacturer specifically for its dealer network. You will find occasional references to these manufacturers' special tools in the text of this manual. Generally, an alternative method of doing the job without the vehicle manufacturers' special tool is given. However, sometimes, there is no alternative to using them. Where this is the case and the relevant tool cannot be bought or borrowed, you will have to entrust the work to a franchised garage.

> Valve spring compressor
> Piston ring compressor
> Balljoint separator
> Universal hub/bearing puller
> Impact screwdriver
> Micrometer and/or vernier gauge
> Dial gauge
> Stroboscopic timing light
> Dwell angle meter/tachometer
> Universal electrical multi-meter
> Cylinder compression gauge
> Lifting tackle
> Trolley jack
> Light with extension lead

## Buying tools

For practically all tools, a tool factor is the best source since he will have a very comprehensive range compared with the average garage or accessory shop. Having said that, accessory shops often offer excellent quality tools at discount prices, so it pays to shop around.

There are plenty of good tools around at reasonable prices, but always aim to purchase items which meet the relevant national safety standards. If in doubt, ask the proprietor or manager of the shop for advice before making a purchase.

## Care and maintenance of tools

Having purchased a reasonable tool kit, it is necessary to keep the tools in a clean serviceable condition. After use, always wipe off any dirt, grease and metal particles using a clean, dry cloth, before putting the tools away. Never leave them lying around after they have been used. A simple tool rack on the garage or workshop wall, for items such as screwdrivers and pliers is a good idea. Store all normal wrenches and sockets in a metal box. Any measuring instruments, gauges, meters, etc, must be carefully stored where they cannot be damaged or become rusty.

Take a little care when tools are used. Hammer heads inevitably become marked and screwdrivers lose the keen edge on their blades from time to time. A little timely attention with emery cloth or a file will soon restore items like this to a good serviceable finish.

## Working facilities

Not to be forgotten when discussing tools, is the workshop itself. If anything more than routine maintenance is to be carried out, some form of suitable working area becomes essential.

It is appreciated that many an owner mechanic is forced by circumstances to remove an engine or similar item, without the benefit of a garage or workshop. Having done this, any repairs should always be done under the cover of a roof.

Wherever possible, any dismantling should be done on a clean, flat workbench or table at a suitable working height.

Any workbench needs a vice: one with a jaw opening of 4 in (100 mm) is suitable for most jobs. As mentioned previously, some clean dry storage space is also required for tools, as well as for lubricants, cleaning fluids, touch-up paints and so on, which become necessary.

Another item which may be required, and which has a much more general usage, is an electric drill with a chuck capacity of at least $\frac{5}{16}$ in (8 mm). This, together with a good range of twist drills, is virtually essential for fitting accessories such as mirrors and reversing lights.

Last, but not least, always keep a supply of old newspapers and clean, lint-free rags available, and try to keep any working area as clean as possible.

## Spanner jaw gap comparison table

| Jaw gap (in) | Spanner size |
|---|---|
| 0.250 | $\frac{1}{4}$ in AF |
| 0.276 | 7 mm |
| 0.313 | $\frac{5}{16}$ in AF |
| 0.315 | 8 mm |
| 0.344 | $\frac{11}{32}$ in AF; $\frac{1}{8}$ in Whitworth |
| 0.354 | 9 mm |
| 0.375 | $\frac{3}{8}$ in AF |
| 0.394 | 10 mm |
| 0.433 | 11 mm |
| 0.438 | $\frac{7}{16}$ in AF |
| 0.445 | $\frac{3}{16}$ in Whitworth; $\frac{1}{4}$ in BSF |
| 0.472 | 12 mm |
| 0.500 | $\frac{1}{2}$ in AF |
| 0.512 | 13 mm |
| 0.525 | $\frac{1}{4}$ in Whitworth; $\frac{5}{16}$ in BSF |
| 0.551 | 14 mm |
| 0.563 | $\frac{9}{16}$ in AF |
| 0.591 | 15 mm |
| 0.600 | $\frac{5}{16}$ in Whitworth; $\frac{3}{8}$ in BSF |
| 0.625 | $\frac{5}{8}$ in AF |
| 0.630 | 16 mm |
| 0.669 | 17 mm |
| 0.686 | $\frac{11}{16}$ in AF |
| 0.709 | 18 mm |
| 0.710 | $\frac{3}{8}$ in Whitworth; $\frac{7}{16}$ in BSF |
| 0.748 | 19 mm |
| 0.750 | $\frac{3}{4}$ in AF |
| 0.813 | $\frac{13}{16}$ in AF |
| 0.820 | $\frac{7}{16}$ in Whitworth; $\frac{1}{2}$ in BSF |
| 0.866 | 22 mm |
| 0.875 | $\frac{7}{8}$ in AF |
| 0.920 | $\frac{1}{2}$ in Whitworth; $\frac{9}{16}$ in BSF |
| 0.938 | $\frac{15}{16}$ in AF |
| 0.945 | 24 mm |
| 1.000 | 1 in AF |
| 1.010 | $\frac{9}{16}$ in Whitworth; $\frac{5}{8}$ in BSF |
| 1.024 | 26 mm |
| 1.063 | $1\frac{1}{16}$ in AF; 27 mm |
| 1.100 | $\frac{5}{8}$ in Whitworth; $\frac{11}{16}$ in BSF |
| 1.125 | $1\frac{1}{8}$ in AF |
| 1.181 | 30 mm |
| 1.200 | $\frac{11}{16}$ in Whitworth; $\frac{3}{4}$ in BSF |
| 1.250 | $1\frac{1}{4}$ in AF |
| 1.260 | 32 mm |
| 1.300 | $\frac{3}{4}$ in Whitworth; $\frac{7}{8}$ in BSF |
| 1.313 | $1\frac{5}{16}$ in AF |
| 1.390 | $\frac{13}{16}$ in Whitworth; $\frac{15}{16}$ in BSF |
| 1.417 | 36 mm |
| 1.438 | $1\frac{7}{16}$ in AF |
| 1.480 | $\frac{7}{8}$ in Whitworth; 1 in BSF |
| 1.500 | $1\frac{1}{2}$ in AF |
| 1.575 | 40 mm; $\frac{15}{16}$ in Whitworth |
| 1.614 | 41 mm |
| 1.625 | $1\frac{5}{8}$ in AF |
| 1.670 | 1 in Whitworth; $1\frac{1}{8}$ in BSF |
| 1.688 | $1\frac{11}{16}$ in AF |
| 1.811 | 46 mm |
| 1.813 | $1\frac{13}{16}$ in AF |
| 1.860 | $1\frac{1}{8}$ in Whitworth; $1\frac{1}{4}$ in BSF |
| 1.875 | $1\frac{7}{8}$ in AF |
| 1.969 | 50 mm |
| 2.000 | 2 in AF |
| 2.050 | $1\frac{1}{4}$ in Whitworth; $1\frac{3}{8}$ in BSF |
| 2.165 | 55 mm |
| 2.362 | 60 mm |

# General repair procedures

Whenever servicing, repair or overhaul work is carried out on the car or its components, it is necessary to observe the following procedures and instructions. This will assist in carrying out the operation efficiently and to a professional standard of workmanship.

### Joint mating faces and gaskets

Where a gasket is used between the mating faces of two components, ensure that it is renewed on reassembly, and fit it dry unless otherwise stated in the repair procedure. Make sure that the mating faces are clean and dry with all traces of old gasket removed. When cleaning a joint face, use a tool which is not likely to score or damage the face, and remove any burrs or nicks with an oilstone or fine file.

Make sure that tapped holes are cleaned with a pipe cleaner, and keep them free of jointing compound if this is being used unless specifically instructed otherwise.

Ensure that all orifices, channels or pipes are clear and blow through them, preferably using compressed air.

### Oil seals

Whenever an oil seal is removed from its working location, either individually or as part of an assembly, it should be renewed.

The very fine sealing lip of the seal is easily damaged and will not seal if the surface it contacts is not completely clean and free from scratches, nicks or grooves. If the original sealing surface of the component cannot be restored, the component should be renewed.

Protect the lips of the seal from any surface which may damage them in the course of fitting. Use tape or a conical sleeve where possible. Lubricate the seal lips with oil before fitting and, on dual lipped seals, fill the space between the lips with grease.

Unless otherwise stated, oil seals must be fitted with their sealing lips toward the lubricant to be sealed.

Use a tubular drift or block of wood of the appropriate size to install the seal and, if the seal housing is shouldered, drive the seal down to the shoulder. If the seal housing is unshouldered, the seal should be fitted with its face flush with the housing top face.

### Screw threads and fastenings

Always ensure that a blind tapped hole is completely free from oil, grease, water or other fluid before installing the bolt or stud. Failure to do this could cause the housing to crack due to the hydraulic action of the bolt or stud as it is screwed in.

When tightening a castellated nut to accept a split pin, tighten the nut to the specified torque, where applicable, and then tighten further to the next split pin hole. Never slacken the nut to align a split pin hole unless stated in the repair procedure.

When checking or retightening a nut or bolt to a specified torque setting, slacken the nut or bolt by a quarter of a turn, and then retighten to the specified setting.

### Locknuts, locktabs and washers

Any fastening which will rotate against a component or housing in the course of tightening should always have a washer between it and the relevant component or housing.

Spring or split washers should always be renewed when they are used to lock a critical component such as a big-end bearing retaining nut or bolt.

Locktabs which are folded over to retain a nut or bolt should always be renewed.

Self-locking nuts can be reused in non-critical areas, providing resistance can be felt when the locking portion passes over the bolt or stud thread.

Split pins must always be replaced with new ones of the correct size for the hole.

### Special tools

Some repair procedures in this manual entail the use of special tools such as a press, two or three-legged pullers, spring compressors etc. Wherever possible, suitable readily available alternatives to the manufacturer's special tools are described, and are shown in use. In some instances, where no alternative is possible, it has been necessary to resort to the use of a manufacturer's tool and this has been done for reasons of safety as well as the efficient completion of the repair operation. Unless you are highly skilled and have a thorough understanding of the procedure described, never attempt to bypass the use of any special tool when the procedure described specifies its use. Not only is there a very great risk of personal injury, but expensive damage could be caused to the components involved.

# Jacking and towing

To change a roadwheel, remove the spare wheel and tool kit from the well in the boot floor. Apply the handbrake and chock the wheel diagonally opposite the one to be changed. Make sure that the car is located on firm level ground. Lever off the wheel trim (photo) and slacken the wheel nuts slightly with the spanner provided. Engage the jack head in the jacking point nearest to the wheel to be changed and raise the car until the tyre is just clear of the ground. Unscrew the wheel nuts and remove the wheel.

Fit the spare wheel to the studs and screw on the wheel nuts. Tighten the wheel nuts initially then lower the car and tighten them again. Refit the wheel trim, stow the wheel and tool kit in the boot and remove the chock.

When jacking up the car with a trolley jack, position the head of the jack under the jacking bracket/towing hook in the centre of the front crossmember (photo), to raise both front wheels. To raise both rear wheels position the jack head under the rear jacking bracket/towing hook (photo). To raise one side of the car at the front or rear, position the jack head under the front or rear jacking points on the side to be raised. In all cases make sure the handbrake is firmly applied and the wheels chocked before raising the car. Always position axle stands or suitable supports under the side jacking points at the front or rear, or under the rear chassis rails situated either side of the spare wheel well, to support the car when it is raised. Never work on, under, or around a raised car unless it is adequately supported in at least two places with axle stands, car ramps or suitable sturdy blocks.

The vehicle may be towed for recovery purposes using the jacking bracket/towing hook on the front crossmember. The rear jacking bracket/towing hook may only be used for towing a light vehicle.

If automatic transmission is fitted the vehicle should only be towed at slow speed for a short distance. If a greater distance must be covered or if there is the possibility of any fault in the transmission, the car must be towed with the front wheels lifted.

Use the tool provided or a screwdriver to lever off the wheel trim

Front jacking bracket/towing hook

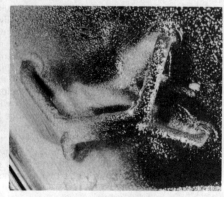

Rear jacking bracket/towing hook

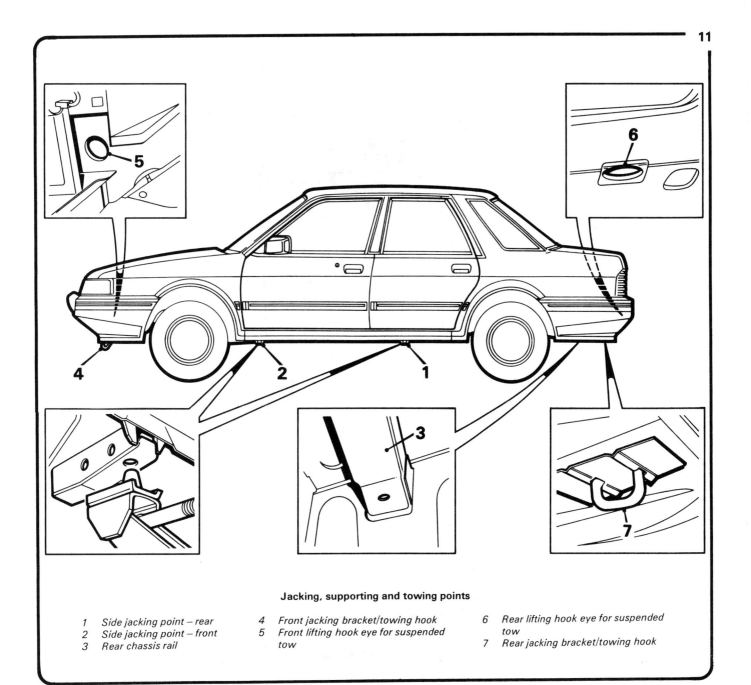

**Jacking, supporting and towing points**

1    Side jacking point – rear
2    Side jacking point – front
3    Rear chassis rail

4    Front jacking bracket/towing hook
5    Front lifting hook eye for suspended
     tow

6    Rear lifting hook eye for suspended
     tow
7    Rear jacking bracket/towing hook

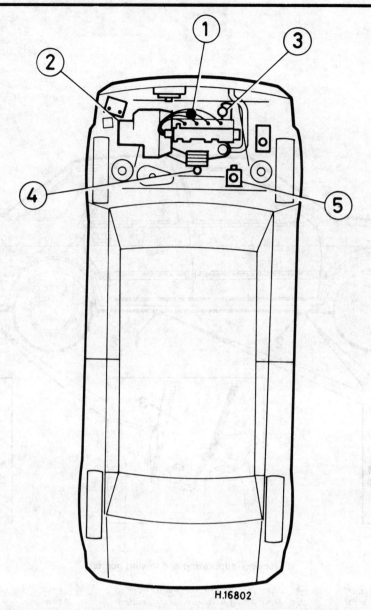

H.16802

# Recommended lubricants and fluids

| Component or system | Lubricant type/specification | Duckhams recommendation |
|---|---|---|
| 1 Engine* | Multigrade engine oil, viscosity SAE 10W/40 (or equivalent multigrade engine oil having a viscosity rating compatible with temperatures in the vehicle operating area – see manufacturer's handbook) | Duckhams QXR, Hypergrade, or 10W/40 Motor Oil |
| 2 Gearbox* | Multigrade engine oil, viscosity SAE 10W/40 (see above) | Duckhams QXR, Hypergrade, or 10W/40 Motor Oil |
| 2 Automatic transmission | Dexron TQ11 type ATF | Duckhams Uni-Matic or D-Matic |
| 3 Power-assisted steering | Dexron TQ11 type ATF | Duckhams Uni-Matic or D-Matic |
| 4 Carburettor piston damper (where applicable) | Multigrade engine oil, viscosity SAE 10W/40 (see above) | Duckhams QXR, Hypergrade, or 10W/40 Motor Oil |
| 5 Brake fluid reservoir | Hydraulic fluid to FMVSS 116 DOT 4 or SAE J1703C | Duckhams Universal Brake and Clutch Fluid |

*Note: *Austin Rover specify a 10W/40 oil to meet warranty requirements for models produced after August 1983. Duckhams QXR and 10W/40 Motor Oil are available to meet these requirements*

# Safety first!

Professional motor mechanics are trained in safe working procedures. However enthusiastic you may be about getting on with the job in hand, do take the time to ensure that your safety is not put at risk. A moment's lack of attention can result in an accident, as can failure to observe certain elementary precautions.

There will always be new ways of having accidents, and the following points do not pretend to be a comprehensive list of all dangers; they are intended rather to make you aware of the risks and to encourage a safety-conscious approach to all work you carry out on your vehicle.

### Essential DOs and DON'Ts

**DON'T** rely on a single jack when working underneath the vehicle. Always use reliable additional means of support, such as axle stands, securely placed under a part of the vehicle that you know will not give way.

**DON'T** attempt to loosen or tighten high-torque nuts (e.g. wheel hub nuts) while the vehicle is on a jack; it may be pulled off.

**DON'T** start the engine without first ascertaining that the transmission is in neutral (or 'Park' where applicable) and the parking brake applied.

**DON'T** suddenly remove the filler cap from a hot cooling system – cover it with a cloth and release the pressure gradually first, or you may get scalded by escaping coolant.

**DON'T** attempt to drain oil until you are sure it has cooled sufficiently to avoid scalding you.

**DON'T** grasp any part of the engine, exhaust or catalytic converter without first ascertaining that it is sufficiently cool to avoid burning you.

**DON'T** allow brake fluid or antifreeze to contact vehicle paintwork.

**DON'T** syphon toxic liquids such as fuel, brake fluid or antifreeze by mouth, or allow them to remain on your skin.

**DON'T** inhale dust – it may be injurious to health (see *Asbestos* below).

**DON'T** allow any spilt oil or grease to remain on the floor – wipe it up straight away, before someone slips on it.

**DON'T** use ill-fitting spanners or other tools which may slip and cause injury.

**DON'T** attempt to lift a heavy component which may be beyond your capability – get assistance.

**DON'T** rush to finish a job, or take unverified short cuts.

**DON'T** allow children or animals in or around an unattended vehicle.

**DO** wear eye protection when using power tools such as drill, sander, bench grinder etc, and when working under the vehicle.

**DO** use a barrier cream on your hands prior to undertaking dirty jobs – it will protect your skin from infection as well as making the dirt easier to remove afterwards; but make sure your hands aren't left slippery. Note that long-term contact with used engine oil can be a health hazard.

**DO** keep loose clothing (cuffs, tie etc) and long hair well out of the way of moving mechanical parts.

**DO** remove rings, wristwatch etc, before working on the vehicle – especially the electrical system.

**DO** ensure that any lifting tackle used has a safe working load rating adequate for the job.

**DO** keep your work area tidy – it is only too easy to fall over articles left lying around.

**DO** get someone to check periodically that all is well, when working alone on the vehicle.

**DO** carry out work in a logical sequence and check that everything is correctly assembled and tightened afterwards.

**DO** remember that your vehicle's safety affects that of yourself and others. If in doubt on any point, get specialist advice.

**IF**, in spite of following these precautions, you are unfortunate enough to injure yourself, seek medical attention as soon as possible.

### Asbestos

Certain friction, insulating, sealing, and other products – such as brake linings, brake bands, clutch linings, torque converters, gaskets, etc – contain asbestos. *Extreme care must be taken to avoid inhalation of dust from such products since it is hazardous to health.* If in doubt, assume that they *do* contain asbestos.

### Fire

Remember at all times that petrol (gasoline) is highly flammable. Never smoke, or have any kind of naked flame around, when working on the vehicle. But the risk does not end there – a spark caused by an electrical short-circuit, by two metal surfaces contacting each other, by careless use of tools, or even by static electricity built up in your body under certain conditions, can ignite petrol vapour, which in a confined space is highly explosive.

Always disconnect the battery earth (ground) terminal before working on any part of the fuel or electrical system, and never risk spilling fuel on to a hot engine or exhaust.

It is recommended that a fire extinguisher of a type suitable for fuel and electrical fires is kept handy in the garage or workplace at all times. Never try to extinguish a fuel or electrical fire with water.

**Note**: *Any reference to a 'torch' appearing in this manual should always be taken to mean a hand-held battery-operated electric lamp or flashlight. It does NOT mean a welding/gas torch or blowlamp.*

### Fumes

Certain fumes are highly toxic and can quickly cause unconsciousness and even death if inhaled to any extent. Petrol (gasoline) vapour comes into this category, as do the vapours from certain solvents such as trichloroethylene. Any draining or pouring of such volatile fluids should be done in a well ventilated area.

When using cleaning fluids and solvents, read the instructions carefully. Never use materials from unmarked containers – they may give off poisonous vapours.

Never run the engine of a motor vehicle in an enclosed space such as a garage. Exhaust fumes contain carbon monoxide which is extremely poisonous; if you need to run the engine, always do so in the open air or at least have the rear of the vehicle outside the workplace.

If you are fortunate enough to have the use of an inspection pit, never drain or pour petrol, and never run the engine, while the vehicle is standing over it; the fumes, being heavier than air, will concentrate in the pit with possibly lethal results.

### The battery

Never cause a spark, or allow a naked light, near the vehicle's battery. It will normally be giving off a certain amount of hydrogen gas, which is highly explosive.

Always disconnect the battery earth (ground) terminal before working on the fuel or electrical systems.

If possible, loosen the filler plugs or cover when charging the battery from an external source. Do not charge at an excessive rate or the battery may burst.

Take care when topping up and when carrying the battery. The acid electrolyte, even when diluted, is very corrosive and should not be allowed to contact the eyes or skin.

If you ever need to prepare electrolyte yourself, always add the acid slowly to the water, and never the other way round. Protect against splashes by wearing rubber gloves and goggles.

When jump starting a car using a booster battery, for negative earth (ground) vehicles, connect the jump leads in the following sequence: First connect one jump lead between the positive ( + ) terminals of the two batteries. Then connect the other jump lead first to the negative (–) terminal of the booster battery, and then to a good earthing (ground) point on the vehicle to be started, at least 18 in (45 cm) from the battery if possible. Ensure that hands and jump leads are clear of any moving parts, and that the two vehicles do not touch. Disconnect the leads in the reverse order.

### Mains electricity and electrical equipment

When using an electric power tool, inspection light etc, always ensure that the appliance is correctly connected to its plug and that, where necessary, it is properly earthed (grounded). Do not use such appliances in damp conditions and, again, beware of creating a spark or applying excessive heat in the vicinity of fuel or fuel vapour. Also ensure that the appliances meet the relevant national safety standards.

### Ignition HT voltage

A severe electric shock can result from touching certain parts of the ignition system, such as the HT leads, when the engine is running or being cranked, particularly if components are damp or the insulation is defective. Where an electronic ignition system is fitted, the HT voltage is much higher and could prove fatal.

# Routine maintenance

*For modifications, and information applicable to later models, see Supplement at end of manual*

Maintenance is essential for ensuring safety and desirable for the purpose of getting the best in terms of performance and economy from your car. Over the years the need for periodic lubrication has been greatly reduced if not totally eliminated. This has unfortunately tended to lead some owners to think that because no such action is required, the items either no longer exist, or will last forever. This is certainly not the case; it is essential to carry out regular visual examination as comprehensively as possible in order to spot any possible defects at an early stage before they develop into major expensive repairs.

The following service schedules are a list of the maintenance requirements and the intervals at which they should be carried out, as recommended by the manufacturers. Where applicable these procedures are covered in greater detail throughout this Manual, near the beginning of each Chapter.

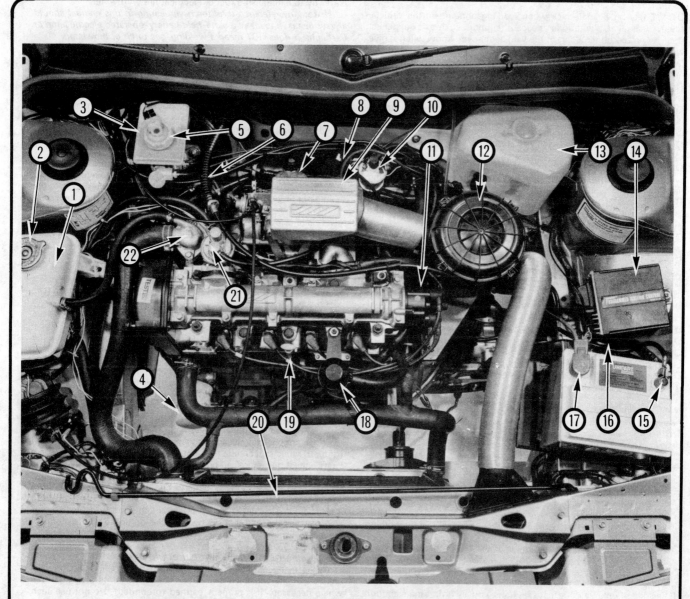

**Engine and under bonnet component locations (models with carburettor induction)**

| | | |
|---|---|---|
| 1 Cooling system expansion tank | 6 Clutch cable self-adjusting mechanism | 11 Distributor cap |
| 2 Cooling system filler cap | 7 Carburettor | 12 Air cleaner |
| 3 Brake master cylinder | 8 Main vacuum line | 13 Washer reservoir |
| 4 Oil filter | 9 Air cleaner plenum chamber | 14 Ignition system electronic control unit |
| 5 Master cylinder reservoir filler cap | 10 Ignition coil | 15 Battery negative terminal |
| | | 16 Fusible links |
| | | 17 Battery positive cable |
| | | 18 Oil filler/breather cap |
| | | 19 Oil dipstick |
| | | 20 Front body panel |
| | | 21 Fuel pump |
| | | 22 Water outlet elbow |

**Front underbody view**

1  Brake caliper
2  Anti-roll bar mounting block
3  Steering tie-rod outer balljoint
4  Suspension lower arm rear mounting
5  Gearbox drain plug
6  Driveshaft inner constant velocity joint
7  Suspension crossmember
8  Engine oil drain plug
9  Anti-roll bar clamp
10  Front jacking point
11  Access panel
12  Front snubber cap
13  Front jacking bracket/ towing hook

**Rear underbody view**

1 Exhaust intermediate silencer
2 Handbrake cable adjuster
3 Handbrake cable connectors
4 Exhaust rear silencer
5 Rear chassis rail
6 Rear suspension strut lower mounting
7 Rear axle mounting pivot bolt
8 Rear brake hose
9 Rear axle transverse member
10 Fuel tank rear mounting bolts
11 Fuel tank front mounting bolts
12 Rear jacking point

**Every 250 miles (400 km) or weekly – whichever occurs first**

## Engine, cooling system and brakes

Check the oil level and top up, if necessary (photo)
Check the coolant level and top up, if necessary (photo)
Check the brake fluid level in the master cylinder and top up, if necessary (photo)

## Lights and wipers

Check the operation of all interior and exterior lights, wipers and washers
Check, and if necessary, top up the washer reservoir (photo), adding a screen wash such as Turtle Wax High Tech Screen Wash

## Tyres

Check the tyre pressures
Visually examine the tyres for wear or damage

**Every 12 000 miles (20 000 km) or 12 months – whichever occurs first**

## Engine

Renew the engine oil and filter (photo)
Visually check the engine for oil leaks and for the security and condition of all related components and attachments

## Cooling system

Check the hoses, hose clips and visible joint gaskets for leaks and any signs of corrosion or deterioration
Check and, if necessary, top up the cooling system
Check the condition of the water pump and alternator drivebelt and renew if worn. Adjust the belt tension

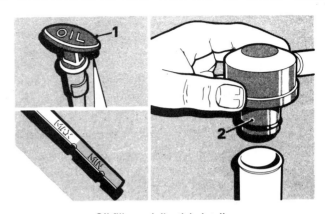

**Oil filler and dipstick details**

*1   Dipstick        2   Oil filler cap*

## Fuel and exhaust system

Renew the air cleaner element
Visually check the fuel pipes and hoses for security, chafing, leaks and corrosion
Check the fuel tank for leaks and any signs of damage or corrosion
Top up the carburettor piston damper (where applicable) – (photo)
Check the operation of the accelerator cable and linkage
Check and, if necessary, adjust the carburettor slow running characteristics (where applicable)
Check the exhaust system for corrosion, leaks and security
Renew fuel filter (if fitted)

Top up the engine oil

Top up the cooling system at the expansion tank

Top up the brake fluid at the master cylinder reservoir

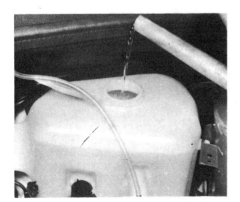

Top up the washer reservoir

Engine oil drain plug location

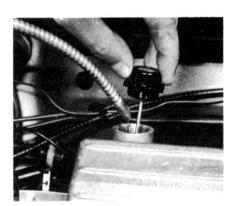

Top up the carburettor piston damper with engine oil

*Ignition system*
Renew the spark plugs
Clean the distributor cap, coil tower and HT leads, and check for tracking

*Clutch*
Check the operation of the clutch and clutch pedal
Check and, if necessary, adjust the clutch cable

*Gearbox*
Visually check for oil leaks around the gearbox joint faces and oil seals
Check and, if necessary, top up the gearbox oil

*Automatic transmission*
Check and, if necessary, top up transmission fluid

*Driveshafts*
Check the driveshaft constant velocity joints for wear or damage and check the rubber gaiters for condition

*Braking system*
Check visually all brake pipes, hoses and unions for corrosion, chafing, leakage and security
Check and, if necessary, top up the brake fluid
Check the operation of the brake warning indicators
Check the brake servo vacuum hose for condition and security
Check the operation of the hand and footbrake
Check the front brake pads for wear, and the discs for condition
Check the rear brake shoes for wear, and the drums for condition

*Electrical system*
Check the condition and security of all accessible wiring connectors, harnesses and retaining clips
Check the operation of all electrical equipment and accessories (lights, indicators, horn, wipers etc)
Check and adjust the operation of the screen washer and, if necessary, top up the reservoir
Clean the battery terminals and smear with petroleum jelly
Have the headlamp alignment checked and, if necessary, adjusted

*Suspension, steering, wheels and tyres*
Check the front and rear suspension struts for fluid leaks
Check the condition and security of the steering gear, steering and suspension joints, and rubber gaiters
Check the front wheel toe setting
Check and adjust the tyre pressures
Check the tyres for damage, tread depth and uneven wear
Inspect the roadwheels for damage
Check the tightness of the wheel nuts
Check the power-assisted steering pump drivebelt. Adjust the tension or renew if worn (where applicable)
Check and, if necessary, top up the power-assisted steering fluid (where applicable)

*Bodywork*
Carefully inspect the paintwork for damage and the bodywork for corrosion
Check the condition of the underseal
Oil all hinges, door locks and the bonnet release mechanism with a few drops of light oil

*Road test*
Check the operation of all instruments and electrical equipment
Check the operation of the seat belts
Check for any abnormalities in the steering, suspension, handling or road feel
Check the performance of the engine, clutch and transmission
Check the operation and performance of the braking system

---

**Every 24 000 miles (40 000 km) or 24 months – whichever occurs first**

---

In addition to all the items in the annual service, carry out the following:

*Engine*
Clean the filter in the oil filler cap
Check and if necessary adjust the valve clearances

*Cooling system*
Flush the cooling system and renew the antifreeze solution
Renew the water pump and alternator drivebelt

*Fuel system*
Check and adjust the carburettor idle speed and mixture settings

*Ignition system*
Check and, if necessary, adjust the ignition timing

*Gearbox*
Renew the gearbox oil

*Automatic transmission*
Renew the transmission fluid (clean the fluid filter if necessary)

*Braking system*
Renew the brake fluid

---

**Every 36 000 miles (60 000 km) or 36 months – whichever occurs first**

---

In addition to the items in the annual service, carry out the following:

*Braking system*
Renew the flexible rubber hoses and the rubber seals in the calipers, wheel cylinders and master cylinder
Renew the air filter in the servo unit

---

**Every 48 000 miles (80 000 km) or 48 months – whichever occurs first**

---

In addition to the items in the annual service, carry out the following:

*Engine*
Renew the timing belt

*Fuel system*
Renew the fuel filter (EFi models)

# Fault diagnosis

## Introduction

The vehicle owner who does his or her own maintenance according to the recommended schedules should not have to use this section of the manual very often. Modern component reliability is such that, provided those items subject to wear or deterioration are inspected or renewed at the specified intervals, sudden failure is comparatively rare. Faults do not usually just happen as a result of sudden failure, but develop over a period of time. Major mechanical failures in particular are usually preceded by characteristic symptoms over hundreds or even thousands of miles. Those components which do occasionally fail without warning are often small and easily carried in the vehicle.

With any fault finding, the first step is to decide where to begin investigations. Sometimes this is obvious, but on other occasions a little detective work will be necessary. The owner who makes half a dozen haphazard adjustments or replacements may be successful in curing a fault (or its symptoms), but he will be none the wiser if the fault recurs and he may well have spent more time and money than was necessary. A calm and logical approach will be found to be more satisfactory in the long run. Always take into account any warning signs or abnormalities that may have been noticed in the period preceding the fault – power loss, high or low gauge readings, unusual noises or smells, etc – and remember that failure of components such as fuses or spark plugs may only be pointers to some underlying fault.

The pages which follow here are intended to help in cases of failure to start or breakdown on the road. There is also a Fault Diagnosis Section at the end of each Chapter which should be consulted if the preliminary checks prove unfruitful. Whatever the fault, certain basic principles apply. These are as follows:

**Verify the fault.** This is simply a matter of being sure that you know what the symptoms are before starting work. This is particularly important if you are investigating a fault for someone else who may not have described it very accurately.

**Don't overlook the obvious.** For example, if the vehicle won't start, is there petrol in the tank? (Don't take anyone else's word on this particular point, and don't trust the fuel gauge either!) If an electrical fault is indicated, look for loose or broken wires before digging out the test gear.

**Cure the disease, not the symptom.** Substituting a flat battery with a fully charged one will get you off the hard shoulder, but if the underlying cause is not attended to, the new battery will go the same way. Similarly, changing oil-fouled spark plugs for a new set will get you moving again, but remember that the reason for the fouling (if it wasn't simply an incorrect grade of plug) will have to be established and corrected.

**Don't take anything for granted.** Particularly, don't forget that a 'new' component may itself be defective (especially if it's been rattling round in the boot for months), and don't leave components out of a fault diagnosis sequence just because they are new or recently fitted. When you do finally diagnose a difficult fault, you'll probably realise that all the evidence was there from the start.

## Electrical faults

Electrical faults can be more puzzling than straightforward mechanical failures, but they are no less susceptible to logical analysis if the basic principles of operation are understood. Vehicle electrical wiring exists in extremely unfavourable conditions – heat, vibration and chemical attack – and the first things to look for are loose or corroded connections and broken or chafed wires, especially where the wires pass through holes in the bodywork or are subject to vibration.

All metal-bodied vehicles in current production have one pole of the battery 'earthed', ie connected to the vehicle bodywork, and in nearly all modern vehicles it is the negative (–) terminal. The various electrical components – motors, bulb holders etc – are also connected to earth, either by means of a lead or directly by their mountings. Electric current flows through the component and then back to the battery via the bodywork. If the component mounting is loose or corroded, or if a good path back to the battery is not available, the circuit will be incomplete and malfunction will result. The engine and/or gearbox are also earthed by means of flexible metal straps to the body or subframe; if these straps are loose or missing, starter motor, generator and ignition trouble may result.

Assuming the earth return to be satisfactory, electrical faults will be due either to component malfunction or to defects in the current supply. Individual components are dealt with in Chapter 9. If supply wires are broken or cracked internally this results in an open-circuit, and the easiest way to check for this is to bypass the suspect wire temporarily with a length of wire having a crocodile clip or suitable connector at each end. Alternatively, a 12V test lamp can be used to verify the presence of supply voltage at various points along the wire and the break can be thus isolated.

If a bare portion of a live wire touches the bodywork or other earthed metal part, the electricity will take the low-resistance path thus formed back to the battery: this is known as a short-circuit. Hopefully a short-circuit will blow a fuse, but otherwise it may cause burning of the insulation (and possibly further short-circuits) or even a fire. This is why it is inadvisable to bypass persistently blowing fuses with silver foil or wire.

## Spares and tool kit

Most vehicles are supplied only with sufficient tools for wheel changing; the *Maintenance and minor repair* tool kit detailed in *Tools*

Carrying a few spares may save you a long walk!

*and working facilities,* with the addition of a hammer, is probably sufficient for those repairs that most motorists would consider attempting at the roadside. In addition a few items which can be fitted without too much trouble in the event of a breakdown should be carried. Experience and available space will modify the list below, but the following may save having to call on professional assistance:

*Spark plugs, clean and correctly gapped*
*HT lead and plug cap – long enough to reach the plug furthest from the distributor*
*Distributor rotor*
*Drivebelt(s) – emergency type may suffice*
*Spare fuses*
*Set of principal light bulbs*
*Tin of radiator sealer and hose bandage*
*Exhaust bandage*
*Roll of insulating tape*
*Length of soft iron wire*
*Length of electrical flex*
*Torch or inspection lamp (can double as test lamp)*
*Battery jump leads*
*Tow-rope*
*Ignition water dispersant aerosol*
*Litre of engine oil*
*Sealed can of hydraulic fluid*
*Emergency windscreen*
*Worm drive clips*

If spare fuel is carried, a can designed for the purpose should be used to minimise risks of leakage and collision damage. A first aid kit and a warning triangle, whilst not at present compulsory in the UK, are obviously sensible items to carry in addition to the above.

When touring abroad it may be advisable to carry additional spares which, even if you cannot fit them yourself, could save having to wait while parts are obtained. The items below may be worth considering:

*Clutch and throttle cables*
*Cylinder head gasket*
*Alternator brushes*
*Tyre valve core*

One of the motoring organisations will be able to advise on availability of fuel etc in foreign countries.

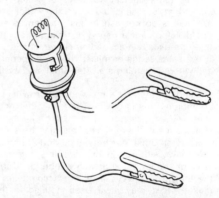

A simple test lamp is useful for checking electrical faults

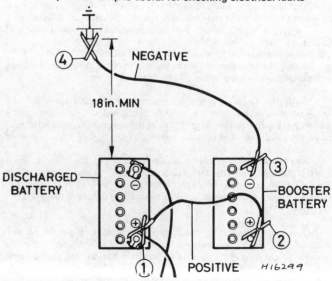

Jump start lead connections for negative earth – connect leads in order shown

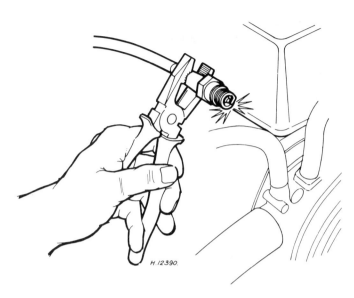

Crank engine and check for spark. Note use of insulated tool to hold plug lead

## Engine will not start

### Engine fails to turn when starter operated
Flat battery (recharge, use jump leads, or push start)
Battery terminals loose or corroded
Battery earth to body defective
Engine earth strap loose or broken
Starter motor (or solenoid) wiring loose or broken
Ignition/starter switch faulty
Major mechanical failure (seizure)
Starter or solenoid internal fault (see Chapter 9)

### Starter motor turns engine slowly
Partially discharged battery (recharge, use jump leads, or push start)
Battery terminals loose or corroded
Battery earth to body defective
Engine earth strap loose
Starter motor (or solenoid) wiring loose
Starter motor internal fault (see Chapter 9)

### Starter motor spins without turning engine
Flat battery
Starter motor pinion sticking on sleeve
Flywheel gear teeth damaged or worn
Starter motor mounting bolts loose

### Engine turns normally but fails to start
Damp or dirty HT leads and distributor cap (crank engine and check for spark) — try moisture dispersant such as Holts Wet Start
No fuel in tank (check for delivery at carburettor)
Excessive choke (hot engine) or insufficient choke (cold engine)
Fouled or incorrectly gapped spark plugs (remove, renew or regap)
Other ignition system fault (see Chapter 4)
Other fuel system fault (see Chapter 3)
Poor compression (see Chapter 1)
Major mechanical failure (eg camshaft drive)

### Engine fires but will not run
Insufficient choke (cold engine)
Air leaks at carburettor or inlet manifold
Fuel starvation (see Chapter 3)
Ballast resistor defective, or other ignition fault (see Chapter 4)

## Engine cuts out and will not restart

### Engine cuts out suddenly — ignition fault
Loose or disconnected LT wires
Wet HT leads or distributor cap (after traversing water splash)
Coil or condenser failure (check for spark)
Other ignition fault (see Chapter 4)

### Engine misfires before cutting out — fuel fault
Fuel tank empty
Fuel pump defective or filter blocked (check for delivery)
Fuel tank filler vent blocked (suction will be evident on releasing cap)
Carburettor needle valve sticking
Carburettor jets blocked (fuel contaminated)
Other fuel system fault (see Chapter 3)

### Engine cuts out — other causes
Serious overheating
Major mechanical failure (eg camshaft drive)

## Engine overheats

### Ignition (no-charge) warning light illuminated
Slack or broken drivebelt — retension or renew (Chapter 2)

### Ignition warning light not illuminated
Coolant loss due to internal or external leakage (see Chapter 2)
Thermostat defective
Low oil level
Brakes binding
Radiator clogged externally or internally
Electric cooling fan not operating correctly
Engine waterways clogged
Ignition timing incorrect or automatic advance malfunctioning
Mixture too weak

**Note**: *Do not add cold water to an overheated engine or damage may result*

## Low engine oil pressure

### Gauge reads low or warning light illuminated with engine running
Oil level low or incorrect grade
Defective gauge or sender unit
Wire to sender unit earthed
Engine overheating
Oil filter clogged or bypass valve defective
Oil pressure relief valve defective
Oil pick-up strainer clogged
Oil pump worn or mountings loose
Worn main or big-end bearings

**Note**: *Low oil pressure in a high-mileage engine at tickover is not necessarily a cause for concern. Sudden pressure loss at speed is far more significant. In any event, check the gauge or warning light sender before condemning the engine.*

## Engine noises

### Pre-ignition (pinking) on acceleration
Incorrect grade of fuel
Ignition timing incorrect
Distributor faulty or worn
Worn or maladjusted carburettor
Excessive carbon build-up in engine

## *Whistling or wheezing noises*

Leaking vacuum hose
Leaking carburettor or manifold gasket
Blowing head gasket

## *Tapping or rattling*

Incorrect valve clearances
Worn valve gear
Worn timing belt
Broken piston ring (ticking noise)

## *Knocking or thumping*

Unintentional mechanical contact (eg fan blades)
Worn fanbelt
Peripheral component fault (generator, water pump etc)
Worn big-end bearings (regular heavy knocking, perhaps less under load)
Worn main bearings (rumbling and knocking, perhaps worsening under load)
Piston slap (most noticeable when cold)

# Chapter 1 Engine

*For modifications, and information applicable to later models, see Supplement at end of manual*

## Contents

## Specifications

### Engine (general)

| | |
|---|---|
| Type ........................................................................................ | 20H overhead camshaft ('O' Series) |
| Number of cylinders ................................................................ | 4 |
| Bore ........................................................................................ | 3.325 in (84.45 mm) |
| Stroke ..................................................................................... | 3.504 in (89.0 mm) |
| Capacity .................................................................................. | 1994 cc (121.68 cu in) |
| Firing order ............................................................................ | 1-3-4-2 |
| Valve operation ...................................................................... | Overhead camshaft |
| Compression ratio .................................................................. | 9.0:1 |

### Crankshaft

| | |
|---|---|
| Main journal diameter ............................................................ | 2.1262 to 2.1270 in (54.005 to 54.026 mm) |
| Crankpin journal diameter ...................................................... | 1.8754 to 1.8759 in (47.635 to 47.647 mm) |
| Crankshaft endthrust .............................................................. | Taken on thrust washers at centre bearing |
| Crankshaft endfloat ................................................................ | 0.001 to 0.005 in (0.025 to 0.14 mm) |

### Main bearings

| | |
|---|---|
| Number and type ..................................................................... | 5, steel-backed thin wall type |
| Width: | |
|     Front, centre and rear ..................................................... | 1.125 in (28.57 mm) |
|     Intermediate ....................................................................... | 0.760 in (19.3 mm) |
| Diametrical clearance ............................................................. | 0.001 to 0.003 in (0.025 to 0.077 mm) |
| Undersizes .............................................................................. | 0.010 in (0.254 mm) |
| | 0.020 in (0.508 mm) |
| | 0.030 in (0.762 mm) |
| | 0.040 in (1.016 mm) |

### Connecting rods

| | |
|---|---|
| Type ........................................................................................ | Horizontal, split big-end, plain small-end offset |
| Length between centres ......................................................... | 5.86 in (149 mm) |

## Big-end bearings
Type ................................................................................................ Steel-backed thin wall
Width .............................................................................................. 0.775 to 0.785 in (19.68 to 19.94 mm)
Diametrical clearance ................................................................... 0.0015 to 0.0032 in (0.038 to 0.081 mm)
Undersizes ..................................................................................... 0.010 in (0.254 mm)
0.020 in (0.508 mm)
0.030 in (0.762 mm)
0.040 in (1.016 mm)

## Gudgeon pin
Type ................................................................................................ Press fit in connecting rod
Fit in piston .................................................................................. Handpush at 16°C (60°F)
Diameter ........................................................................................ 0.8125 to 0.8127 in (20.638 to 20.643 mm)

## Pistons
Type ................................................................................................ Duotherm, solid skirt with combustion chamber in crown
Clearance in cylinder:
  Below oil control groove ........................................................ 0.0008 to 0.0024 in (0.02 to 0.06 mm)
  Bottom of skirt ......................................................................... 0.0004 to 0.0014 in (0.01 to 0.038 mm)
Number of rings ............................................................................ 3 (2 compression, 1 oil control)
Width of ring grooves:
  Top and second ....................................................................... 0.070 to 0.071 in (1.78 to 1.80 mm)
  Oil control ................................................................................. 0.157 to 0.158 in (4.00 to 4.02 mm)
Gudgeon pin bore ........................................................................ 0.8128 to 0.8130 in (20.646 to 20.651 mm)
  Offset from centre ................................................................... 0.059 in (1.5 mm)

## Piston rings
Compression rings:
  Type:
    Top ...................................................................................... Plain, chrome faced
    Second ................................................................................ Stepped scraper
  Ring-to-groove clearance ....................................................... 0.0015 to 0.0027 in (0.04 to 0.07 mm)
  Fitted gap ................................................................................. 0.012 to 0.020 in (0.3 to 0.5 mm)
Oil control ring:
  Type .......................................................................................... Two chrome faced rings with butted expander
  Fitted gap ................................................................................. 0.013 to 0.055 in (0.33 to 1.4 mm)

## Camshaft
Journal diameters ........................................................................ 1.888 to 1.889 in (47.96 to 47.97 mm)
Number of bearings ...................................................................... 3
Bearing type ................................................................................. Direct in cylinder head and cover
Diametrical clearance ................................................................... 0.0017 to 0.0037 in (0.043 to 0.094 mm)
Endthrust ...................................................................................... Taken on rear cover
Endfloat ......................................................................................... 0.003 to 0.007 in (0.07 to 0.18 mm)
Drive .............................................................................................. Toothed belt from crankshaft sprocket
Timing belt tension:
  New belt .................................................................................... 13 lbf (58N)
  Used belt .................................................................................. 11 lbf (49N)

## Tappets
Type ................................................................................................ Bucket with flat base
Outside diameter .......................................................................... 1.2491 to 1.2498 in (31.729 to 31.745 mm)
Adjustment .................................................................................... Selective shim

## Valves
Face angle ..................................................................................... 45° 30'
Seat angle ..................................................................................... 45°
Head diameter:
  Inlet ........................................................................................... 1.575 in (40 mm)
  Exhaust ..................................................................................... 1.339 in (34 mm)
Stem diameter:
  Inlet ........................................................................................... 0.2917 to 0.2921 in (7.41 to 7.42 mm)
  Exhaust ..................................................................................... 0.2909 to 0.2917 in (7.39 to 7.41 mm)
Stem-to-guide clearance:
  Inlet ........................................................................................... 0.001 to 0.002 in (0.027 to 0.053 mm)
  Exhaust ..................................................................................... 0.0015 to 0.0028 in (0.04 to 0.073 mm)
Cam lift ......................................................................................... 0.375 in (9.525 mm)

## Valve guides
Length ........................................................................................... 1.532 in (38.90 mm)
Outside diameter .......................................................................... 0.474 to 0.475 in (12.04 to 12.06 mm)
Inside diameter ............................................................................ 0.293 to 0.2937 in (7.45 to 7.46 mm)
Fitted height above head ............................................................. 0.394 in (10 mm)
Interference fit in head ................................................................ 0.0015 to 0.003 in (0.04 to 0.09 mm)

## Valve springs
Free length ................................................................................... 1.646 in (41.81 mm)
Fitted length ................................................................................. 1.375 in (34.92 mm)
Number of working coils .............................................................. 4.5

## Valve timing
Inlet valve:

 Opens .................................................................................... 19° BTDC

 Closes ................................................................................... 41° ABDC

Exhaust valve:

 Opens .................................................................................... 61° BBDC

 Closes ................................................................................... 15° ATDC

## Valve clearance (cold) ................................................. 0.011 to 0.013 in (0.28 to 0.33 mm)
Adjust only if less than ..................................................... 0.008 in (0.20 mm)

## Lubrication system
Oil type/specification* ........................................................ Multigrade engine oil, viscosity SAE 10W/40 (or equivalent multigrade engine oil having a viscosity rating compatible with the vehicle operating area – see manufacturer's handbook) (Duckhams QXR, Hypergrade, or 10W/40 Motor Oil)

Oil filter type ...................................................................... Champion B101

Oil pump:

 Type .................................................................................... Bi-rotor

 Outer rotor to body clearance ......................................... 0.007 to 0.009 in (0.18 to 0.23 mm)

 Rotor lobe clearance ........................................................ 0.010 to 0.011 in (0.25 to 0.28 mm)

 Outer rotor endfloat ......................................................... 0.001 to 0.003 in (0.03 to 0.07 mm)

System pressure:

 Idling ................................................................................. 15 lbf/in$^2$ (1.0 bar)

 Running .............................................................................. 60 lbf/in$^2$ (4.1 bar)

Pressure relief valve spring free length ........................... 1.525 in (38.7 mm)

*Note: *Austin Rover specify a 10W/40 oil to meet warranty requirements for models produced after August 1983. Duckhams QXR and 10W/40 Motor Oil are available to meet these requirements*

## Torque wrench settings

| | lbf ft | Nm |
|---|---|---|
| Main bearing cap bolts | 75 | 100 |
| Big-end bearing cap nuts | 33 | 45 |
| Gearbox adaptor plate: | | |
|  8 mm bolts | 22 | 30 |
|  10 mm bolts | 37 | 51 |
| Flywheel retaining bolts | 42 | 58 |
| Oil pump backplate bolts | 2 | 3 |
| Oil pressure switch | 9 | 12 |
| Camshaft sprocket bolt | 48 | 66 |
| Cylinder head bolts: | | |
|  Stage 1 | 33 | 45 |
|  Stage 2 | 60 | 80 |
|  Stage 3 | Further 60° or to 80 lbf ft (108 Nm) – whichever comes first | |
| Crankshaft pulley bolt | 62 | 85 |
| Camshaft cover bolts | 13 | 18 |
| Knock sensor to cylinder block | 9 | 12 |
| Oil pump housing bolts | 8 | 11 |
| Oil separator to cylinder block | 22 | 30 |
| Sump drain plug | 27 | 37 |
| Sump pan bolts: | | |
|  Stage 1 | 2 | 3 |
|  Stage 2 | 7 | 10 |
| Bolts not specified into cast iron: | | |
|  M6 | 8 | 11 |
|  M8 | 22 | 30 |
|  M10 | 37 | 50 |
| Bolts not specified into aluminium: | | |
|  M6 | 8 | 11 |
|  M8 | 18 | 25 |
| Engine mountings to engine, gearbox and body members | 34 | 45 |
| Engine mounting through bolts | 66 | 90 |
| Front snubber bracket to gearbox bellhousing | 66 | 90 |
| Snubber cup bolts | 34 | 45 |

## 1 General description

The engine is a water-cooled, four-cylinder four-stroke petrol engine of overhead camshaft configuration and 1994 cc capacity.

The combined crankcase and cylinder block is of cast iron construction and houses the pistons, connecting rods and crankshaft. The solid skirt cast aluminium alloy pistons are retained on the connecting rods by gudgeon pins which are an interference fit in the connecting rod small-end bore. The connecting rods are attached to the crankshaft by renewable shell type big-end bearings.

The forged steel crankshaft is carried in five main bearings also of the renewable shell type. Crankshaft endfloat is controlled by thrust washers which are located on either side of the centre main bearing.

The camshaft is located in the cylinder head and is retained in position by an aluminium cover which is bolted to the upper face of the cylinder head. The camshaft is supported by three bearing journals machined directly in the head and camshaft cover. Drive to the camshaft is by a toothed composite rubber timing belt from a sprocket on the front end of the crankshaft. A spring-loaded tensioner is fitted to eliminate backlash and prevent slackness of the belt.

Two valves per cylinder are mounted vertically in the aluminium cylinder head. The valves are operated by bucket type tappets acted upon directly by the lobes of the camshaft. Valve/tappet clearance adjustment is by selective shims.

Engine lubrication is by the conventional forced feed system and a detailed description of its operation will be found in Section 25.

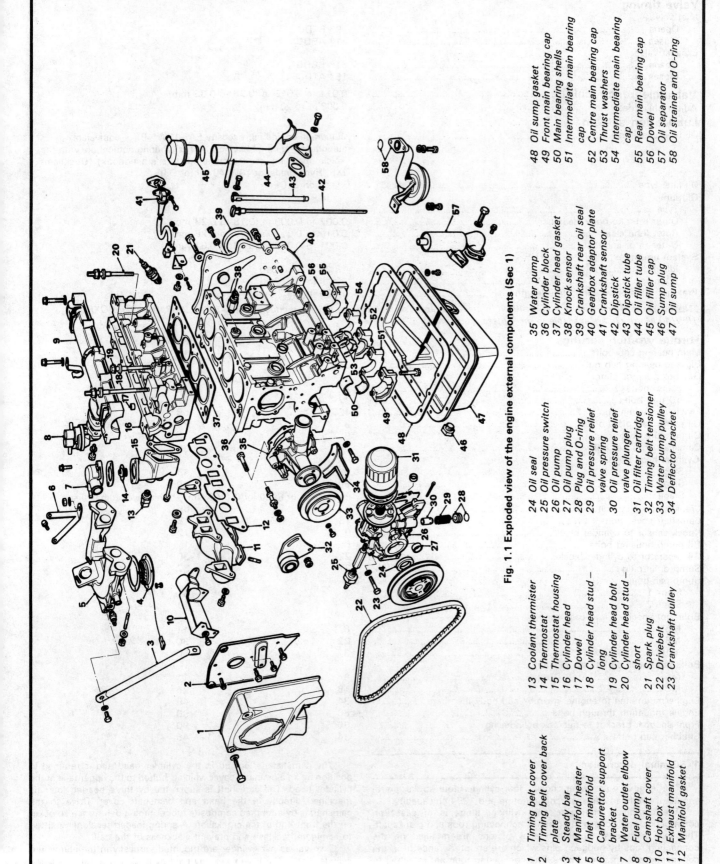

**Fig. 1.1 Exploded view of the engine external components (Sec 1)**

1  Timing belt cover
2  Timing belt cover back plate
3  Steady bar
4  Manifold heater
5  Inlet manifold
6  Carburettor support bracket
7  Water outlet elbow
8  Fuel pump
9  Camshaft cover
10 Hot box
11 Exhaust manifold
12 Manifold gasket

13 Coolant thermister
14 Thermostat
15 Thermostat housing
16 Cylinder head
17 Dowel
18 Cylinder head stud – long
19 Cylinder head bolt
20 Cylinder head stud – short
21 Spark plug
22 Drivebelt
23 Crankshaft pulley

24 Oil seal
25 Oil pressure switch
26 Oil pump
27 Oil pump plug
28 Plug and O-ring
29 Oil pressure relief valve spring
30 Oil pressure relief valve plunger
31 Oil filter cartridge
32 Timing belt tensioner
33 Water pump pulley
34 Deflector bracket

35 Water pump
36 Cylinder block
37 Cylinder head gasket
38 Knock sensor
39 Crankshaft rear oil seal
40 Gearbox adaptor plate
41 Crankshaft sensor
42 Dipstick
43 Dipstick tube
44 Oil filler tube
45 Oil filler cap
46 Sump plug
47 Oil sump

48 Oil sump gasket
49 Front main bearing cap
50 Main bearing shells
51 Intermediate main bearing cap
52 Centre main bearing cap
53 Thrust washers
54 Intermediate main bearing cap
55 Rear main bearing cap
56 Dowel
57 Oil separator
58 Oil strainer and O-ring

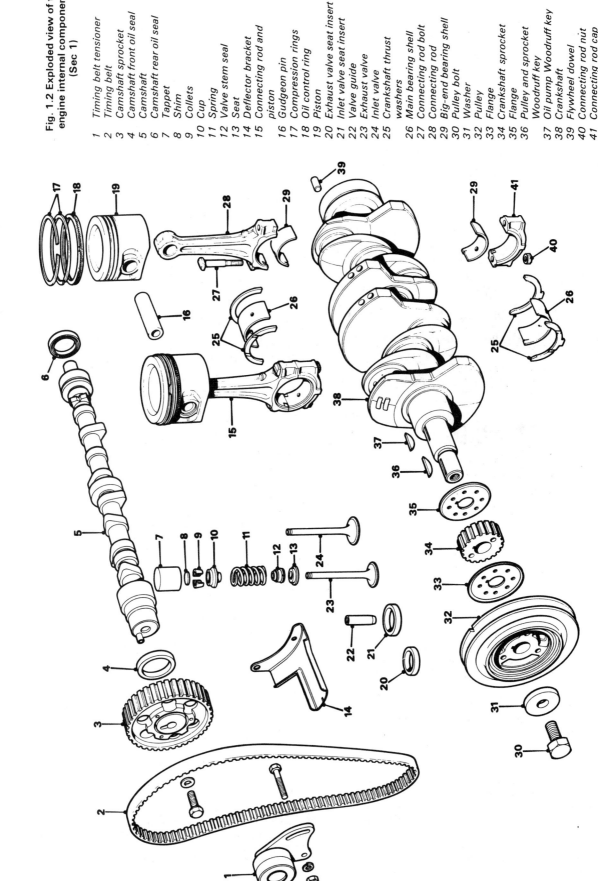

**Fig. 1.2 Exploded view of the engine internal components (Sec 1)**

1 Timing belt tensioner
2 Timing belt
3 Camshaft sprocket
4 Camshaft front oil seal
5 Camshaft
6 Camshaft rear oil seal
7 Tappet
8 Shim
9 Collets
10 Cup
11 Spring
12 Valve stem seal
13 Seat
14 Deflector bracket
15 Connecting rod and piston
16 Gudgeon pin
17 Compression rings
18 Oil control ring
19 Piston
20 Exhaust valve seat insert
21 Inlet valve seat insert
22 Valve guide
23 Exhaust valve
24 Inlet valve
25 Crankshaft thrust washers
26 Main bearing shell
27 Connecting rod bolt
28 Connecting rod
29 Big-end bearing shell
30 Pulley bolt
31 Washer
32 Pulley
33 Flange
34 Crankshaft sprocket
35 Flange
36 Pulley and sprocket Woodruff key
37 Oil pump Woodruff key
38 Crankshaft
39 Flywheel dowel
40 Connecting rod nut
41 Connecting rod cap

## 2 Maintenance and inspection

1   At the intervals given in Routine Maintenance at the beginning of this Manual, carry out the following operations on the engine.
2   Visually inspect the engine joint faces, gaskets and seals for any sign of oil or water leaks. Pay particular attention to the areas around the camshaft cover, cylinder head, crankshaft front oil seal and sump joint faces. Rectify any leaks by referring to the appropriate Sections of this Chapter.
3   Place a suitable container beneath the oil drain plug located on the rear right-hand side of the sump. Unscrew the plug using a spanner or socket and allow the oil to drain. Inspect the condition of the drain plug sealing washer, and renew it if necessary. Refit and tighten the plug after draining.
4   Move the bowl to the front of the engine under the oil filter.
5   Using a strap wrench or filter removal tool, slacken the filter and unscrew it from the engine and discard (photo).

2.5 Removing the oil filter with a filter removal tool

6   Wipe the filter mating face with a rag and then lubricate the seal of a new filter using clean engine oil.
7   Screw the filter into position and tighten it by hand only, do not use any tools.
8   Refill the engine using the correct grade of oil through the filler tube at the front of the engine. Fill until the level reaches the MAX mark on the dipstick – 0.5 litre will raise the level from MIN to MAX.
9   With the engine running, check for leaks around the filter seal.
10   At the less frequent service intervals given in Routine Maintenance check and if necessary adjust the valve clearances as described in Section 12.
11   Remove the oil filler cap from the filler tube, detach the breather hose and clean the filter in the cap. Check the condition of the crankcase ventilation hoses and renew any that are perished or show signs of deterioration.
12   Renew the timing belt at the specified service interval using the procedure described in Section 11.

## 3 Major operations possible with the engine in the car

The following operations can be carried out without having to remove the engine from the car:

(a)   Removal and refitting of the camshaft and tappets
(b)   Removal and refitting of the timing belt and tensioner
(c)   Removal and refitting of the cylinder head
(d)   Removal and refitting of the sump
(e)   Removal and refitting of the big-end bearings

(f)   Removal and refitting of the piston and connecting rod assemblies
(g)   Removal and refitting of the oil pump
(h)   Removal and refitting of the engine mountings

## 4 Major operations requiring engine removal

Strictly speaking it is only necessary to remove the engine if the crankshaft or main bearings require attention. However, owing to the possibility of dirt entry, and to allow greater working access, it is preferable to remove the engine if working on the piston and connecting rod assemblies, or when carrying out any major engine overhaul or repair.

## 5 Methods of engine removal

The engine and gearbox assembly can be lifted from the car as a complete unit, as described in the following Section, or the gearbox may first be removed, as described in Chapter 6. It is not possible to remove the engine on its own owing to space restrictions in the engine bay.

## 6 Engine and gearbox assembly – removal and refitting (all models except MG EFi)

1   Drain the cooling system as described in Chapter 2, the gearbox oil as described in Chapter 6 and the engine oil as described in Section 2 of this Chapter.
2   Disconnect the battery negative and positive terminals then remove the clamp bolt and lift out the battery.
3   Undo the nuts and bolts securing the battery tray to the valance noting the earth wire locations on the front retaining bolt (photo). Remove the battery tray from the engine compartment (photo).
4   Refer to Chapter 3 and remove the air cleaner assembly.
5   Slacken the screw securing the accelerator cable to the connector on the carburettor linkage. Withdraw the cable from its support bracket and move it aside.
6   Disconnect the two wires at the fuel shut off valve solenoid on the carburettor and the multi-plug at the carburettor stepper motor.
7   Disconnect the vacuum hoses at the manifold take off, at the carburettor and at the carburettor vacuum switch.
8   Undo the servo vacuum hose banjo union bolt, withdraw the union and recover the two copper washers (photo). Move the vacuum hose to one side.
9   Disconnect the fuel inlet hose at the fuel pump and plug the end of the hose to prevent fuel spillage.
10   Disconnect the heater hoses at the two heater pipes and at the inlet manifold (photo).
11   Disconnect the radiator top hose and expansion tank hose at the water outlet elbow (photo).
12   Release the top hose retaining clip on the timing belt cover (photo) and move the hoses clear.
13   Disconnect the radiator bottom hose at the water pump outlet.
14   Disconnect the clutch cable from the operating lever and engine mounting bracket as described in Chapter 5.
15   Disconnect the coolant thermister wiring multi-plug from the thermister on the thermostat housing.
16   Disconnect the inlet manifold heater wire and the two wires at the inlet manifold temperature sensor.
17   Disconnect the multi-plug from the knock sensor on the cylinder block and the two reversing lamp wires at their connectors (photos).
18   Note the locations of the wires at the starter solenoid and disconnect them from their spade and stud terminals.
19   Undo the retaining bolt and withdraw the engine earth strap from the gearbox breather bracket.
20   Undo the bolt securing the speedometer cable retaining plate to the gearbox. Withdraw the plate and lift out the speedometer cable and pinion assembly.
21   Disconnect the crankshaft sensor wiring multi-plug at the rear of the engine.
22   Disconnect the ignition coil HT lead from the centre of the distributor cap.

6.3A Note the earth wires at the battery tray retaining bolt ...

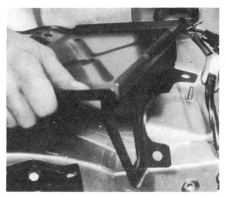

6.3B ... remove the bolts and the tray

6.8 Remove the servo vacuum hose banjo union

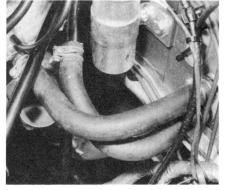

6.10 Disconnect the heater hoses

6.11 Disconnect the hoses at the water outlet elbow

6.12 Top hose retaining clip secured to timing belt cover bolt

6.17A Disconnect the knock sensor multi-plug ...

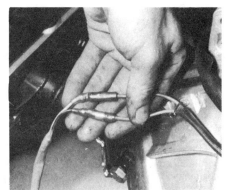

6.17B ... and the reversing lamp wires

6.24 Remove the wheel arch access panels

23 Jack up the front of the car and support it on axle stands. Remove both front roadwheels.

24 Undo the retaining screws and remove the access panels from both front wheel arches (photo).

25 If power-assisted steering is fitted, slacken the pump adjustment and pivot bolts, move the pump towards the engine and remove the drivebelt. Release the fluid hoses from the sump bracket and undo the bolts securing the pump mounting bracket to the cylinder block. Lower the pump, with hoses still connected, and suspend it using string or wire from the front jacking bracket towing hook.

26 Undo the two nuts and bolts each side securing the front suspension struts to the swivel hubs.

27 Using a stout screwdriver or flat bar, ease the driveshaft inner

constant velocity joints out of the final drive assembly. Once the joints have been released, tip the swivel hub outwards at the top, as far as possible without straining the brake hoses unduly, then withdraw the inner joints fully from their locations.

28 Undo the two bolts securing the alternator heat shield to the cylinder block and remove the shield.

29 Release the retaining wire and disconnect the alternator multi-plug. Disconnect the oil pressure switch wire at the pump.

30 Undo the retaining nuts and separate the exhaust downpipe flange from the manifold. Recover the flange gasket.

31 Remove the cover band then tap out the roll pin securing the gearchange rod to the gearchange shaft (photos). Undo the small bolt and remove the dished washer, steady bar and flat washer from the

6.31A Remove the gearchange rod cover band (arrowed) ...

6.31B ... and tap out the roll pin (arrowed)

6.32 Rear mounting to gearbox retaining bolt (arrowed)

6.33 Front snubber bracket and snubber cup

6.37 Undo the mounting bracket bolts

6.39 Undo the right-hand engine mounting bolts accessible from above (arrowed)

gearbox case. Slide the gearchange rod rearwards and off the shaft.

32 Undo and remove the two bolts each side securing the rear engine mounting to the gearbox (photo).

33 Undo the two bolts securing the engine front snubber bracket to the gearbox, and the two bolts and eccentric flat washers securing the snubber cup to the chassis member (photo). Remove the snubber cup and bracket assembly.

34 Refer to Chapter 11 if necessary and remove the bonnet.

35 Attach a suitable hoist to the engine using chains or rope slings. Preferably attach the chains to home made brackets fastened to the cylinder head studs. Raise the hoist to just take the weight of the engine.

36 Undo the bolts securing the left-hand engine mounting bracket to the gearbox, and the through bolt and nut securing the mounting to the bracket on the chassis member. Lower the engine slightly and remove the mounting. Recover the two flat washers located each side of the mounting rubber.

37 Undo the bolts securing the mounting bracket to the chassis member and remove the bracket (photo).

38 Undo the screw securing the cooling system expansion tank to its support bracket and the bolts securing the bracket to the chassis member. Remove the bracket.

39 Undo the bolts securing the right-hand engine mounting to the chassis member. The bolts are accessible from above and from under the wheel arch (photo).

40 Make a final check that everything attaching the engine and gearbox to the car have been disconnected and moved well clear.

41 Raise the engine and gearbox carefully, moving them slightly to clear any protrusions. When the unit has been raised sufficiently, draw the hoist forward to bring the engine and gearbox over the front body panel then lower the assembly to the floor (photo).

42 Refitting is a straightforward reverse of the removal sequence, bearing in mind the following points:

(a) Refit all the engine mounting bolts loosely and then tighten them in the sequence – right-hand mounting, left-hand mounting, rear mounting, snubber bracket. When all the mountings have been tightened, centralise the snubber cup around the snubber rubber and tighten the cup bolts

(b) If power-assisted steering is fitted, adjust the pump drivebelt as described in Chapter 10

(c) Reconnect the clutch cable with reference to Chapter 5

(d) Adjust the accelerator cable as described in Chapter 3

(e) Refill the cooling system as described in Chapter 2 and the gearbox with oil as described in Chapter 6. Refill the engine with oil as described in Section 2 of this Chapter

6.41 Removing the engine and gearbox assembly

## 7 Engine and gearbox assembly – removal and refitting (MG EFi models)

**Note:** *Refer to Chapter 3 for all operations relating to the electronic fuel injection system.*

1   Drain the cooling system as described in Chapter 2, the gearbox oil as described in Chapter 6 and the engine oil as described in Section 2 of this Chapter.

2   Depressurise the fuel system then disconnect the battery terminals.

3   Undo the battery clamp retaining bolt and lift out the battery.

4   Undo the nuts and bolts securing the battery tray to the valance, noting the earth wire locations on the front retaining bolt. Remove the battery tray from the engine compartment.

5   Remove the air cleaner and mounting bracket assembly as described in Chapter 3.

6   Slacken the retaining clip and detach the intake tube from the throttle housing. Remove the intake tube and air flow meter assembly from the engine compartment.

7   Disconnect the clutch cable from the operating lever and engine mounting bracket as described in Chapter 5.

8   Disconnect the earth cable at the gearbox breather bracket.

9   Disconnect the wiring multi-plugs from the fuel injectors and fuel temperature switch.

10   Disconnect the wiring plug at the end of the throttle potentiometer lead and the multi-plug at the air valve stepper motor.

11   Make a note of the wiring locations at the starter solenoid and disconnect them.

12   Detach the air and vacuum hoses at the rear of the throttle housing.

13   Undo the servo vacuum hose banjo union bolt on the inlet manifold. Remove the two copper washers and move the hose aside.

14   Slacken the screw securing the accelerator cable to the connector on the throttle housing linkage. Withdraw the cable from the connector and support bracket and move it well clear.

15   Undo the bolt, lift off the retaining plate and withdraw the speed transducer and pinion assembly from the gearbox.

16   Disconnect the crankshaft sensor multi-plug from the sensor lead.

17   Disconnect the multi-plug at the knock sensor on the front of the cylinder block and the two wires at the reversing lamp switch.

18   Disconnect the coil HT lead from the centre of the distributor cap.

19   Disconnect the two wires at the oil level sensor, if fitted, on the front of the sump.

20   Undo the union nut and disconnect the fuel inlet hose at the fuel rail. Plug the hose after removal.

21   Undo the union nut and disconnect the fuel return hose at the fuel pressure regulator. Plug the hose after removal.

22   Disconnect the heater hoses at the heater pipes and at the throttle housing.

23   Disconnect the radiator top hose and expansion tank hose at the water outlet elbow.

24   Release the top hose retaining clip on the timing belt cover and move the hoses clear.

25   Disconnect the radiator bottom hose at the water pump outlet.

26   Disconnect the coolant thermister wiring multi-plug from the thermister on the thermostat housing.

27   Release the fuel filter and bracket from their location and place the assembly to one side.

28   The remainder of the removal procedure is the same as described in Section 6, paragraphs 23 to 41 inclusive.

29   Refitting is a straightforward reverse of the removal sequence, bearing in mind the following points:

(a)   *Refit all the engine mounting bolts loosely and then tighten them in the sequence – right-hand mounting, left-hand mounting, rear mounting, snubber bracket. When all the mountings have been tightened, centralise the snubber cup around the snubber rubber and tighten the cup bolts*

(b)   *If power-assisted steering is fitted, adjust the pump drivebelt as described in Chapter 10*

(c)   *Reconnect the clutch cable with reference to Chapter 5*

(d)   *Adjust the accelerator cable as described in Chapter 3*

(e)   *Refill the cooling system as described in Chapter 2 and the gearbox oil as described in Chapter 6. Refill the engine with oil as described in Section 2 of this Chapter*

(f)   *Reconnect the fuel relay circuit as described in Chapter 3 before reconnecting the battery*

## 8 Engine – separation from gearbox

1   With the engine and gearbox removed from the car, undo the starter motor retaining bolts and nuts and remove the unit from the gearbox bellhousing (photo).

2   Noting the various lengths of the retaining bolts, undo and remove all the fixings securing the gearbox to the engine adaptor plate.

3   Withdraw the gearbox off the adaptor plate after releasing the dowels (photo).

8.1 Removing the starter motor

8.3 Separating the engine and gearbox

## 9 Engine dismantling – general

1   If possible mount the engine on a stand for the dismantling procedure, but failing this, support it in an upright position with blocks of wood placed under each side of the sump or crankcase.

2   Drain the oil into a suitable container before cleaning the engine or major dismantling, if this has not already been done.

3   Cleanliness is most important, and if the engine is dirty it should be cleaned with paraffin or a suitable solvent while keeping it in an upright position.

4   Avoid working with the engine directly on a concrete floor, as grit presents a real source of trouble.

5   As parts are removed, clean them in a paraffin bath. However, do not immerse parts with internal oilways in paraffin as it is difficult to remove, usually requiring a high pressure hose. Clean oilways with nylon pipe cleaners.

6   It is advisable to have suitable containers to hold small items, as this will help when reassembling the engine and also prevent possible losses.

7   Always obtain complete sets of gaskets when the engine is being dismantled, but retain the old gaskets with a view to using them as a pattern to make a replacement if a new one is not available.

8   When possible, refit nuts, bolts and washers in their location after being removed, as this helps to protect the threads and will also be helpful when reassembling the engine.

9   Retain unserviceable components in order to compare them with the new parts supplied.

## 10  Ancilliary components – removal

1   With the engine separated from the gearbox the externally mounted ancillary components, as given in the following list can be removed with reference to the relevant Chapters of this Manual, where necessary. The removal sequence need not necessarily follow the order given:

    *Thermostat housing and heater pipes – photo*
    *Inlet and exhaust manifolds and carburettor (Chapter 3) – not MG EFi models*
    *Inlet and exhaust manifolds, fuel rail and fuel pressure regulator (Chapter 3) – MG EFi models*
    *Alternator (Chapter 9) – photo*
    *Alternator and engine mounting bracket – photo*
    *Alternator adjustment arm bracket – photo*
    *Fuel pump (Chapter 3) – not MG EFi models*
    *Spark plugs (Chapter 4)*
    *Distributor cap, rotor arm and shield – photos*
    *Oil filler tube – photo*
    *Crankshaft sensor (Chapter 4) – photo*
    *Knock sensor (Chapter 4) – photo*
    *Oil filter (Section 2 of this Chapter)*
    *Clutch assembly (Chapter 5)*
    *Dipstick*

10.1B Remove the alternator

10.1C Remove the engine mounting bracket

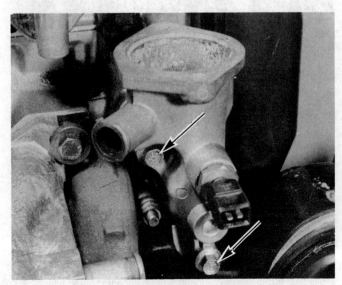

10.1A Thermostat housing retaining bolts (arrowed)

10.1D Remove the alternator adjustment arm bracket

# Are your plugs trying to tell you something?

**Normal.**
Grey-brown deposits, lightly coated core nose. Plugs ideally suited to engine, and engine in good condition.

**Heavy Deposits.**
A build up of crusty deposits, light-grey sandy colour in appearance.
Fault: Often caused by worn valve guides, excessive use of upper cylinder lubricant, or idling for long periods.

**Lead Glazing.**
Plug insulator firing tip appears yellow or green/yellow and shiny in appearance.
Fault: Often caused by incorrect carburation, excessive idling followed by sharp acceleration. Also check ignition timing.

**Carbon fouling.**
Dry, black, sooty deposits.
Fault: over-rich fuel mixture.
Check: carburettor mixture settings, float level, choke operation, air filter.

**Oil fouling.**
Wet, oily deposits. Fault: worn bores/piston rings or valve guides; sometimes occurs (temporarily) during running-in period.

**Overheating.**
Electrodes have glazed appearance, core nose very white – few deposits. Fault: plug overheating. Check: plug value, ignition timing, fuel octane rating (too low) and fuel mixture (too weak).

**Electrode damage.**
Electrodes burned away; core nose has burned, glazed appearance. Fault: pre-ignition. Check: for correct heat range and as for 'overheating'.

**Split core nose.**
(May appear initially as a crack). Fault: detonation or wrong gap-setting technique. Check: ignition timing, cooling system, fuel mixture (too weak).

# WHY DOUBLE COPPER IS BETTER FOR YOUR ENGINE.

Unique Trapezoidal Copper Cored Earth Electrode — 50% Larger Spark Area

Copper Cored Centre Electrode

Champion Double Copper plugs are the first in the world to have copper core in both centre <u>and</u> earth electrode. This innovative design means that they run cooler by up to 100°C – giving greater efficiency and longer life. These double copper cores transfer heat away from the tip of the plug faster and more efficiently. Therefore, Double Copper runs at cooler temperatures than conventional plugs giving improved acceleration response and high speed performance with no fear of pre-ignition.

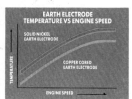

Champion Double Copper plugs also feature a unique trapezoidal earth electrode giving a 50% increase in spark area. This, together with the double copper cores, offers greatly reduced electrode wear, so the spark stays stronger for longer.

 **FASTER COLD STARTING**

 **FOR UNLEADED OR LEADED FUEL**

 **ELECTRODES UP TO 100°C COOLER**

 **BETTER ACCELERATION RESPONSE**

 **LOWER EMISSIONS**

 **50% BIGGER SPARK AREA**

**THE LONGER LIFE PLUG**

**Plug Tips/Hot and Cold.**
Spark plugs must operate within well-defined temperature limits to avoid cold fouling at one extreme and overheating at the other.
Champion and the car manufacturers work out the best plugs for an engine to give optimum performance under all conditions, from freezing cold starts to sustained high speed motorway cruising.
Plugs are often referred to as hot or cold. With Champion, the higher the number on its body, the hotter the plug, and the lower the number the cooler the plug.

**Plug Cleaning**
Modern plug design and materials mean that Champion no longer recommends periodic plug cleaning. Certainly don't clean your plugs with a wire brush as this can cause metal conductive paths across the nose of the insulator so impairing its performance and resulting in loss of acceleration and reduced m.p.g.
However, if plugs are removed, always carefully clean the area where the plug seats in the cylinder head as grit and dirt can sometimes cause gas leakage.
Also wipe any traces of oil or grease from plug leads as this may lead to arcing.

This photographic sequence shows the steps taken to repair the dent and paintwork damage shown above. In general, the procedure for repairing a hole will be similar; where there are substantial differences, the procedure is clearly described and shown in a separate photograph.

First remove any trim around the dent, then hammer out the dent where access is possible. This will minimise filling. Here, after the large dent has been hammered out, the damaged area is being made slightly concave.

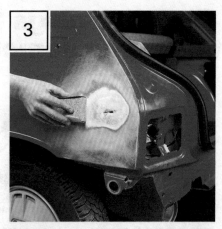

Next, remove all paint from the damaged area by rubbing with coarse abrasive paper or using a power drill fitted with a wire brush or abrasive pad. 'Feather' the edge of the boundary with good paintwork using a finer grade of abrasive paper.

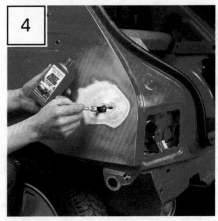

Where there are holes or other damage, the sheet metal should be cut away before proceeding further. The damaged area and any signs of rust should be treated with Turtle Wax Hi-Tech Rust Eater, which will also inhibit further rust formation.

*For a large dent or hole* mix Holts Body Plus Resin and Hardener according to the manufacturer's instructions and apply around the edge of the repair. Press Glass Fibre Matting over the repair area and leave for 20-30 minutes to harden. Then ...

... brush more Holts Body Plus Resin and Hardener onto the matting and leave to harden. Repeat the sequence with two or three layers of matting, checking that the final layer is lower than the surrounding area. Apply Holts Body Plus Filler Paste as shown in Step 5B.

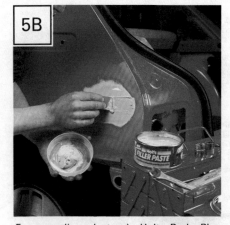

*For a medium dent*, mix Holts Body Plus Filler Paste and Hardener according to the manufacturer's instructions and apply it with a flexible applicator. Apply thin layers of filler at 20-minute intervals, until the filler surface is slightly proud of the surrounding bodywork.

*For small dents and scratches* use Holts No Mix Filler Paste straight from the tube. Apply it according to the instructions in thin layers, using the spatula provided. It will harden in minutes if applied outdoors and may then be used as its own knifing putty.

Use a plane or file for initial shaping. Then, using progressively finer grades of wet-and-dry paper, wrapped round a sanding block, and copious amounts of clean water, rub down the filler until glass smooth. 'Feather' the edges of adjoining paintwork.

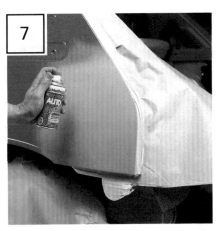

Protect adjoining areas before spraying the whole repair area and at least one inch of the surrounding sound paintwork with Holts Dupli-Color primer.

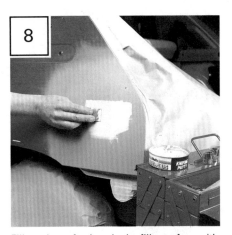

Fill any imperfections in the filler surface with a small amount of Holts Body Plus Knifing Putty. Using plenty of clean water, rub down the surface with a fine grade wet-and-dry paper – 400 grade is recommended – until it is really smooth.

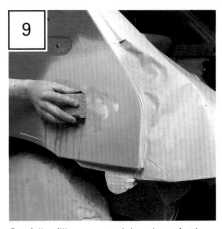

Carefully fill any remaining imperfections with knifing putty before applying the last coat of primer. Then rub down the surface with Holts Body Plus Rubbing Compound to ensure a really smooth surface.

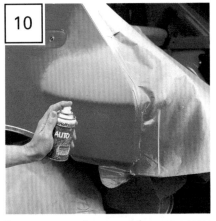

Protect surrounding areas from overspray before applying the topcoat in several thin layers. Agitate Holts Dupli-Color aerosol thoroughly. Start at the repair centre, spraying outwards with a side-to-side motion.

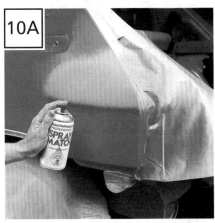

If the exact colour is not available off the shelf, local Holts Professional Spraymatch Centres will custom fill an aerosol to match perfectly.

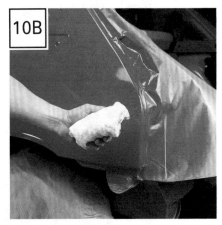

To identify whether a lacquer finish is required, rub a painted unrepaired part of the body with wax and a clean cloth.

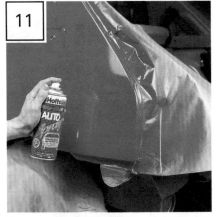

If *no* traces of paint appear on the cloth, spray Holts Dupli-Color clear lacquer over the repaired area to achieve the correct gloss level.

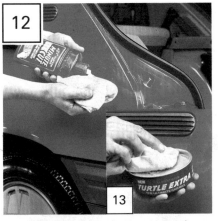

The paint will take about two weeks to harden fully. After this time it can be 'cut' with a mild cutting compound such as Turtle Wax Minute Cut prior to polishing with a final coating of Turtle Wax Extra.

When carrying out bodywork repairs, remember that the quality of the finished job is proportional to the time and effort expended.

10.1E Remove the distributor cap and rotor arm ...

10.1H Remove the crankshaft sensor ...

10.1F ... followed by the shield

10.1I ... and knock sensor

10.1G Remove the oil filler tube

## 11 Timing belt – removal, refitting and adjustment

1 Disconnect the battery negative terminal.

2 If power-assisted steering is fitted, refer to Chapter 10 and remove the pump drivebelt.

3 Refer to Chapter 2 and remove the water pump and alternator drivebelt.

4 Undo the bolts securing the radiator top hose clip and timing belt cover in position. Note the spacer behind the hose clip and lift off the timing belt cover.

5 Using a spanner or socket on the crankshaft pulley bolt, turn the crankshaft until the dimple in the rear edge of the camshaft sprocket is in line with the hole in the timing belt cover backplate. The notch on the crankshaft pulley should also be aligned with the corresponding notch in the deflector bracket below the water pump (photo). In this position the crankshaft is at 90° BTDC with No 1 piston on its compression stroke.

6 Slacken the timing belt tensioner retaining nuts (photo) and move the tensioner away from the engine.

7 Slip the belt off the two sprockets and remove it from the engine.

8 If the timing belt is to be re-used, mark it with an arrow using chalk to indicate its running direction and store it on edge while it is off the engine.

11.5 Pulley timing notch aligned with deflector bracket notch

11.6 Timing belt tensioner retaining nuts (arrowed)

11.9 Dimple on camshaft sprocket aligned with hole in backplate (arrows)

9   Check that the crankshaft pulley notch is still aligned with the deflector bracket notch and the dimple on the camshaft sprocket is still in line with the hole in the timing belt cover backplate (photo).

10  Slip the belt over the two sprockets, push the tensioner against the belt and tighten the two nuts.

11  Turn the crankshaft through two complete revolutions in the normal direction of rotation then realign the pulley and deflector bracket notches.

12  Check that the camshaft sprocket dimple and backplate hole are also still aligned. If not, slacken the tensioner, move the belt one tooth either way as necessary on the camshaft sprocket then retighten the tensioner. Repeat the above procedure until the belt is positioned correctly.

13  The tension of the timing belt can now be accurately adjusted as follows.

14  Attach a spring balance to the timing belt using a suitably shaped length of stiff wire or thin rod, so that the spring balance is running in line with the water pump inlet (Fig. 1.3).

15  Pull the spring balance until the belt is deflected as far as the cast line on the water pump inlet. Observe the reading on the spring balance and compare this with the timing belt tension figures given in the Specifications.

16  If adjustment is necessary, slacken the two tensioner retaining nuts and reposition the tensioner slightly as required. Re-tighten the nuts and check the tension with the spring balance again. Repeat this procedure until the correct tension is obtained.

17  Refit the timing belt cover and top hose support clip.

18  Refit the water pump and alternator drivebelt as described in Chapter 2.

19  If power-assisted steering is fitted, refit the pump drivebelt as described in Chapter 10.

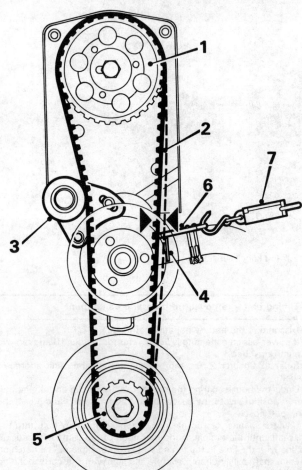

**Fig. 1.3 Timing belt tension adjustment (Sec 11)**

1   Camshaft sprocket
2   Timing belt
3   Timing belt tensioner
4   Cast line on water pump inlet
5   Crankshaft sprocket
6   Shaped wire
7   Spring balance

### 12 Valve clearances – checking and adjustment

**Note:** *To retain the camshaft in position during the checking operation, BL special tool 18G 1301 will be required. If this tool is not available it is possible to make up suitable alternatives as described below and shown in the photos. Read through the entire section first to familiarize yourself with the procedure, and then obtain or make up the special tools before proceeding.*

1   Disconnect the battery negative terminal.

2   Undo the bolts securing the radiator top hose support clip and the timing belt cover in position. Note the spacer behind the clip then lift off the cover.

3   Undo the small bolt securing the timing belt cover backplate to the camshaft cover. Use a socket and extension bar inserted through one of the holes in the camshaft sprocket to undo the bolt.

4   Disconnect the coil HT lead from the centre of the distributor cap then undo the two screws and move the cap aside.

5   Using an Allen key, undo the rotor arm retaining screw and withdraw the rotor arm and shield from the camshaft.

6   If BL special tool 18G 1301 is not available, make up three camshaft retaining brackets to the dimensions shown in Fig. 1.4, using suitable scrap metal and angle iron.

7   Progressively slacken the nine camshaft cover retaining bolts then remove them from their locations.

8   Lift off the camshaft cover and move it clear of the camshaft, but avoid stretching the fuel hoses to the fuel pump. Make sure that the

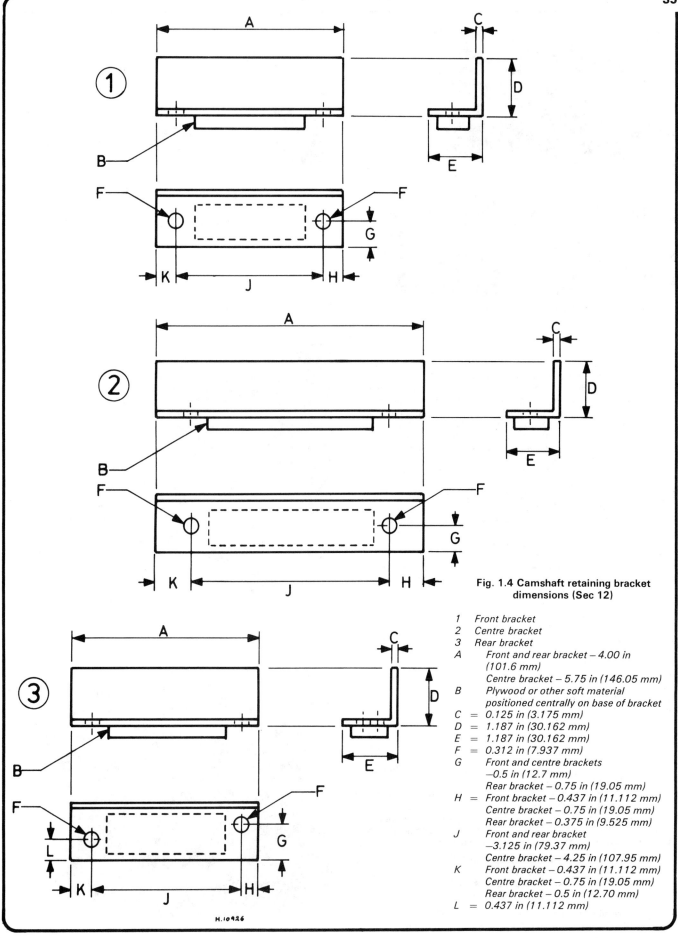

Fig. 1.4 Camshaft retaining bracket
dimensions (Sec 12)

1   Front bracket
2   Centre bracket
3   Rear bracket
A     Front and rear bracket – 4.00 in
      (101.6 mm)
      Centre bracket – 5.75 in (146.05 mm)
B     Plywood or other soft material
      positioned centrally on base of bracket
C  =  0.125 in (3.175 mm)
D  =  1.187 in (30.162 mm)
E  =  1.187 in (30.162 mm)
F  =  0.312 in (7.937 mm)
G     Front and centre brackets
      –0.5 in (12.7 mm)
      Rear bracket – 0.75 in (19.05 mm)
H  =  Front bracket – 0.437 in (11.112 mm)
      Centre bracket – 0.75 in (19.05 mm)
      Rear bracket – 0.375 in (9.525 mm)
J     Front and rear bracket
      –3.125 in (79.37 mm)
      Centre bracket – 4.25 in (107.95 mm)
K     Front bracket – 0.437 in (11.112 mm)
      Centre bracket – 0.75 in (19.05 mm)
      Rear bracket – 0.5 in (12.70 mm)
L  =  0.437 in (11.112 mm)

H.10926

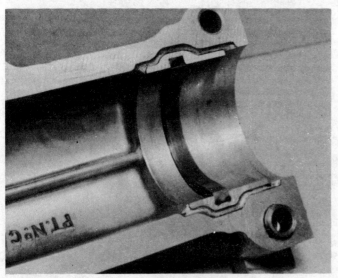

12.8 Camshaft thrust plate location in camshaft cover

12.9 Camshaft retaining brackets in position

camshaft thrust plate is secure in the cover rear journal and is not likely to drop out. If so remove the plate and store it safely (photo).

9 Fit tool 18G 1301 or the three homemade brackets in position over the camshaft bearings and secure with the cover retaining bolts (photo). When in position the brackets should be exerting a light pressure only on the camshaft bearing journals; sufficient to hold the camshaft firmly in place while still allowing it to be rotated.

10 Turn the crankshaft using a socket or spanner on the pulley bolt until Nos 1 and 3 cam lobes are vertical with their heels above the valves. Tighten the bracket retaining bolts so that the camshaft bearings are fully seated in their journals.

11 Using feeler gauges, measure and record the clearance between Nos 1 and 3 cam lobes and their respective tappets (photo).

12 Slacken the bracket retaining bolts, turn the crankshaft until Nos 4 and 7 cam lobes are vertical then tighten the brackets again. Measure and record the clearance then repeat the procedure for Nos 2 and 5 and Nos 6 and 8 cam lobes.

13 Valve clearance adjustment is only necessary if any of the measured clearances are greater than the clearance settings given in the Specifications, or are less than 0.008 in (0.20 mm).

14 If the measured clearances are within tolerance proceed to paragraph 24. If adjustment is necessary, proceed as follows.

15 Slacken the bracket retaining bolts and turn the crankshaft until the dimple on the edge of the camshaft sprocket rear face is aligned with the hole in the timing belt cover backplate. The notch in the crankshaft pulley should now align with the notch in the deflector plate below the water pump. In this position the crankshaft is at 90° BTDC with No 1 piston on its compression stroke.

16 Slacken the timing belt tensioner retaining nuts and slip the belt off the camshaft sprocket.

17 Undo the sprocket retaining bolt and withdraw the sprocket from the camshaft (photo).

18 Undo the retaining bolts and remove the timing belt cover backplate (photo).

19 Remove tool 18G 1301, or the homemade brackets, and lift out the camshaft with its two oil seals.

20 Remove each tappet requiring adjustment in turn and extract the shim from the valve spring cup (or from inside the tappet) (photos). Measure and record the thickness of the shim with a micrometer.

21 If the measured clearance was too small then a thinner shim is required. If the clearance was too large a thicker shim is needed, ie

12.11 Valve clearance measurement

| A+B−C | | = Shim thickness required |
|---|---|---|
| Where: | A = | Measured clearance |
| | B = | Thickness of existing shim |
| | C = | Specified valve clearance |

Shims are available in the thickness range from 0.091 in (2.32 mm) to 0.125 in (3.17 mm), in increments of 0.002 in (0.05 mm).

12.17 Undo the camshaft sprocket retaining bolt

12.18 Timing belt cover backplate retaining bolts (arrowed)

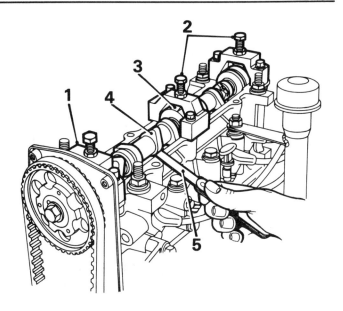

**Fig. 1.5 Valve clearance adjustment (Sec 12)**

1 *Tool 18G 1301*
2 *Tool compression bolt*
3 *Tool pressure pad*

4 *Camshaft lobe vertical for clearance checking*
5 *Feeler gauge*

22 Having selected each required new shim, place it in the spring cup on top of the valve and refit the tappet in its original bore.
23 Liberally lubricate the camshaft bearing journals with engine oil and place the camshaft with its oil seals in position.
24 Remove all traces of old sealant from the camshaft cover and cylinder head and ensure that the mating faces are clean and dry.
25 Apply a bead of RTV silicone sealant to the camshaft cover mating face. Ensure that the camshaft thrust plate is in place then fit the cover and retaining bolts. Tighten the bolts progressively and in a diagonal sequence to the specified torque.
26 Refit the timing belt cover backplate and the camshaft sprocket. Align the sprocket dimple with the hole in the backplate.
27 Ensure that the crankshaft is still at the 90° BTDC position then slip the timing belt over the camshaft sprocket.
28 Adjust the tension of the timing belt as described in Section 11.
29 Refit the rotor arm and shield, the distributor cap and the coil HT lead.
30 Refit the timing belt cover and top hose support clip then reconnect the battery.

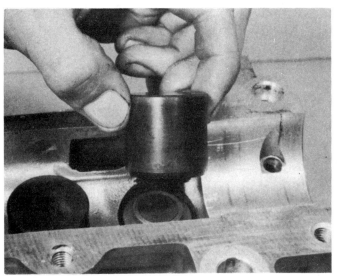

12.20A Lift out the tappet ...

**13 Camshaft and tappets – removal and refitting**

1 Disconnect the battery negative terminal.
2 Disconnect the coil HT lead from the centre of the distributor cap, undo the two screws and move the cap to one side.
3 Using an Allen key, undo the rotor arm retaining screw and withdraw the rotor arm and shield from the camshaft.
4 Undo the bolts securing the radiator top hose support clip and the timing belt cover in position. Note the spacer behind the clip then lift off the cover.
5 Turn the crankshaft using a socket or spanner on the pulley bolt until the dimple on the edge of the camshaft sprocket rear face is aligned with the hole in the timing belt cover backplate. The notch on the crankshaft pulley should now be in line with the corresponding notch in the deflector plate below the water pump. In this position the crankshaft is at 90° BTDC with No 1 piston on its compression stroke.
6 Slacken the two timing belt tensioner retaining nuts and slip the belt off the camshaft sprocket.
7 Undo the camshaft sprocket retaining bolt and withdraw the sprocket from the camshaft.

12.20B ... and extract the shim

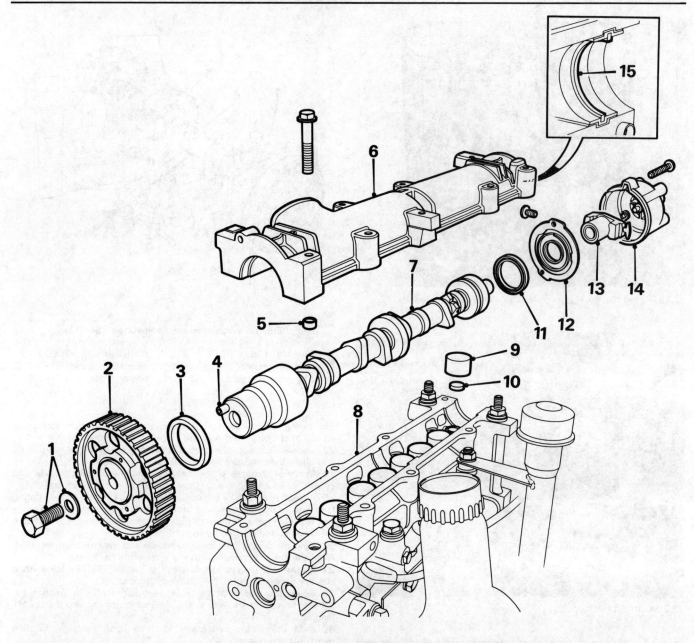

**Fig. 1.6 Exploded view of the camshaft and tappets (Sec 13)**

| | | | | | |
|---|---|---|---|---|---|
| 1 | Sprocket retaining bolt and washer | 4 | Dowel | 8 | Cylinder head |
| 2 | Camshaft sprocket | 5 | Cover dowel | 9 | Tappet |
| 3 | Oil seal | 6 | Camshaft cover | 10 | Shim |
| | | 7 | Camshaft | 11 | Oil seal |

| | |
|---|---|
| 12 | Rotor arm shield |
| 13 | Rotor arm |
| 14 | Distributor cap |
| 15 | Camshaft thrust plate |

8   Undo the timing belt cover backplate retaining bolts and remove the backplate.

9   Progressively slacken the nine camshaft cover retaining bolts then remove them from their locations.

10   Lift off the camshaft cover and move it clear of the camshaft, but avoid stretching the fuel hoses to the fuel pump. Make sure that the camshaft thrust plate is secure in the cover rear journal and is not likely to drop out. If so remove the plate and store it safely.

11   Lift out the camshaft and slide off the oil seals.

12   Lift out each tappet with its adjusting shim and keep them in strict order of removal.

13   To refit the camshaft, liberally lubricate the tappet bores and the camshaft journals.

14   Place each shim in its respective valve spring cup then place the tappet over it.

15   Lubricate the oil seal lips and carefully slide them into place on the

camshaft with their open face inwards. Locate the camshaft with its oil seals in the cylinder head.

16   The valve clearances should now be checked and if necessary adjusted using the procedure described in Section 12, which also includes the camshaft refitting procedures.

## 14 Cylinder head – removal and refitting (all models except MG EFi)

1   Disconnect the battery negative terminal.

2   Jack up the front of the car and support it on axle stands.

3   Drain the cooling system as described in Chapter 2 and remove the air cleaner assembly as described in Chapter 3.

4   If power-assisted steering is fitted, slacken the pump adjustment

and pivot bolts, move the pump towards the engine and remove the drivebelt. Release the fluid hoses from the sump bracket and undo the bolts securing the pump mounting bracket to the cylinder head. Lower the pump, with hoses still attached and suspend it using string or wire from the front jacking bracket towing hook.

5 Slacken the screw securing the accelerator cable to the connector on the carburettor linkage. Withdraw the cable from its support bracket and move it aside.

6 Disconnect the two wires at the fuel shut off valve on the carburettor and the multi-plug at the carburettor stepper motor.

7 Disconnect the vacuum hoses at the manifold take off, at the carburettor and at the carburettor vacuum switch.

8 Undo the servo vacuum hose banjo union bolt, withdraw the union and recover the two copper washers. Move the vacuum hose to one side.

9 Disconnect the fuel inlet hose at the fuel pump and plug the end of the hose to prevent fuel spillage.

10 Disconnect the heater hoses at the two heater pipes and at the inlet manifold.

11 Disconnect the hoses at the thermostat housing end of the heater pipes, remove the pipe support bracket retaining bolts and withdraw the heater pipe assembly.

12 Disconnect the radiator top hose and expansion tank hose at the water outlet elbow. Release the top hose support clip on the timing belt cover and move the hoses clear.

13 Disconnect the coolant thermister wiring multi-plug from the thermister on the thermostat housing.

14 Disconnect the inlet manifold heater wire and the two wires at the inlet manifold temperature sensor.

15 Disconnect the ignition coil HT lead from the centre of the distributor cap.

16 Disconnect the float chamber overflow hose at the carburettor and undo the bolt securing the overflow pipe to the inlet manifold. Remove the pipe and hose.

17 Disconnect the crankcase breather hose at the carburettor.

18 Undo the retaining nuts and bolts and remove the oil filler tube from the cylinder block and head.

19 Undo the bolt securing the dipstick tube to the cylinder head.

20 Slip the clutch cable out of its steady bracket, undo the retaining bolt and remove the bracket.

21 From under the car undo the nuts securing the exhaust downpipes to the manifold, separate the flange joint and recover the gasket.

22 Undo the retaining bolts and remove the heat shield from behind the alternator. Release the clip and disconnect the alternator multi-plug.

23 Disconnect the oil pressure switch wire at the switch on the oil pump housing.

24 From above, lift up the complete engine wiring harness and move it clear of its clips and away from the cylinder head. Undo the manifold steady bar upper retaining bolts.

25 Remove the timing belt cover, slacken the two tensioner retaining nuts and slip the timing belt off the camshaft sprocket.

26 Progressively slacken the cylinder head retaining bolts in the reverse sequence of that shown in Fig. 1.7. Remove the bolts when all have been slackened.

27 Lift the cylinder head, complete with manifolds and carburettor off the engine and recover the cylinder head gasket. If the head is stuck, tap it upwards with a hide or plastic mallet to free it. Do not lever between the head and block face or the mating surface may be damaged.

28 Before refitting the cylinder head, turn the crankshaft so that the notch in the pulley is aligned with the corresponding notch in the deflector bracket below the water pump. Similarly turn the camshaft sprocket so that the dimple on the edge of the sprocket rear face is aligned with the hole in the timing belt cover backplate.

29 Ensure that the cylinder block and head mating surfaces are perfectly clean and dry. Where there has been an oil leak problem from the cylinder head joint face, reference should be made to Section 4 of Chapter 12, *Cylinder head gasket (all models) – oil leak*.

30 Locate a new cylinder head gasket on the cylinder block. *Do not use any form of jointing compound on the cylinder head gasket.* Lower the cylinder head unit onto the gasket and locate the retaining bolts. Working in the sequence shown in Fig. 1.7, initially tighten the bolts to the specified Stage 1 torque setting, then to the Stage 2 torque setting. Finally tighten through a further 60° (1/6 of a turn), or to the Stage 3 torque setting, whichever comes first.

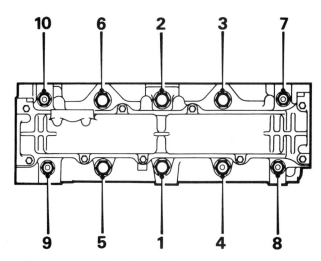

Fig. 1.7 Cylinder head bolt tightening sequence (Sec 14)

31 Refit all the wiring, pipes, hoses and components to the cylinder head using the reverse sequence of removal. Adjust the timing belt tension as described in Section 11.

32 Adjust the power-assisted steering pump drivebelt, if fitted, as described in Chapter 10. Adjust the accelerator cable as described in Chapter 3.

33 Refit the air cleaner as described in Chapter 3 and refill the cooling system as described in Chapter 2.

34 After refitting, run the engine for at least 15 minutes or drive the car for approximately 5 miles (8 km). Allow the engine to cool and then working in the sequence shown in Fig. 1.7, slacken each cylinder head bolt in turn and then re-tighten to the specified torque wrench setting.

## 15 Cylinder head – removal and refitting (MG EFi models)

**Note:** *Refer to Chapter 3 for details of all operations relating to the electronic fuel injection system.*

1 Depressurise the fuel system then disconnect the battery negative terminal.

2 Jack up the front of the car and support it on axle stands.

3 Drain the cooling system as described in Chapter 2.

4 If power-assisted steering is fitted, slacken the pump adjustment and pivot bolts, move the pump towards the engine and remove the drivebelt. Release the fluid hoses from the sump bracket and undo the bolts securing the pump mounting bracket to the cylinder head. Lower the pump, with hoses still attached and suspend it using string or wire from the front jacking bracket towing hook.

5 Undo the brake servo vacuum hose banjo union bolt at the inlet manifold. Withdraw the bolt and union, recover the two copper washers and move the hose aside.

6 Slacken the clip and disconnect the intake air hose from the throttle housing.

7 Disconnect the air and vacuum hoses at the rear of the throttle housing and the water and crankcase breather hoses from below it.

8 Disconnect the air valve stepper motor and throttle potentiometer wiring multi-plugs.

9 Slacken the screw securing the accelerator cable to the connector on the throttle housing linkage. Withdraw the cable from its support bracket and place aside.

10 Disconnect the wiring multi-plugs at the fuel injectors and at the fuel temperature switch.

11 Disconnect the fuel feed hose at the fuel rail and the fuel return hose at the pressure regulator. Plug the hoses after removal.

12 Disconnect the heater hoses at the heater pipes.

13 Disconnect the radiator top hose and expansion tank hose at the water outlet elbow. Release the top hose support clip on the timing belt cover and move the hoses clear.

14 Disconnect the ignition coil HT lead from the centre of the distributor cap.

15 The remainder of the procedure is the same as for vehicles with carburettor induction. Refer to Section 14 and carry out the procedures described in paragraphs 18 to 34.

## 16 Cylinder head – overhaul

1　With the cylinder head on the bench, remove the fuel pump, manifolds, spark plugs, distributor cap, rotor arm, camshaft and tappets as described in the relevant Sections and Chapters of this Manual.

2　To remove the valves, compress each spring in turn with a universal valve spring compressor until the two halves of the collet can be removed. Release the compressor and lift away the spring top cup, valve spring, oil seal, valve spring seat and the valve (photos).

3　If, when the valve spring compressor is screwed down, the valve spring top cup refuses to free and expose the split collet, do not continue to screw down on the compressor, but gently tap the top of the tool directly over the cup with a light hammer. This should free the cup. To avoid the compressor jumping off the valve retaining cup when it is tapped, hold the compressor firmly in position.

4　It is essential that the valves are kept in their correct order unless they are so badly worn or burnt that they are to be renewed. If they are going to be refitted, place them in their correct sequences along with the tappets and shims previously removed. Also keep the valve springs, cups and collets in the same order.

5　With the valves removed from the cylinder head, scrape away all traces of carbon from the valves and the combustion chambers and ports in the cylinder head using a knife and suitable scraper.

6　Examine the heads of the valves for signs of cracking, burning away or pitting of the valve face or the edge of the valve head. The valve seats in the cylinder head should also be examined for the same signs. Usually it is the valve that deteriorates first, but if a bad valve is not rectified the seat will suffer, and this is more difficult to repair. If the valve face and seat are deeply pitted, or if the valve face is concave where it contacts the seat, it will be necessary to have the valve refaced and the seat recut by a BL dealer or motor engineering specialist. It is worth considering having this work done in any case if the engine has covered a high mileage. A little extra time and money spent ensuring that the cylinder head and valve gear are in first class

condition will make a tremendous difference to the performance and economy of the engine after overhaul. If any of the valves are cracked or burnt away they must be renewed. Any similar damage that may have occurred to the valve seats can be repaired by renewing the seat. However, this is a job that can only be carried out by a specialist.

7　Another form of valve wear can occur on the stem where it runs in the guide in the cylinder head. This can be detected by trying to rock the valve from side to side. if there is any movement at all it is an indication that the valve stem or guide is worn. Check the stem first with a micrometer at points along and around its length. If they are not within the specified size, new valves will probably solve the problem. If the guides are worn, however, they will need renewing. The valve seats will also need recutting to ensure they are concentric with the stems. This work should be given to your BL dealer or local engineering works.

8　Assuming that the valve faces and seats are only lightly pitted, or that new valves are to be fitted, the valves should be lapped into their seats. This may be done by placing a smear of fine carborundum paste on the edge of the valve and, using a suction type valve holder, lapping the valve in situ. This is done with a semi-rotary action, rotating the handle of the valve holder between the hands and lifting it occasionally to redistribute the traces of paste. As soon as a matt grey unbroken line appears on both the valve face and seat, the valve is 'ground in'.

9　When all work on the cylinder head and valves is complete, it is essential that all traces of carbon dust and grinding paste are removed. This should be done by thoroughly washing the components in paraffin, or a suitable engine cleaner and blowing out with a jet of air. If particles of carbon or grinding paste should work their way into the engine, they would cause havoc with bearings or cylinder walls.

10　With the valves and valve seats suitably prepared and in their correct order, start with No 1 cylinder.

11　Lubricate the valve stem with oil and insert the valve into its guide.

12　Fit a new oil seal well lubricated with engine oil.

13　Fit the valve spring seat, valve spring and valve spring cup over the valve stem.

14　Using a valve spring compressor tool, compress the valve spring

16.2A Compress the valve spring and remove the collets

16.2B Remove the spring top cup ...

16.2C ... followed by the spring ...

16.2D ... oil seal

16.2E ... and spring seat ...

16.2F ... then withdraw the valve

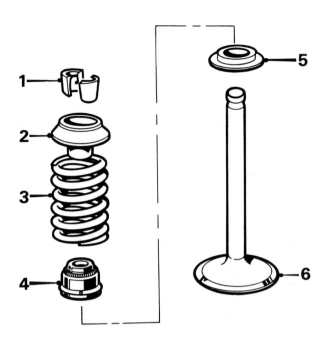

**Fig. 1.8 Valve components (Sec 16)**

| | |
|---|---|
| 1   Collets | 4   Oil seal |
| 2   Spring top cup | 5   Spring seat |
| 3   Spring | 6   Valve |

until the split collets can be slid into position, then carefully release the valve spring compressor in order not to displace the collets.

15  Refit the other valves in the same way. When they are all fitted, tap the end of each valve stem using a plastic-faced hammer or a hammer with a block of hardwood interposed. This will settle the valve components ready for checking the valve clearances.

16  Refit the components listed in paragraph 1 using the reverse sequence of removal and with reference to the applicable Sections and Chapters of this manual.

## 17  Sump – removal and refitting

1   Jack up the front of the car and support it on axle stands.

2   Disconnect the battery negative terminal then drain the engine oil as described in Section 2.

3   Disconnect the crankcase breather hose from the oil separator.

4   If power-assisted steering is fitted, release the fluid pipes from the support brackets on the sump. On MG EFi models, disconnect the sump oil level sensor wires at their connectors (photo).

5   Undo the sump retaining bolts and recover the carburettor float chamber drain tube bracket (photo).

6   Withdraw the sump from the crankcase, tapping it from side to side with a hide or plastic mallet if it is stuck.

7   If required, undo the two bolts securing the oil pick-up pipe to the crankcase and the bolt securing the pipe bracket to the main bearing cap (photo). Recover the O-ring from the pick-up pipe flange.

8   Clean the sump thoroughly and remove all traces of old gasket from the sump and crankcase mating faces.

9   If removed, refit the pick-up pipe using a new O-ring and tighten the bolts to the specified torque.

10  Apply jointing compound to the sump and crankcase faces, place a new gasket in position and refit the sump. Tighten the retaining bolts progressively to the specified torque.

11  Where applicable reconnect the oil level sensor and refit the power-assisted steering fluid pipes to their support brackets.

12  Make sure that the oil drain plug is tight then lower the car to the ground.

13  Refit the crankcase breather hose and fill the engine with oil.

17.4 Disconnect the sump oil level sensor wires

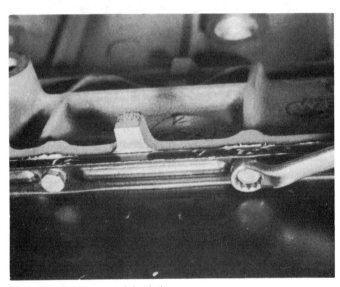

17.5 Undo the sump retaining bolts

17.7 Oil pick-up pipe attachments (arrowed)

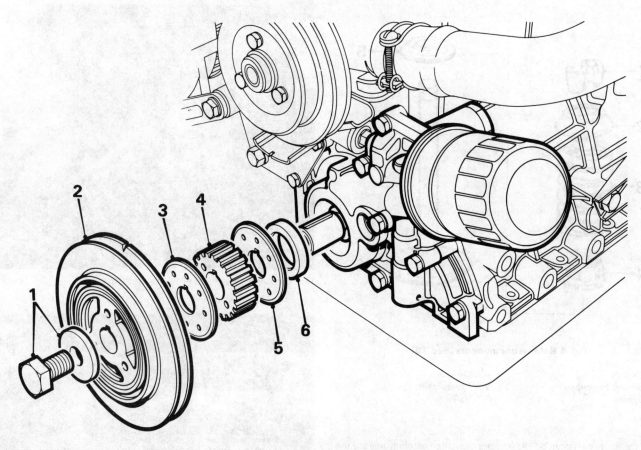

**Fig. 1.9 Crankshaft sprocket details (Sec 18)**

| 1 | Pulley bolt and washer | 3 | Outer flange | 5 | Inner flange |
|---|---|---|---|---|---|
| 2 | Crankshaft pulley | 4 | Sprocket | 6 | Oil seal |

## 18 Oil pump and housing – removal and refitting

1    Disconnect the battery negative terminal.

2    Put the transmission in gear and firmly engage the handbrake. Using a socket and long handle, slacken the crankshaft pulley retaining bolt. If the engine is not in the car, engage a small strip of angle iron between the flywheel teeth and one of the adaptor plate dowels to prevent rotation of the crankshaft. Return the transmission to neutral.

3    Refer to Section 11 and remove the timing belt.

4    Unscrew the crankshaft pulley bolt and withdraw the pulley. Carefully lever it off using two screwdrivers if it is tight (photo).

5    Withdraw the outer flange, crankshaft sprocket and inner flange (photos). If the sprocket is tight, lever between the sprocket and inner flange carefully using two screwdrivers.

6    Disconnect the oil pressure switch wire at the switch on the front of the oil pump housing.

7    Undo the pump housing retaining bolts (photo) and withdraw the assembly from the crankshaft and crankcase.

8    If required the oil filter and oil pressure switch can be removed by unscrewing them from the pump housing.

9    Ensure that all traces of old gasket are removed from the pump housing and crankcase then place a new gasket in position (photo).

10   Apply some grease to the lips of the crankshaft front oil seal and refit the pump housing. Ease the seal carefully over the crankshaft.

11   Fit the retaining bolts and tighten them to the specified torque.

12   Refit the oil pressure switch and a new oil filter, if removed, and reconnect the switch wire.

13   Refit the inner flange, crankshaft sprocket and outer flange, followed by the pulley and retaining bolt. Tighten the bolt finger tight only at this stage.

14   Refit and tension the timing belt using the procedure described in Section 11.

15   Using the same method as for removal, tighten the pulley retaining bolt to the specified torque.

16   Finally reconnect the battery negative terminal.

18.4 Undo the crankshaft pulley bolt and remove the pulley

18.5A Withdraw the outer flange ...

18.7 Undo the pump housing retaining bolts

18.5B ... sprocket ...

18.9 Fit the housing using a new gasket

18.5C ... and inner flange

## 19  Oil pump – overhaul

1    With the pump housing removed from the engine, scribe a line on the pump housing and pump backplate to ensure that the backplate is refitted in its original position.

2    Using a 3 mm Allen key, remove the backplate securing bolts and lift off the backplate (photo).

3    Remove the inner and outer rotors (photos).

4    Remove the split pin that retains the pressure relief valve, then remove the plug from the housing by pushing down on the relief valve spring with a screwdriver through the oil return hole. Collect the spring and valve plunger. Discard the plug O-ring seal (photo).

5    Remove the front crankshaft oil seal from the oil pump housing (photo).

6    Thoroughly clean all the component parts then check the rotor endfloat and lobe clearances as follows.

7    Fit the outer rotor in the housing and, using a feeler gauge, check the clearance between the outer rotor and the pump body (photo).

8    Fit the inner rotor and measure the inner and outer rotor lobe clearances (photo).

9    Measure the outer rotor endfloat between rotor and housing.

10   If any of the clearances exceed the limits given in the Specifications, the pump must be renewed.

19.2 Remove the backplate securing bolts

19.3A Remove the inner ...

19.3B ... and outer rotors

19.4 Oil pressure relief valve components

19.5 Remove the crankshaft front oil seal

19.7 Checking outer rotor to body clearance ...

19.8 ... and rotor lobe clearance

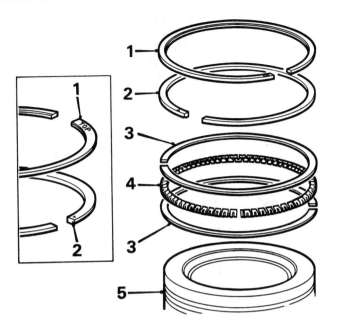

**Fig. 1.10 Piston ring fitting details (Sec 20)**

*1   Top compression ring*          *4   Expander*
*2   Second compression ring*       *5   Piston*
*3   Oil control rails*

11 Check the free length of the pressure relief valve spring. If its length differs from the dimension given in the Specifications, renew the spring.
12 Reassembly is the reverse of the dismantling procedure. Liberally lubricate the oil pump components with engine oil, fit a new O-ring on the pressure relief valve plug and a new crankshaft front oil seal in the housing.

## 20 Pistons and connecting rods – removal and refitting

1   Remove the cylinder head, the sump and the oil pick-up pipe, as described in earlier Sections of this Chapter.
2   Note that the pistons are marked on the top with the word FRONT, or by an arrow or a groove which must face towards the crankshaft pulley end of the engine.
3   Check that the big-end bearing caps and connecting rods have identification marks. If not, suitably mark them to ensure that the correct end caps are fitted to the correct connecting rods and the correct connecting rods are fitted in their respective cylinder bores.
4   Remove the big-end cap securing nuts and put them to one side in the order in which they were removed.
5   Remove the big-end caps, taking care to keep them in the right order. Ensure that the bearing shells are kept with their respective big-end caps unless the bearings are to be renewed.
6   If the big-end caps are difficult to remove, they may be gently tapped with a soft-faced hammer.
7   To remove the shell bearings, press the bearing opposite the notch in both the connecting rod and its cap, and the bearing shell will slide out easily.
8   Push the piston and connecting rod assemblies upwards and withdraw them from the top of the cylinder block.
9   If the pistons are to be separated from the connecting rods, leave this work to a BL dealer or engineering works. The gudgeon pin is an interference fit in the connecting rod and special tools including a press are needed for removal.
10 To remove the piston rings, slide them carefully over the top of the piston, taking care not to scratch the aluminium alloy of the piston. Never slide them off the bottom of the piston skirt. It is very easy to break piston rings if they are pulled off roughly so this operation should be done with extreme caution. It is helpful to use an old 0.020 in (0.5 mm) feeler gauge to facilitate their removal as follows.
11 Lift one end of the piston ring to be removed out of its groove and insert the end of the feeler gauge under it.
12 Turn the feeler gauge slowly round the piston; as the ring comes out of its groove it rests on the land above. It can then be eased off the

piston with the feeler gauge stopping it from slipping into any empty grooves.
13 If new pistons or piston rings are being fitted to old bores, it is essential to roughen the cylinder bore walls slightly with medium grit emery cloth to allow the rings to bed in. Do this with a circular up-and-down action to produce a criss-cross pattern on the cylinder bore walls. Make sure that the bearing journal on the crankshaft is protected with masking tape during this operation. Thoroughly clean the bores with a paraffin-soaked rag and dry with a lint-free cloth. Remove the tape from the crankshaft journals and clean them also.
14 Check that the piston ring grooves and oilways are thoroughly clean and unblocked. Piston rings must always be fitted over the head of the piston and *never* from the bottom.
15 The easiest method to use when fitting rings is to position two feeler blades on either side of the piston and slide the rings down over the blades. This will stop the rings from dropping into a vacant ring groove. When the ring is adjacent to its correct groove, slide out the feeler blades and the ring will drop in.
16 The procedure for fitting the rings is as follows. Start by sliding the bottom rail of the oil control ring down the piston and position it below the bottom piston ring groove. Fit the oil control expander into the bottom ring groove and then move the bottom oil control ring rail into the bottom groove. Now fit the top oil control ring rail into the bottom groove. Make sure that the ends of the expander are butting together and not overlapping Position the gaps of the two rails and the expander at 90° to each other.
17 Fit the second compression ring to its groove with the step towards the gudgeon pin and the word TOP or the letter T facing the top of the piston.
18 Fit the chrome plated top compression ring to its groove with the word TOP or the letter T facing the top of the piston.
19 When all the piston rings are in place, set the ring gaps of the compression rings at 90° to each other and away from the thrust side of the piston.
20 Wipe the cylinder bores clean with a clean rag.
21 The pistons, complete with connecting rods, must be fitted to their bores from the top of the block.
22 Check that the two upper piston ring gaps are spaced at 90° to each other.
23 Check that the piston is the correct one for the cylinder bore and

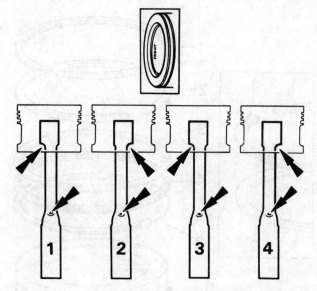

**Fig. 1.11 Connecting rod offset and oil squirt hole arrangement (Sec 20)**

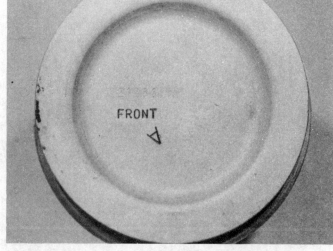

20.26B ... with the word 'front' towards the crankshaft pulley

that the small-end offset of the connecting rod is correct for the bore; see Fig. 1.11.

24  Lubricate the cylinder bore and piston with clean engine oil.

25  Fit a universal piston ring compressor over the rings. A large diameter worm-drive hose clip will serve as a ring compressor if the proper tool is not available.

26  Insert the first piston into its bore making sure that the front of the piston (marked on top with the word FRONT or an arrow) is towards the crankshaft pulley end of the engine (photos).

27  Slide the assembly down the bore until the bottom of the piston ring compressor rests on the cylinder block face. Now gently, but firmly, tap the piston through the compressor using a block of wood or the handle of a mallet.

28  Wipe the connecting rod and bearing shell then fit the shell to the rod with its tag engaged with the notch in the rod.

29  Lubricate the crankpin journals with engine oil and draw the rod down onto its journal.

30  Fit the bearing shell to the connecting rod cap then refit the cap to the rod. Note that the bearing shell notches in the cap and rod must both be on the same side (photos).

31  Refit the cap retaining nuts and tighten them to the specified torque (photo).

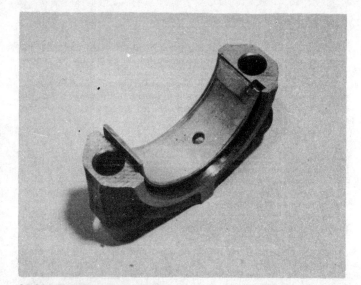

20.30A Fit the bearing shell ...

20.26A Insert the piston and connecting rod into its bore ...

20.30B ... then refit the cap to the rod

20.31 Tighten the nuts to the specified torque

21.14 Fit the thrust washers to the main bearing

32  Repeat the fitting procedure for the remaining piston and connecting rod assemblies.
33  Refit the oil pick-up strainer, sump and cylinder head as described in earlier Sections of this Chapter.

## 21  Crankshaft and main bearings – removal and refitting

1    With the engine removed from the car as described in Sections 6 and 7 and with all the components removed from it, as described in earlier Sections, the crankshaft can be removed as follows.
2    Note that the main bearing caps Nos 2, 3 and 4 have their numbers, together with arrows, cast on the front face of the caps.
3    Undo by one turn at a time the bolts that secure the main bearing caps.
4    Lift away each main bearing cap and the bottom half of each bearing shell, taking care to keep the bearing shell with the right cap.
5    When removing the front and rear main bearing caps, note the cork sealing joints.
6    When removing the centre main bearing cap, note the bottom semi-circular halves of the thrust washers, one half located on each side of the main bearing. Lay them, with the centre bearing cap, along the correct side.
7    Lift out the crankshaft followed by the bearing shell upper halves and the thrustwashers.
8    Ensure that the crankcase and crankshaft are thoroughly clean and that all oilways are clear. If possible blow the drillings out with compressed air and then inject clean engine oil through them to ensure they are clear.
9    Fit the five upper halves of the main bearing shells to their location in the crankcase after wiping the location clean.
10  Note that on the back of each bearing is a tab which engages in locating grooves in either the crankcase or the main bearing cap housings.
11  If new bearings are being fitted, carefully clean away all traces of the protective grease with which they are coated.
12  With the five upper bearing shells securely in place, wipe the lower bearing cap housings and fit the five lower shell bearings to their caps, ensuring that the right shell goes into the right cap if the old bearings are being refitted.
13  Wipe the recesses either side of the centre main bearing which locate the upper halves of the thrust washers.
14  Introduce the upper halves of the thrust washers (the halves without tabs) into their grooves either side of the centre main bearing with their oil grooves facing outwards (photo).
15  Generously lubricate the crankshaft journals and the upper and lower main bearing shells and carefully lower the crankshaft into position. Make sure that it is the right way round (photos).

21.15A Lubricate the bearing shells ...

21.15B ... and lower the crankshaft into place

16 Fit the main bearing caps into position, ensuring that they locate properly. The mating surfaces must be spotlessly clean or the caps will not seat correctly.

17 When refitting the centre main bearing cap, ensure that the thrust washers, generously lubricated, are fitted with their oil grooves facing outwards and the locating tab of each washer is in the slot in the bearing cap (photo).

21.20 Tighten the bolts to the specified torque

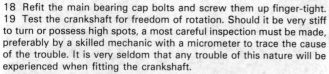

21.17 Fit the centre main bearing cap and thrustwashers

18 Refit the main bearing cap bolts and screw them up finger-tight.
19 Test the crankshaft for freedom of rotation. Should it be very stiff to turn or possess high spots, a most careful inspection must be made, preferably by a skilled mechanic with a micrometer to trace the cause of the trouble. It is very seldom that any trouble of this nature will be experienced when fitting the crankshaft.
20 Tighten the main bearing cap bolts to the specified torque and recheck the crankshaft for freedom of rotation (photo).
21 Using a screwdriver between one crankshaft web and main bearing cap, lever the crankshaft forwards and check the endfloat using feeler gauges. This should be as shown in the Specifications. If excessive, new thrust washers must be fitted (photo).

21.21 Checking crankshaft endfloat

## 22 Flywheel – removal and refitting

1 With the engine removed from the car and separated from the gearbox, remove the clutch assembly as described in Chapter 5.
2 Knock back the tab of the locking plate and undo the six flywheel retaining bolts. To prevent the flywheel turning, lock the ring gear teeth using a small strip of angle iron engaged in the teeth and against the adaptor plate dowel (photos).
3 Lift off the locking plate then withdraw the flywheel from the crankshaft (photo).
4 Refitting is the reverse sequence of removal. Tighten the bolts to the specified torque then bend over the tabs of a new locking plate.

## 23 Gearbox adaptor plate – removal and refitting

1 Remove the flywheel as described in Section 22.
2 Undo the two bolts securing the crankshaft sensor to the adaptor plate and withdraw the sensor.
3 Undo the bolts securing the adaptor plate to the cylinder block noting that four of these are the socket-headed TORX type (photo).
4 Remove the adaptor plate from the cylinder block.
5 The crankshaft rear oil seal should be renewed as a matter of course. Tap out the old seal and fit a new one with its open side

22.2A Knock back the locking plate tabs

22.2B Angle iron strip to lock the flywheel ring gear

22.3 Remove the locking plate

23.3 Gearbox adaptor plate bolts

towards the engine. Tap the seal into place using a block of wood or the old seal.

6    Refitting the adaptor plate is the reverse sequence of removal, but tighten the bolts to the specified torque.

## 24 Engine components – examination and renovation

### Crankshaft

Inspect the main bearing journals and crankpins. If there are any scratches or score marks then the shaft will need regrinding. Such conditions will nearly always be accompanied by similar deterioration in the matching bearings shells. ·

Each bearing journal should also be round and can be checked with a micrometer or caliper gauge around the periphery at several points. If there is more than 0.001 in (0.025 mm) of ovality, regrinding is necessary.

A main BL agent or motor engineering specialist will be able to decide to what extend regrinding is necessary, and also supply the special undersize shell bearing to match whatever may need grinding off.

Before taking the crankshaft for regrinding, also check the cylinder bores and pistons, as it may be advantageous to have the whole engine done at the same time.

### Main and big-end bearings

With careful servicing and regular oil and filter changes, bearings will last for a very long time, but they can still fail for unforeseen reasons. With big-end bearings the indication is a regular rhythmic loud knocking from the crankcase. The frequency depends on engine speed and is particularly noticeable when the engine is under load. This symptom is accompanied by a fall in oil pressure, although this is not normally noticeable unless an oil pressure gauge is fitted. Main bearing failure is usually indicated by serious vibration, particularly at higher engine revolutions, accompanied by a more significant drop in oil pressure and a rumbling noise.

Bearing shells in good condition have bearing surfaces with a smooth even matt silver/grey colour all over. Worn bearings will show patches of a different colour where the bearing metal has worn away and exposed the underlay. Damaged bearings will be pitted or scored. It is always well worthwhile fitting new shells as their cost is relatively low. If the crankshaft is in good condition it is merely a question of obtaining another set of standard size. A reground crankshaft will need new bearing shells as a matter of course.

### Cylinder bores

A new cylinder is perfectly round and the walls parallel throughout its length. The action of the piston tends to wear the walls at right angles to the gudgeon pin due to side thrust. This wear takes place principally on that section of the cylinder swept by the piston rings.

It is possible to get an indication of bore wear by removing the cylinder head with the engine still in the car. With the piston down in the bore, first signs of wear can be seen and felt just below the top of the bore where the top piston ring reaches, and there will be a noticeable lip. If there is no lip it is fairly reasonable to assume that bore wear is not severe and any lack of compression or excessive oil consumption is due to worn or broken piston rings or pistons.

If it is possible to obtain a bore measuring micrometer, measure the bore in the thrust plane below the lip and again at the bottom of the cylinder in the same plane. If the difference is more than 0.006 in (0.15 mm), a rebore is necessary. Similarly, a difference of 0.006 in (0.15 mm) or more between two measurements of the bore diameter taken at right angles to each other is a sign of excessive ovality, calling for a rebore.

Any bore which is significantly scratched or scored will need reboring. This symptom usually indicates that the piston or rings are also damaged. Even in the event of only one cylinder in need of reboring, it will still be necessary for all four to be bored and fitted with new oversize pistons and rings. A motor engineering specialist will be able to rebore the cylinders and supply the necessary matched pistons. If the crankshaft is also undergoing regrinding, it is a good idea to let the same firm renovate and reassemble the crankshaft and pistons to the block. A reputable firm normally gives a guarantee for such work.

## Pistons and piston rings

If the old pistons are to be refitted, carefully remove the piston rings and then thoroughly clean them. Take particular care to clean out the piston ring grooves. Do not scratch the aluminium in any way. If new rings are to be fitted to the old pistons, then the top ring should be of the stepped type so as to clear the ridge left above the previous top ring. If a normal but oversize new ring is fitted, it will hit the ridge and break, because the new ring will not have worn in the same way as the old.

Before fitting the rings on the pistons, each should be inserted approximately 3 in (76 mm) down the cylinder bore and the gap measured with a feeler gauge. This should be between the limits given in the Specifications at the beginning of this Chapter. It is essential that the gap is measured at the bottom of the ring travel, for if it is measured at the top of a worn bore and gives a perfect fit, it could easily seize at the bottom. If the ring gap is too small, rub down the ends of the ring with a very fine file until the gap is correct when fitted. To keep the rings square in the bore for measurement, line each one up in turn with an old piston inserted in the bore upside down, and use the piston to push the ring down about 3 in (76 mm). Remove the piston and measure the piston ring gap.

The groove clearance of the new rings in old pistons should be within the tolerance given in the Specifications. If it is not enough, the rings could stick in the piston grooves causing loss of compression. The ring grooves in the piston in this case will need machining out to accept the new rings.

Before fitting new rings onto an old piston, clean out the grooves with a piece of broken ring.

If new pistons are obtained, the rings will be included, so it must be emphasised that the top ring be stepped if fitted to a cylinder bore that has not been rebored or has not had the top ridge removed.

## Camshaft and tappets

Check the camshaft journals and lobes for scoring and wear. If there are very slight scoring marks these can be removed with emery cloth or a fine oil stone. The greatest care must be taken to keep the cam profiles smooth.

Examine the camshaft bearing surfaces in the cylinder head and camshaft cover, if these are scored and worn it means a new cylinder head and camshaft cover will be required.

Check for scoring or pitting of the tappets and check the fit of the tappets in their respective bores. Renew any that show signs of wear.

## Timing belt and tensioner

Check the belt for tension and for any sign of cracks or splits in the belt particularly around the roots of the teeth. Renew the belt if the wear is obvious, if there are signs of oil contamination or if the belt has exceeded its service life (see Routine Maintenance). Also renew the sprockets if they show any signs of wear or chipping of the teeth.

Spin the tensioner and ensure that there is no roughness or harshness in the bearing. Also check that the endfloat does not exceed the maximum value stated on the tensioner face. Renew the tensioner if worn.

## Cylinder head and pistons – decarbonising

This can be carried out with the engine either in or out of the car. With the cylinder head removed, carefully use a wire brush and blunt scraper to clean all traces of carbon deposits from the combustion spaces and the ports. The valve head stems and valve guides should also be freed from any carbon deposits. Wash the combustion spaces and ports down with petrol and scrape the cylinder head surface free of any foreign matter with the side of a steel rule or similar article.

If the engine is installed in the car, clean the pistons and the top of the cylinder bores. If the pistons are still in the block, then it is essential that great care is taken to ensure that no carbon gets into the cylinder bores as this could scratch the cylinder walls or cause damage to the piston and rings. To ensure this does not happen, first turn the crankshaft so that two of the pistons are at the top of their bores. Stuff

rag into the other two bores or seal them off with paper and masking tape. The waterways should also be covered with small pieces of masking tape to prevent particles of carbon entering the cooling system and damaging the water pump.

Press a little grease into the gap between the cylinder walls and the two pistons which are to be worked on. With a blunt scraper carefully scrape away the carbon from the piston crown, taking great care not to scratch the aluminium. Also scrape away the carbon from the surrounding lip of the cylinder wall. When all carbon has been removed, scrape away the grease which will now be contaminated with carbon particles, taking care not to press any into the bores. To assist prevention of carbon build-up the piston crown can be polished with a metal polish. Remove the rags or masking tape from the other two cylinders and turn the crankshaft so that the two pistons which were at the bottom are now at the top. Place rag or masking tape in the cylinders which have been decarbonised and proceed as just described. Decarbonising is now complete.

## 25  Lubrication system – description

The pressed steel sump is attached to the underside of the crankcase and acts as a reservoir for the engine oil. The oil pump draws oil through a strainer attached to the pick-up pipe and submerged in the oil. The pump passes the oil along a short passage and into the full-flow filter which is screwed onto the pump housing. The freshly filtered oil flows from the filter and enters the main gallery to the five main bearings. Oil passes from the main bearings, through drillings in the crankshaft to the big-end bearings.

When the crankshaft is rotating, oil is squirted from the hole in each connecting rod and splashes the thrust side of the piston and cylinder bore.

Further drillings connect the main oil gallery to the camshaft in order to lubricate the bearings, cam lobes and tappets. The oil then drains back into the sump via large drillings in the cylinder head and cylinder block.

A pressure relief valve is incorporated in the oil pump housing to maintain the oil pressure within specified limits.

## 26  Engine mountings – removal and refitting

### Right-hand mounting

1  On MG EFi models refer to Chapter 3 and remove the inlet and exhaust manifolds.
2  On all models remove the alternator as described in Chapter 9.
3  Place a jack with interposed block of wood beneath the sump and just take the weight of the engine.
4  Undo the bolts securing the mounting to the cylinder block.
5  From above and from under the wheel arch undo the bolts securing the mounting to the body. On MG EFi models move the fuel filter support bracket to gain access.
6  Remove the mounting from the car and, if necessary, undo the through-bolt and separate the two parts.
7  Refitting is the reverse sequence of removal.

### Left-hand mounting

8  Disconnect the battery and support the gearbox on a jack.
9  Undo the nut and through-bolt securing the mounting to the bracket on the chassis member.
10  Undo the two bolts securing the mounting to the gearbox and withdraw the mounting. Recover the two spacer washers.
11  Refitting is the reverse sequence of removal.

### Rear mounting

12  Jack up the front of the car and support it on axle stands.
13  Undo the four bolts securing the mounting to the gearbox and the bolts securing the mounting bracket to the crossmember. Withdraw

the mounting and bracket. Undo the nuts and remove the mounting from the bracket.

14 Refitting is the reverse sequence of removal.

## Front snubber

15 Undo the two bolts securing the snubber bracket to the gearbox bellhousing and the two bolts and eccentric washers securing the snubber cup to the front chassis member. Remove the snubber and snubber cup.

16 Refitting is the reverse sequence of removal, but centralise the snubber cup around the snubber rubber then tighten the retaining bolts.

## 27 Ancillary components – refitting

Refer to Section 10 and refit the listed components with reference to the Sections and Chapters specified where applicable.

## 28 Engine – attachment to gearbox

Refer to Section 8 and attach the gearbox to the engine using the reverse of the removal sequence. Tighten the retaining bolts to the specified torque.

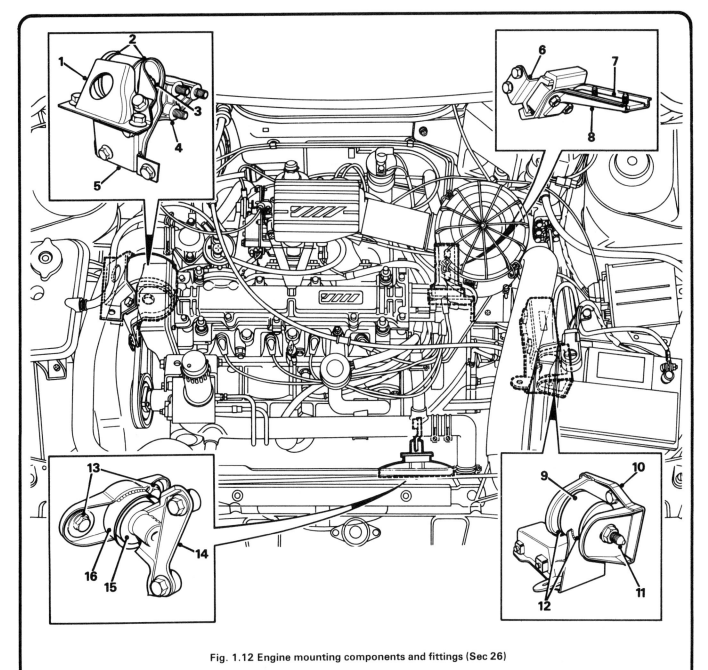

Fig. 1.12 Engine mounting components and fittings (Sec 26)

| | | |
|---|---|---|
| 1 Right-hand mounting and upper body bolts | 5 Right-hand mounting to body lower bolts | 9 Left-hand mounting |
| 2 Flat washers | 6 Rear mounting | 10 Left-hand mounting chassis member bracket |
| 3 Through-bolt | 7 Mounting bracket | 11 Through-bolt |
| 4 Right-hand mounting to engine bolts | 8 Retaining bolts | 12 Flat washers |

13 Snubber cup bolts
14 Snubber bracket
15 Snubber rubber
16 Snubber cup

## 29 Fault diagnosis – engine

| Symptom | Reason(s) |
| --- | --- |
| Engine fails to start | Discharged battery |
| | Loose battery connection |
| | Loose or broken ignition leads |
| | Moisture on spark plugs, distributor cap, or HT leads |
| | Incorrect spark plug gaps |
| | Cracked distributor cap or rotor |
| | Other ignition system fault |
| | Dirt or water in carburettor |
| | Empty fuel tank |
| | Faulty fuel pump |
| | Other fuel system fault |
| | Faulty starter motor |
| | Low cylinder compressions |
| Engine idles erratically | Inlet manifold air leak |
| | Leaking cylinder head gasket |
| | Worn camshaft lobes |
| | Faulty fuel pump |
| | Incorrect valve clearances |
| | Loose crankcase ventilation hoses |
| | Carburettor adjustment incorrect |
| | Uneven cylinder compressions |
| Engine misfires | Spark plugs worn or incorrectly gapped |
| | Dirt or water in carburettor |
| | Carburettor adjustment incorrect |
| | Burnt out valve |
| | Leaking cylinder head gasket |
| | Distributor cap cracked |
| | Incorrect valve clearances |
| | Uneven cylinder compressions |
| | Worn carburettor or other fuel system fault |
| Engine stalls | Carburettor adjustment incorrect |
| | Inlet manifold air leak |
| | Ignition timing incorrect |
| Excessive oil consumption | Worn pistons, cylinder bores or piston rings |
| | Valve guides and valve stem seals worn |
| | Oil seal or gasket leakage |
| Engine backfires | Carburettor adjustment incorrect |
| | Ignition timing incorrect |
| | Incorrect valve clearances |
| | Inlet manifold air leak |
| | Sticking valve |

# Chapter 2 Cooling system

*For modifications, and information applicable to later models, see Supplement at end of manual*

## Contents

## Specifications

| | |
|---|---|
| **System type** | Pressurized, water pump assisted thermo-syphon with front mounted radiator and electric cooling fan |

**Thermostat**

| | |
|---|---|
| Type | Wax |
| Start to open temperature | 169 to 176°F (76 to 80°C) |
| Fully open temperature | 190°F (88°C) |
| Lift height | 0.32 in (8.1 mm) |

| | |
|---|---|
| **Expansion tank cap pressure** | 15 lbf/in$^2$ (1.0 bar) |
| **Drivebelt tension** | 0.3 to 0.5 in (7.0 to 12.0 mm) deflection midway between crankshaft and alternator pulleys under load of 10 lbf (44.4N) |
| **System capacity (including heater)** | 15.0 Imp pints (8.5 litres) |
| **Antifreeze type/specification** | Ethylene glycol-based antifreeze with non-phosphate corrosion inhibitors, suitable for mixed-metal engines, containing no methanol and meeting specifications BS6580 and BS5117 (Duckhams Universal Antifreeze and Summer Coolant) |

**Antifreeze properties and quantities**

33% antifreeze (by volume):

| | |
|---|---|
| Commences freezing | −2°F (−19°C) |
| Frozen solid | −33°F (−36°C) |

| **Antifreeze** | **Water** |
|---|---|
| 5.0 Imp pints (2.8 litres) | 10.0 Imp pints (5.7 litres) |

Quantities (system refill)

50% antifreeze (by volume):

| | |
|---|---|
| Commences freezing | −33°F (−36°C) |
| Frozen solid | −53°F (−48°C) |

| **Antifreeze** | **Water** |
|---|---|
| 7.5 Imp pints (4.3 litres) | 7.5 Imp pints (4.3 litres) |

Quantities (system refill)

**Torque wrench settings**

| | lbf ft | Nm |
|---|---|---|
| Coolant temperature thermister | 5 | 7 |
| Water outlet elbow | 9 | 12 |
| Water pump bolts | 8 | 11 |
| Water pump pulley bolts | 9 | 12 |

## 1   General description

The cooling system is of the pressurized, pump-assisted thermo-syphon type. The system consists of the radiator, water pump, thermostat, electric cooling fan, expansion tank and associated hoses. The impeller type water pump is mounted on the right-hand end of the engine and, in conjunction with the alternator, is driven by the crankshaft pulley by a drivebelt.

The system functions as follows. Cold coolant in the bottom of the radiator left-hand tank passes through the hose to the water pump where it is pumped around the cylinder block and head passages. After cooling the cylinder bores, combustion surfaces and valve seats, the coolant reaches the underside of the thermostat, which is initially closed and is diverted through the heater inlet hose to the heater. After passing through the heater the coolant travels through the water jacket of the inlet manifold before returning to the water pump inlet hose. When the engine is cold the thermostat remains closed and the coolant only circulates around the engine, heater and inlet manifold. When the coolant reaches a predetermined temperature the thermostat opens and the coolant passes through the top hose to the radiator right-hand tank. As the coolant circulates around the radiator it is cooled by the inrush of air when the car is in forward motion. Air flow is supplemented by the action of the electric cooling fan when necessary. Upon reaching the left-hand side of the radiator, the coolant is now cooled and the cycle is repeated.

When the engine is at normal operating temperature the coolant expands and some of it is displaced into the expansion tank. This coolant collects in the tank and is returned to the radiator when the system cools.

The electric cooling fan mounted in front of the radiator is controlled by a thermostatic switch located in the radiator side tank. At a predetermined coolant temperature the switch contacts close, thus actuating the fan.

On models equipped with automatic transmission, a fluid cooler is mounted to the transmission casing by a hollow centre bolt. The cooler is connected to the main cooling system, engine coolant being the transmission fluid cooling medium.

## 2   Maintenance and inspection

1   Check the coolant level in the system weekly and, if necessary, top up with a water and antifreeze mixture until the level is up to the indicator in the expansion tank. With a sealed type cooling system, topping-up should only be necessary at very infrequent intervals. If this is not the case and frequent topping-up is required, it is likely there is a leak in the system. Check all hoses and joint faces for any staining, or actual wetness, and rectify if necessary. If no leaks can be found it is advisable to have the system pressure tested, as the leak could possibly be internal. It is a good idea to keep a check on the engine oil level as a serious internal leak can often cause the level in the sump to rise, thus confirming suspicions.

2   At the service intervals given in Routine Maintenance at the beginning of this Manual carefully inspect all the hoses, hose clips and visible joint gaskets for cracks, corrosion, deterioration or leakage. Renew any hoses and clips that are suspect, and also renew any gaskets, if necessary.

3   At the same service interval check the condition of the water pump and alternator drivebelt and renew it if there is any sign of

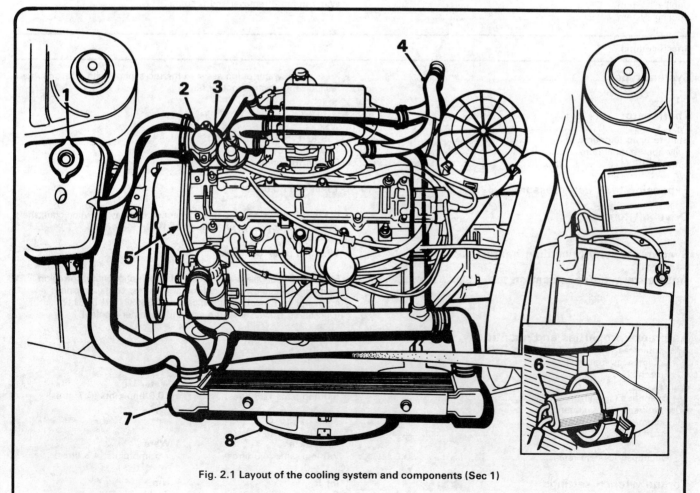

**Fig. 2.1 Layout of the cooling system and components (Sec 1)**

| | | | |
|---|---|---|---|
| 1 *Expansion tank filler cap* | 3 *Thermostat housing* | 5 *Water pump* | 7 *Radiator* |
| 2 *Water outlet elbow* | 4 *Heater hose* | 6 *Cooling fan thermostatic switch* | 8 *Electric cooling fan and cowl* |

cracking or fraying. Check and adjust the tension of the belt as described in Section 12.

4   At the less frequent service intervals indicated, drain, flush and refill the cooling system using fresh antifreeze, as described in Sections 3, 4, and 5 respectively.

5   Also, at this service interval the drivebelt should be renewed using the procedure described in Section 12.

## 3   Cooling system – draining

1   It is preferable to drain the cooling system when the engine is cold. *If the engine is hot the pressure in the cooling system must be released before attempting to drain the system.* Place a cloth over the pressure cap of the expansion tank and turn the cap anti-clockwise until it reaches its stop.

2   Wait until the pressure has escaped, then press the cap downwards and turn it further in an anti-clockwise direction. Release the downward pressure on the cap very slowly and, after making sure that all the pressure in the system has been relieved, remove the cap.

3   Place a suitable container beneath the left-hand side of the radiator. Slacken the hose clip and carefully ease the bottom hose off the radiator outlet. Allow the coolant to drain into the container.

4   If the cooling system is to be flushed after draining, see the next Section otherwise refit the bottom hose then fill the system as described in Section 5.

## 4   Cooling system – flushing

1   With time, the cooling system may gradually lose its efficiency as the radiator core becomes choked with rust, scale deposits from the water, and other sediment.

2   To flush the system, first drain the coolant as described in the previous Section.

3   Disconnect the top hose at the water outlet elbow and leave the bottom hose disconnected at the radiator outlet.

4   Insert a garden hose into the top hose and allow water to circulate through the radiator until it runs clear from the outlet.

5   Disconnect the heater inlet hose from the thermostat housing. Insert the hose and allow water to circulate through the heater and manifold and out through the bottom hose until clear.

6   In severe cases of contamination the system should be reverse flushed. To do this remove the radiator as described in Section 7, invert it and insert the garden hose in the bottom outlet. Continue flushing until clear water runs from the top hose outlet.

7   The engine should also be flushed. To do this remove the thermostat, as described in Section 8, and insert the hose into the cylinder head. Flush the system until clear water runs from the bottom hose.

8   If, after a reasonable period the water still does not run clear, the radiator can be flushed with a good proprietary cleaning agent such as Holts Radflush or Holts Speedflush.

## 5   Cooling system – filling

1   If the system has been flushed, refit any hoses or components that were removed for this purpose.

2   Fill the system through the expansion tank with the appropriate mixture of water and antifreeze (see Section 6) until the tank is half full. Do not refit the expansion tank cap at this stage.

3   Start the engine and run it at a fast idle for approximately one minute. During this time compress the top hose several times to release any air pockets in the system.

4   Stop the engine, top up the expansion tank to the indicated level then refit the filler cap.

## 6   Antifreeze mixture

1   The antifreeze should be renewed at regular intervals (see Routine Maintenance). This is necessary not only to maintain the antifreeze properties, but also to prevent corrosion which would otherwise occur as the corrosion inhibitors become progressively less effective.

Fig. 2.2 Expansion tank level indicator (Sec 5)

2   Use an antifreeze, in the specified concentration, which conforms to the standard given in the Specifications at the beginning of this Chapter. Note that Austin Rover recommend the use of Unipart Superplus antifreeze, and state that no other antifreeze should be mixed with this type, such as when topping up is required. If another type of antifreeze is to be used, the system must be drained and flushed prior to its use.

3   Before adding antifreeze the cooling system should be completely drained and flushed, and all hoses checked for condition and security.

4   The quantity of antifreeze and levels of protection are indicated in the Specifications.

5   After filling with antifreeze, a label should be attached to the radiator stating the type and concentration of antifreeze used and the date installed. Any subsequent topping-up should be made with the same type and concentration of antifreeze.

6   Do not use engine antifreeze in the screen washer system, as it will cause damage to the vehicle paintwork. Use a screen wash such as Turtle Wax High Tech Screen Wash in the washer system.

## 7   Radiator – removal, inspection, cleaning and refitting

1   Disconnect the battery negative terminal.

2   Drain the cooling system, as described in Section 3. Leave the bottom radiator hose disconnected.

3   Slacken the retaining clip and detach the radiator top hose.

4   Detach the electrical connectors from the radiator cooling fan thermostatic switch (photo).

5   Release the upper retaining clips, tip the radiator grille forward at the top and lift it to release the lower mounting lugs from their locations (photo).

7.4 Electrical connections at the cooling fan thermostatic switch

7.5 Grille lower mounting lug and location in body panel

6    Undo and remove the three bolts each side securing the front body panel in position. Release the air cleaner cold air intake hose from the panel then withdraw the panel and place it to one side leaving the bonnet release cable still attached (photos).
7    Release the two plastic panel support clips from the left-hand side of the radiator.
8    Disconnect the cooling fan motor wiring multi-plug (photo) and carefully lift out the radiator.

9    Radiator repair is best left to a specialist, but minor leaks may be sealed using a proprietary coolant additive such as Holts Radweld with the radiator and associated ancillaries *in situ*.
10    Reverse flush the radiator, as described in Section 4. Renew the top and bottom hoses and clips if they are damaged or have deteriorated.
11    Refitting the radiator is the reverse sequence of removal, but ensure that the lower mounting lugs engage in the rubber grommets. Fill the cooling system as described in Section 5 after fitting.

## 8    Thermostat – removal, testing and refitting

1    Remove the expansion tank filler cap. *If the engine is hot, place a cloth over the cap and turn it slowly anti-clockwise until the first stop is reached.* Wait until all the pressure has been released and then remove the cap completely.
2    Place a suitable container beneath the radiator bottom hose outlet. Disconnect the bottom hose and drain approximately 4 pints (2.3 litres) of the coolant. Reconnect the bottom hose and tighten the clip.
3    Slacken the clips and detach the radiator top hose, and expansion tank overflow hose from the water outlet elbow.
4    Undo and remove the two bolts securing the water outlet elbow to the thermostat housing and lift off the elbow (photo). After removal recover the gasket.
5    Withdraw the thermostat from its seat in the housing (photo).
6    To test whether the unit is serviceable, suspend it on a string in a saucepan of cold water together with a thermometer. Do not allow the thermostat or thermometer to touch the bottom of the pan. Heat the water and note the temperature at which the thermostat begins to open. Continue heating the water until the thermostat is fully open, note the temperature, then remove the unit from the water.
7    The temperature at which the thermostat should start to open and be fully open are given in the Specifications. The fully open temperature is also stamped on the wax capsule at the base of the

7.6A Air cleaner cold air intake hose fitment in front body panel

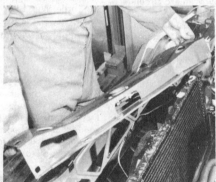

7.6B Front body panel removal

7.8 Cooling fan motor wiring multi-plug

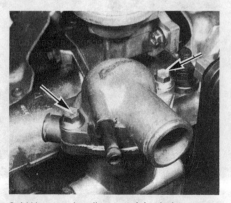

8.4 Water outlet elbow retaining bolts (arrowed)

8.5 Remove the thermostat from the housing

thermostat. If the unit does not start to open or fully open at the specified temperatures, or if it does not close when removed from the water, then it must be discarded and a new one fitted. Under no circumstances should the car be used without a thermostat, as uneven cooling of the cylinder walls and head passages will occur, causing distortion and possible seizure of the engine internal components.

8   Refitting the thermostat is the reverse sequence of removal. Ensure that all traces of old gasket are removed from the mating faces of the water outlet elbow and thermostat housing. Use a new gasket lightly smeared with jointing compound and tighten the retaining bolts to the specified torque. On completion top up the cooling system with reference to Section 5.

## 9   Water pump – removal and refitting

**Note:** *Water pump failure is indicated by water leaking from the gland at the front of the pump, or by rough and noisy operation. This is usually accompanied by excessive play of the pump spindle which can be checked by moving the pulley from side to side. Repair or overhaul of a faulty pump is not possible, as internal parts are not available separately. In the event of failure a replacement pump must be obtained.*

1   Disconnect the battery negative terminal.
2   Drain the cooling system as described in Section 3.
3   Remove the drivebelt as described in Section 12.
4   Refer to Chapter 1 and remove the timing belt.
5   From under the right-hand wheel arch, undo the three retaining bolts and remove the pulley from the water pump spindle (photo).
6   Slacken the retaining clip and remove the water inlet hose from the pump.
7   Undo the two nuts, remove the washers and withdraw the timing belt tensioner from the pump studs.
8   Undo the water pump retaining bolts noting the location of the special stud which retains the timing belt cover, and the two bolts which secure the water deflector bracket (photo).
9   Have a container handy to catch any remaining coolant then withdraw the pump from the engine.
10   If necessary carefully tap the pump body with a soft-faced mallet to free it.
11   Remove the gasket, scraping away any remaining traces with a sharp knife.
12   Before refitting wipe the pump and cylinder block mating faces and ensure that they are dry and perfectly clean.
13   Position the two timing belt tensioner bolts through the rear of the pump (photo) then fit the pump using a new gasket lightly smeared with jointing compound. Refit the pump retaining bolts, but apply jointing compound to the threads of the bolts which locate in holes open through to the water jacket. Ensure that the water deflector bracket is fitted to the two lower bolts. Tighten all the bolts progressively to the specified torque.
14   Refit the timing belt tensioner, but only tighten the retaining bolts finger tight at this stage.
15   Refit the water inlet hose and tighten the clip securely.

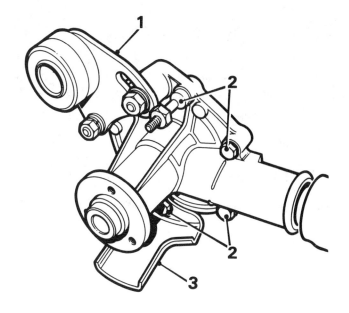

**Fig. 2.3 Water pump attachments (Sec 9)**

1   Timing belt tensioner
2   Pump retaining bolts
3   Water deflector bracket

16   Refit the water pump pulley and tighten the bolts to the specified torque.
17   Refit the access panel, and the front roadwheel then lower the car to the ground.
18   Refer to Chapter 1 and refit the timing belt.
19   Refit the alternator drivebelt as described in Section 12 then refill the cooling system as described in Section 5.

## 10   Cooling fan assembly – removal and refitting

1   Disconnect the battery negative terminal.
2   Release the upper retaining clips, tip the radiator grille forward at the top and lift it to release the lower mounting lugs from their locations.
3   Undo and remove the three bolts each side securing the front body panel in position. Release the air cleaner cold air intake hose from the panel then lift the panel upwards and place it to one side leaving the bonnet release cable still attached (photos 7.6A and 7.6B).
4   Disconnect the fan motor wiring multi-plug.

9.5 Water pump pulley retaining bolts

9.8 Water pump retaining bolts (arrowed)

9.13 Fit the timing belt tensioner bolts to the rear of the pump

10.5 Cooling fan cowl lower retaining nuts (arrowed)

11.4 Cooling fan thermostatic switch and retaining plate

5    Undo the three nuts securing the fan cowl to the radiator and withdraw the assembly from its location (photo).
6    If required the fan may be removed by pulling it off the motor shaft. Mark the outer face to ensure correct refitment. To remove the fan motor, drill off the rivet heads, tap out the rivets securing the motor to the cowl and lift away the motor.
7    Refitting is the reverse sequence of removal.

## 11 Cooling fan thermostatic switch – testing, removal and refitting

1    The radiator cooling fan is operated by a thermostatic switch located on the right-hand side of the radiator. When the coolant exceeds a predetermined temperature the switch contacts close and the fan is activated.
2    If the operation of the fan or switch is suspect, run the engine until normal operating temperature is reached and then allow it to idle. If the fan does not cut-in within a few minutes, switch off the engine and disconnect the wiring plug from the thermostatic switch. Bridge the two terminals in the wiring plug with a length of wire and switch on the ignition. If the fan now operates, the thermostatic switch is faulty and must be renewed. If the fan still fails to operate check that battery voltage is present at the plug terminals. If not check for a blown fuse or wiring fault. If voltage is present the fan motor is faulty.
3    To remove the switch disconnect the battery negative terminal then drain the cooling system as described in Section 3.
4    Disconnect the wiring plug, release the retaining plate and withdraw the switch and seal from the radiator (photo).
5    Refitting is the reverse sequence of removal, but use a new seal on the switch and refill the cooling system as described in Section 5.

## 12 Drivebelt – removal, refitting and adjustment

1    The drivebelt should be checked and re-tensioned at regular intervals (see Routine Maintenance). It should be renewed at these service intervals if it shows any sign of fraying, deterioration or oil contamination.
2    On vehicles equipped with power-assisted steering it will first be necessary to remove the steering pump drivebelt as described in Chapter 10.
3    Jack up the front of the car and support it on axle stands. Remove the right-hand front roadwheel and the access panel from under the wheel arch (photo).
4    Slacken the alternator mounting nuts and the adjustment arm bolt (photo). Move the alternator towards the engine and slip the belt off the three pulleys.

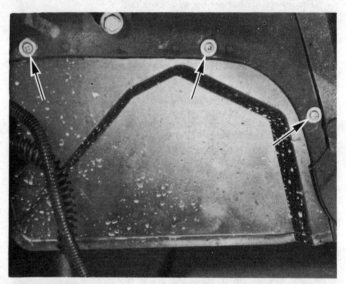

12.3 Access panel retaining screw locations (arrowed) under front wheel arch

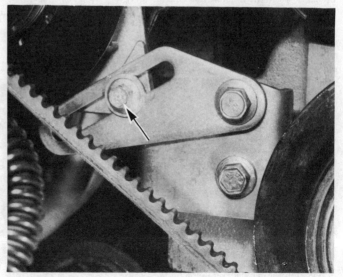

12.4 Alternator adjustment arm bolt (arrowed)

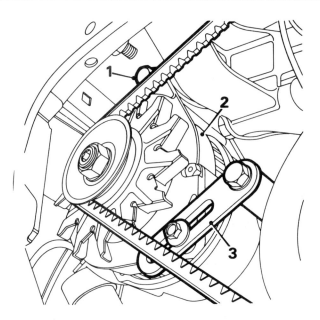

**Fig. 2.4 Drivebelt adjustment points (Sec 12)**

1  *Mounting nut*              3  *Adjustment arm*
2  *Alternator*

5   Fit the new drivebelt over the pulleys then lever the alternator away from the engine until the drivebelt is moderately tight. The alternator must only be levered with care at the drive end bracket.
6   Hold the alternator in this position and tighten the adjustment arm bolt, followed by the mounting nuts.
7   Run the engine at a fast idle for approximately five minutes then switch off. Readjust the drivebelt to the correct tension as given in the Specifications.
8   Refit the access panel and roadwheel then lower the car to the ground. If power-assisted steering is fitted, refit the pump drivebelt as described in Chapter 10.

## 13  Coolant temperature thermister – removal and refitting

1   The thermister contains an element, the resistance of which alters according to coolant temperature. The unit controls the operation of the temperature gauge and is used by the fuel and ignition system control units to determine engine temperature.
2   Remove the expansion tank filler cap. *If the engine is hot place a cloth over the cap and release it slowly to allow the pressure to escape.*
3   Place a suitable container beneath the radiator bottom hose outlet. Disconnect the hose, drain approximately 4 pints (2.3 litres) of the coolant then refit the hose.
4   Disconnect the wiring multi-plug then unscrew the thermister from its location in the thermostat housing.
5   Refitting is the reverse sequence of removal. Tighten the thermister to the specified torque and top up the cooling system with reference to Section 5.

## 14  Fault diagnosis – cooling system

| Symptom | Reason(s) |
| --- | --- |
| Overheating | Low coolant level (this may be the result of overheating for other reasons) |
| | Drivebelt slipping or broken |
| | Radiator blockage (internal or external), or grille restricted |
| | Thermostat defective |
| | Carburettor or fuel injection system incorrectly adjusted |
| | Ignition system fault |
| | Faulty cooling fan thermostatic switch |
| | Faulty cooling fan |
| | Blown cylinder head gasket (combustion gases in coolant) |
| | Water pump defective |
| | Expansion tank pressure cap faulty |
| | Brakes binding |
| Overcooling | Thermostat missing, defective or wrong heat range |
| Water loss – external | Loose hose clips |
| | Perished or cracked hoses |
| | Radiator core leaking |
| | Heater matrix leaking |
| | Expansion tank pressure cap leaking |
| | Boiling due to overheating |
| | Water pump or thermostat housing leaking |
| | Core plug leaking |
| Water loss – internal | Cylinder head gasket blown |
| | Cylinder head cracked or warped |
| | Cylinder block cracked |
| Corrosion | Infrequent draining and flushing |
| | Incorrect antifreeze mixture or inappropriate type |
| | Combustion gases contaminating coolant |

# Chapter 3 Fuel and exhaust systems

*For modifications, and information applicable to later models, see Supplement at end of manual*

## Contents

## Specifications

### Part A: Conventional carburettor induction

**Air cleaner element** ........................................................... Champion W114

**Fuel pump**

Type ....................................................................................... Mechanical, operated by eccentric on camshaft
Make ...................................................................................... SU AUF 800
Delivery pressure ................................................................ 4.0 lbf/in² (0.3 bar)

**Fuel grade** ....................................................................... See Chapter 12, Section 6

**Carburettor**

Type ....................................................................................... SU HIF 44E variable choke with electronic mixture control
Specification number ......................................................... FZX 1438
Piston spring colour ........................................................... Yellow
Jet size .................................................................................. 0.100 in
Needle identification .......................................................... BFU
Adjustment data:
    Fast idle rod minimum clearance .............................. 0.005 in (0.13 mm)
    Throttle lever lost motion gap .................................... 0.060 to 0.080 in (1.52 to 2.03 mm)
    Float height (see text) ................................................ 0.040 to 0.060 in (1.00 to 1.52 mm)
Idling speed ......................................................................... 600 to 700 rpm
Fast idle speed .................................................................... 1050 to 1150 rpm
CO mixture ........................................................................... 1.5 to 3.5%
Piston damper oil type/specification ............................... Multigrade engine oil, viscosity SAE 10W/40 (Duckhams QXR, Hypergrade, or 10W/40 Motor Oil)

## Torque wrench settings

| | lbf ft | Nm |
|---|---|---|
| Carburettor retaining nuts | 19 | 25 |
| Fuel pump retaining nuts | 16 | 22 |
| Manifold retaining nuts and bolts | 18 | 24 |

### *Part B: Electronic fuel injection system*

**System type** ...................................... Lucas indirect multi-point injection with microprocessor control

**Air cleaner element (all models)** ............. Champion W114

### Fuel pump
Type ........................................................ Electric, self-priming centrifugal
Make ....................................................... Lucas 4FP
Maximum delivery pressure ........................ 60 lbf/in² (4.1 bar)
Regulated pressure range .......................... 36 to 26 lbf/in² (2.5 to 1.8 bar)

**Fuel injectors** ..................................... Lucas 8NJ

**Fuel grade** .......................................... See Chapter 12, Section 6

**Fuel filter type** ................................... Champion L206

### Fuel pressure regulator
Make ....................................................... Lucas 4RV vacuum controlled
Pressure range at maximum vacuum ........... 36 to 26 lbf/in² (2.5 to 1.8 bar)

### General
Fuel temperature switch closing temperature ...... 194°F (90°C)
Throttle potentiometer ............................... Lucas 215SA
Fuel ECU ................................................. Lucas 11CU – 84399
Idle speed adjustment setting (see text):
   Ambient temperature below 77°F (25°C) ....... 670 rpm
   Ambient temperature above 77°F (25°C) ....... 650 rpm
CO mixture .............................................. 1.0 to 1.5%

## Part A: Conventional carburettor induction (all models except MG)

### 1 General description

The fuel system on models with carburettor induction consists of a centrally mounted fuel tank, camshaft operated mechanical fuel pump and SU variable choke side draught carburettor with electronic mixture control system.

The air cleaner is of the automatic air temperature control type and contains a disposable paper element.

The exhaust system is in three sections; the front section incorporates the twin downpipe and is fitted with ball and socket type flexible joints, the intermediate section incorporates the front and intermediate silencers and the rear section incorporates the rear silencer and tailpipe. The exhaust system is suspended on rubber mountings at the rear and is bolted to the cast iron manifold at the front.

**Warning**: *Many of the procedures in this Chapter entail the removal of fuel pipes and connections which may result in some fuel spillage. Before carrying out any operation on the fuel system refer to the precautions given in Safety First! at the beginning of this Manual and follow them implicitly. Petrol is a highly dangerous and volatile liquid and the precautions necessary when handling it cannot be overstressed.*

### 2 Maintenance and inspection

1 At the intervals given in Routine Maintenance at the beginning of this Manual, carry out the following service operations to the fuel system components.
2 With the car over a pit, raised on a vehicle lift, or securely supported on axle stands, carefully inspect the fuel pipes, hoses and unions for chafing, leaks and corrosion. Renew any pipes that are severely pitted with corrosion or in any way damaged. Renew any hoses that show signs of cracking or other deterioration.
3 Examine the fuel tank for leaks, particularly around the fuel gauge sender unit, and for signs of corrosion or damage.
4 Check condition of exhaust system, as described in Section 14.

5 From within the engine compartment, check the security of all fuel hose attachments and inspect the fuel hoses and vacuum hoses for kinks, chafing or deterioration.
6 Renew the air cleaner element and check the operation of the air cleaner automatic temperature control, as described in Section 3.
7 Check the operation of the accelerator linkage and lubricate the linkage, cable and pedal pivot with a few drops of engine oil.
8 Top up the carburettor piston damper to the top of the hollow piston rod with engine oil (photo). Note that although this operation is only considered necessary by the manufacturers at the annual major service interval, practical experience has shown that the damper oil may require topping-up more frequently.
9 Check and if necessary adjust the carburettor idle, fast idle and mixture settings as described in Section 10.

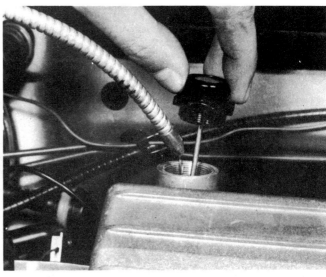

2.8 Topping up the carburettor piston damper

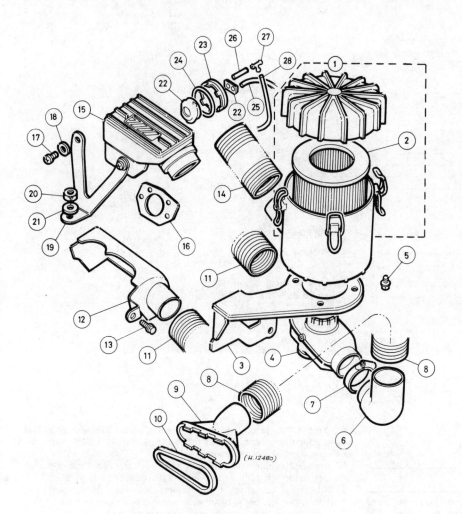

**Fig. 3.1 Exploded view of the air cleaner assembly and related components (Sec 3)**

1  Air cleaner assembly
2  Filter element
3  Mounting bracket
4  Thermac unit
5  Bolt
6  Intake tube elbow
7  Retaining clip
8  Cold air intake tube
9  Intake tube adaptor
10  Rubber seal
11  Hot air intake tube
12  Hot air box
13  Bolt
14  Plenum chamber intake tube
15  Plenum chamber
16  Gasket
17  Screw
18  Washer
19  Support bracket
20  Nut
21  Spring washer
22  Thermac switch
23  Grommet
24  Seal
25  Large diameter vacuum hose
26  Small diameter vacuum hose
27  T-piece connector
28  Vacuum hose – inlet manifold to T-piece

### 3  Air cleaner and element – removal and refitting

1  To renew the air cleaner element, spring back the retaining clips and lift off the air cleaner cover (photo).
2  Withdraw the element and wipe clean the inside of the air cleaner body and cover.
3  Fit a new element, locate the cover on the air cleaner body and secure with the retaining clips.
4  To remove the air cleaner assembly from the engine, release the cold air intake tube from the front body panel (photo), and withdraw the tube from the air cleaner body (photo).

5  Undo the bolts securing the air cleaner mounting bracket to the engine, lift up the unit and release it from the hot air and plenum chamber intake tubes (photos).
6  Detach the vacuum hose from the thermac unit at the base of the air cleaner and remove the air cleaner from the car.
7  To remove the plenum chamber, undo the two screws and one nut securing the support bracket to the plenum chamber, carburettor and cylinder head. Lift away the support bracket and withdraw the plenum chamber from the carburettor. Recover the gasket (photo).
8  Detach the vacuum hoses at the thermac switch and remove the plenum chamber from the car.
9  To test the operation of the air cleaner air temperature control

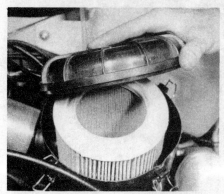

3.1 Lift off the air cleaner cover to renew the element

3.4A Cold air intake tube attachment at the front body panel ...

3.4B ... and at the air cleaner body

3.5A Undo the air cleaner bracket retaining bolts (arrowed) ...

3.5B ... then withdraw the assembly from the hot air tube and plenum chamber

3.7 Removing the plenum chamber from the carburettor

system, reconnect the vacuum hose at the air cleaner thermac unit and at the thermac switch on the plenum chamber. Reconnect the other hose to the thermac switch, but disconnect it at the T-piece connector.

10  Observe the position of the air flap at the base of the air cleaner which should be set to admit air from the cold air intake into the air cleaner.

11  Apply suction to the thermac switch vacuum hose and check that the flap moves to admit hot air into the air cleaner and closes off the cold air intake.

12  Using a domestic hair dryer, warm the thermac switch in the plenum chamber whilst still applying suction to the hose. When the temperature reaches approximately 85°F (30°C) the thermac switch should start to close the vacuum supply to the thermac unit, and the flap should start to move back to the cold air delivery position, closing off the hot air intake.

13  If the system does not operate as described, check the condition of the hoses. If these are satisfactory remove the thermac unit to thermac switch hose at the switch and apply suction. If the flap moves, the thermac switch is faulty and should be renewed. If the flap still fails to operate, renew the thermac unit.

14  Refit the plenum chamber and air cleaner using the reverse sequence of removal. The hose connections at the thermac unit are as follows:

*Large diameter hose – thermac switch to thermac unit*
*Small diameter hose – thermac switch to manifold vacuum hose T-piece connector*

## 4  Fuel pump – removal, testing and refitting

**Note:** *Refer to the warning note in Section 1 before proceeding.*

1  The mechanical fuel pump is bolted to the rear facing side of the camshaft cover and is driven by an eccentric lobe on the camshaft. The pump is of sealed construction and in the event of faulty operation must be renewed as a complete unit.

2  To remove the pump, first disconnect the battery negative terminal.

3  Note the location of the fuel inlet hose and fuel outlet hose and then disconnect the hoses from the pump (photo).

4  Undo the two bolts securing the pump to the camshaft cover, withdraw the pump and recover the insulator block and gaskets (photos).

5  To test the pump operation, refit the fuel inlet hose to the pump inlet and hold a wad of rag near the outlet. Operate the pump lever by hand and if the pump is in a satisfactory condition a strong jet of fuel should be ejected from the pump outlet as the lever is released. If this is not the case, check that fuel will flow from the inlet hose when it is held below the tank level, if so the pump is faulty.

6  Before refitting the pump, clean all traces of old gasket from the pump flange, insulating block and cover face. Use a new gasket on each side of the insulator block and refit the pump using the reverse sequence to removal. Ensure that the pump lever locates over the top of the camshaft eccentric as the pump is installed. If this proves difficult, turn the engine over until the large offset of the eccentric is facing downward.

## 5  Fuel tank – removal, servicing and refitting

**Note:** *Refer to the warning note in Section 1 before proceeding.*

1  Disconnect the battery negative terminal. Remove the fuel tank filler cap.

2  A drain plug is not provided and it will therefore be necessary to syphon, or hand pump, all the fuel from the tank before removal.

3  Having emptied the tank jack up the rear of the car and support it securely on axle stands.

4.3 Disconnect the fuel inlet and outlet hoses at the pump

4.4A Undo the two retaining bolts (arrowed) ...

4.4B ... then withdraw the pump and recover the gaskets and insulator block

4   Disconnect the electrical leads from the fuel gauge sender unit and release the clips securing the leads to the tank (photo).

5   Using pliers, release the fuel feed hose retaining clip and disconnect the hose from the outlet on the sender unit.

6   Unscrew the retaining clip securing the fuel filler hose to the tank and the screw securing the filler neck bracket to the wheel arch panel. Disconnect the filler hose from the tank.

7   Support the tank on blocks, or with a jack, and undo the two rear and two front retaining bolts and stiffener plates (photo).

5.4 Electrical leads (A) and fuel feed hose (B) at the fuel gauge sender unit

5.7 Fuel tank front retaining bolts and stiffener plates (arrowed)

8   Move the tank to the left to clear the brake pipe and lower it sufficiently to allow the breather hose to be detached. Now lower the tank to the ground and withdraw it from under the car.

9   If the tank is contaminated with sediment or water, remove the sender unit, as described in Section 6, and swill the tank out with clean fuel. If the tank is damaged, or leaks, it should be repaired by a specialist or, alternatively, renewed. **Do not** under any circumstances solder or weld the tank.

10   Refitting the tank is the reverse sequence to removal.

## 6   Fuel gauge sender unit – removal and refitting

**Note:** Refer to the warning note in Section 1 before proceeding.

1   Follow the procedure given in Section 5, paragraphs 1 to 5 inclusive.

2   Engage a screwdriver, flat bar or other suitable tool with the lugs of the locking ring, and turn the ring anti-clockwise to release it.

3   Withdraw the locking ring, seal and sender unit.

4   Refitting is the reverse sequence to removal, but always use a new seal.

## 7   Accelerator cable – removal and refitting

1   Slacken the screw to release the inner cable from the connector on the carburettor linkage.

2   Release the outer cable from the support bracket and withdraw the cable from the carburettor.

3   Working inside the car, prise the retaining clip from the top of the accelerator pedal and disconnect the inner cable.

4   Release the accelerator cable from the engine compartment bulkhead and withdraw the complete cable from the car.

5   Refitting is the reverse sequence of removal, but adjust the

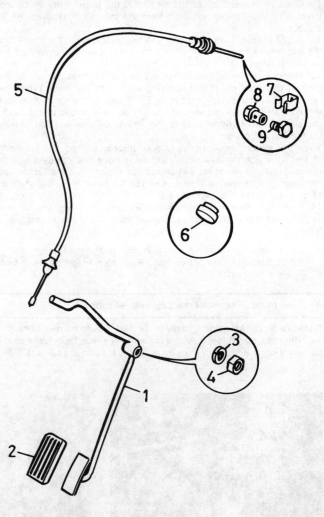

**Fig. 3.2 Accelerator cable and pedal components (Secs 7 and 8)**

| | |
|---|---|
| 1   Accelerator pedal | 6   Pedal stop |
| 2   Pedal pad | 7   Cable retaining clip |
| 3   Washer | 8   Cable connector |
| 4   Retaining nut | 9   Connector screw |
| 5   Accelerator cable | |

position of the inner cable in its connector to give a small amount of free play with the throttle closed. Check that with the accelerator pedal fully depressed, the linkage is fully open, and with the pedal released the linkage closes fully.

## 8 Accelerator pedal – removal and refitting

1 Disconnect the battery negative terminal.
2 Prise the retaining clip from the top of the accelerator pedal and disconnect the accelerator cable from the pedal arm.
3 Unhook the pedal return spring, undo the retaining nut and washer and slide the pedal off the shaft.
4 Refitting is the reverse sequence of removal. After fitting, check the accelerator cable adjustment as described in Section 7.

## 9 SU carburettor – description and operation

The SU HIF (Horizontal Integral Float chamber) carburettor is of the variable choke, constant depression type incorporating a sliding piston which automatically controls the mixture of air and fuel supplied to the engine with respect to the throttle valve position and engine speed. In addition the carburettor is equipped with an electronically operated mixture control device. This alters the mixture strength and engine speed when starting and during slow running, and also controls the operation of a fuel shut-off valve when decelerating or descending a hill.

The carburettor functions as follows. When the engine is started and is allowed to idle, the throttle valve passes a small amount of air. Because the piston is in a low position, it offers a large restriction, and the resultant pressure reduction draws fuel from the jet, and atomisation occurs to provide a combustible mixture. Since the inside section of the tapered needle is across the mouth of the jet, a relatively small amount of fuel is passed.

When the throttle valve is opened, the amount of air passing through the carburettor is increased, which causes a greater depression beneath the sliding piston. An internal passageway connects this depression with the suction chamber above the piston, which now rises. The piston offers less of a restriction and the depression is reduced, with the result that a point is reached where the forces of depression, gravity, and spring tension balance out. The tapered needle has now been raised, and more fuel passes from the jet.

Incorporated in the jet adjusting (mixture) screw mechanism is a bi-metal strip which alters the position of the jet to compensate for varying fuel densities resulting from varying fuel temperatures.

Fuel enrichment for cold starting is by an internal valve which admits more fuel into the airstream passing through the carburettor.

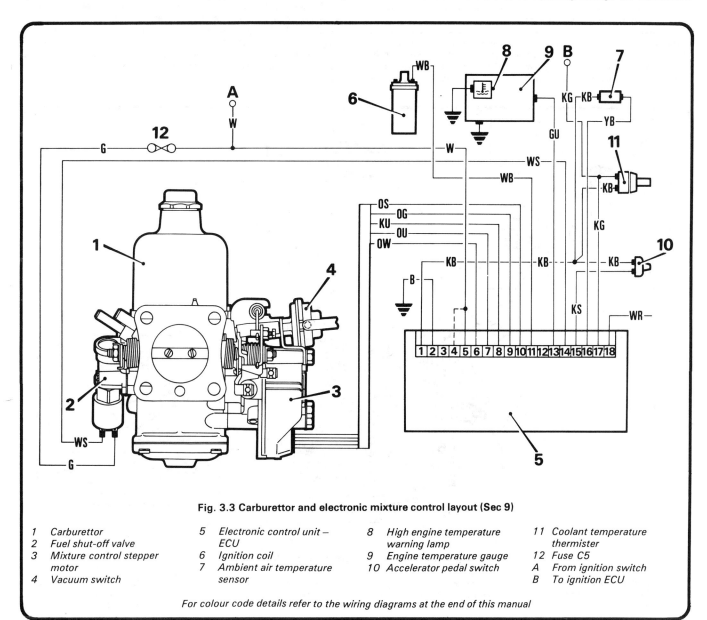

**Fig. 3.3 Carburettor and electronic mixture control layout (Sec 9)**

| | | | |
|---|---|---|---|
| 1 Carburettor | 5 Electronic control unit – ECU | 8 High engine temperature warning lamp | 11 Coolant temperature thermister |
| 2 Fuel shut-off valve | 6 Ignition coil | 9 Engine temperature gauge | 12 Fuse C5 |
| 3 Mixture control stepper motor | 7 Ambient air temperature sensor | 10 Accelerator pedal switch | A From ignition switch |
| 4 Vacuum switch | | | B To ignition ECU |

*For colour code details refer to the wiring diagrams at the end of this manual*

This valve is operated by a stepper motor which also controls the engine idling speed. An electronic control unit (ECU) which is a small microprocessor receives inputs from the coolant thermister, ambient air temperature sensor, accelerator pedal switch and ignition coil, and adjust the engine idle speed and mixture accordingly. The ECU also controls the operation of a fuel shut-off valve which comes into operation when decelerating or descending a hill. If, during these conditions, the engine speed is in excess of 1300 rpm, the ambient air temperature and engine temperature are above a predetemined value, and the accelerator pedal switch is closed (pedal released), the valve will be opened and closed at half second intervals. This introduces a partial vacuum to the top of the float chamber thus weakening the mixture. The fuel shut-off circuit is de-activated if the engine speed suddenly drops, ie when declutching. When accelerating, a vacuum operated switch acts upon the mixture control, allowing more fuel to be drawn through, resulting in the necessary richer mixture.

The overall effect of this type of carburettor is that it will remain in tune during the lengthy service intervals and also under varying operating conditions and temperature changes. The design of the unit and its related systems ensures a fine degree of mixture control over the complete throttle range, coupled with enhanced engine fuel economy.

## 10 SU carburettor – adjustments

**Note:** *Before carrying out any carburettor adjustments, ensure that the spark plug gaps and valve clearances are correctly set and that the ignition system is operating correctly. To carry out the following adjustments an accurate tachometer will be required. The use of an* exhaust gas analyser (CO meter) is also preferable, although not essential.

1   Begin by removing the plenum chamber as described in Section 3.
2   Unscrew and remove the piston damper from the suction chamber.
3   Undo and remove the three securing screws and lift off the suction chamber, complete with piston and piston spring. After removal avoid rotating the piston in the suction chamber.
4   Invert the suction chamber assembly and drain the oil from the hollow piston rod. Check that the needle guide is flush with the piston face and is secure. Do not be concerned that the needle appears loose in the guide, this is perfectly normal.
5   Observe the position of the jet in relation to the jet guide located in the centre of the carburettor venturi. The jet will probably be slightly below the top face of the jet guide. Turn the mixture adjusting screw until the top of the jet is flush with the top of the jet guide. Now turn the adjusting screw two complete turns clockwise. If the mixture adjusting screw is covered by a small blue or red tamperproof plug, hook this out with a small screwdriver and discard it.
6   Refit the piston and suction chamber assembly, taking care not to turn the piston in the suction chamber any more than is necessary to align the piston groove with its guide. If the piston is turned excessively the spring will be wound up and the assembly will have to be dismantled, as described in Section 12.
7   Check that the piston is free to move in the suction chamber by lifting it and allowing it to drop under its own weight. A definite metallic click should be heard as the piston falls and contacts the bridge in the carburettor body. If this is not the case, dismantle and clean the suction chamber, as described in Section 12. If satisfactory, top up the damper oil with engine oil to the top of the hollow piston rod, and refit the damper.

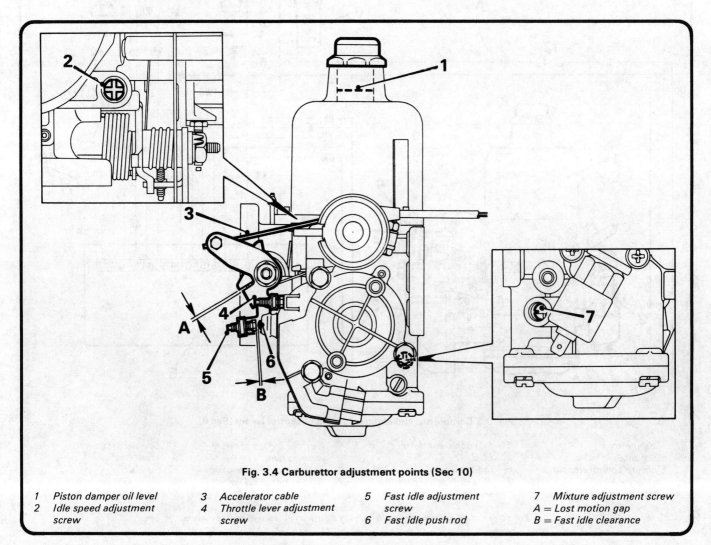

**Fig. 3.4 Carburettor adjustment points (Sec 10)**

| | | | |
|---|---|---|---|
| 1   Piston damper oil level | 3   Accelerator cable | 5   Fast idle adjustment screw | 7   Mixture adjustment screw |
| 2   Idle speed adjustment screw | 4   Throttle lever adjustment screw | 6   Fast idle push rod | A = Lost motion gap |
| | | | B = Fast idle clearance |

8 Check that the throttle linkage operates smoothly and that there is a small amount of free play in the accelerator cable.

9 Connect a tachometer to the engine in accordance with the manufacturer's instructions, and also a CO meter if this is to be used.

10 Reconnect the vacuum hoses to the plenum chamber and lay the unit alongside the carburettor.

11 Start the engine and run it at a fast idle speed until it reaches its normal operating temperature. Continue to run the engine for a further five minutes before commencing adjustment.

12 Increase the engine speed to 2500 rpm for 30 seconds and repeat this at three minute intervals during the adjustment procedure. This will ensure that any excess fuel is cleared from the inlet manifold.

13 Disconnect the coolant thermister wiring plug (see Chapter 2 if necessary) and join the two plug terminals together using a suitable length of wire. This will ensure that the mixture control stepper motor is not actuated during adjustment.

14 If the cooling fan is running, wait until it stops then turn the idle speed adjustment screw as necessary until the engine is idling at the specified speed.

15 Switch off the engine.

16 Check the clearance between the fast idle push rod and fast idle adjustment screw using feeler gauges (position B in Fig. 3.4). Turn the fast idle adjustment screw as necessary to obtain the specified clearance.

17 Check the throttle lever lost motion gap using feeler gauges (position A in Fig. 3.4) and, if necessary, turn the throttle lever adjustment screw to obtain the specified clearance.

18 Start the engine and slowly turn the mixture adjustment screw clockwise (to enrich) or anti-clockwise (to weaken), until the fastest idling speed which is consistent with smooth even running is obtained. If a CO meter is being used, adjust the mixture screw to obtain the specified idling exhaust gas content.

19 Reset the idling speed if necessary using the idle speed adjustment screw then switch off the engine once more.

20 To adjust the fast idle speed disconnect the two wires at the ambient air temperature sensor located in the engine compartment behind the left-hand headlamp (photo). Remove the sensor and join the two wires together using a male-to-male connector or suitable length of wire.

21 Remove the wire connecting the coolant thermister wiring plug terminals together, but leave the plug disconnected.

22 Start the engine again. The mixture control stepper motor should move the fast idle pushrod to the fast idle position. Compare the engine fast idle speed with the specified setting and if necessary adjust by turning the fast idle adjustment screw as required.

23 Switch off the engine, reconnect the ambient air temperature sensor and coolant thermister wiring plug. **Note**: *after carrying out this adjustment, ensure that the specified minimum clearance still exists between pushrod and screw as described in paragraph 16. Adjust the screw if the clearance is less than specified.*

24 Make a final check that the idling speed and mixture are correct after refitting the plenum chamber then switch off the engine and disconnect the instruments.

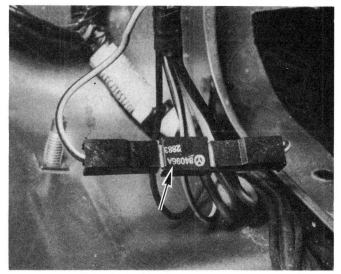

10.20 Ambient air temperature sensor (arrowed) located behind left-hand headlamp

## 11 SU carburettor – removal and refitting

1 Disconnect the battery negative terminal.

2 Remove the air cleaner and the plenum chamber as described in Section 3.

3 Slacken the retaining screw and release the accelerator inner cable from the connector on the carburettor linkage.

4 Release the outer cable from the support bracket and place the cable to one side.

5 Detach the small diameter vacuum hose at the carburettor vacuum take off connector (photo) and at the connector on the vacuum switch. Move the vacuum hose to one side.

6 Disconnect the wiring multi-plug at the stepper motor and the two wires at the fuel shut off valve solenoid (photos).

7 Detach the fuel inlet hose, crankcase breather hose and float chamber vent hose from their carburettor connections.

8 Undo the two nuts securing the carburettor to the inlet manifold and withdraw the unit off the studs (photo). Recover the gaskets and insulating block.

9 Refitting is the reverse sequence of removal. Ensure that all mating faces are perfectly clean and use new gaskets, one each side of the insulating block, if the old ones are in any way damaged. After refitting connect the accelerator cable with reference to Section 7, and adjust the carburettor as described in Section 10.

11.5 Detach the vacuum hose at the carburettor and vacuum switch (arrowed)

11.6A Disconnect the carburettor stepper motor multi-plug (arrowed) ...

11.6B ... and the two wires (arrowed) at the fuel shut off valve solenoid

11.8 Removing the carburettor from the manifold studs

12.3 Removing the carburettor suction chamber and piston assembly

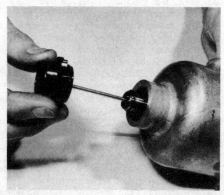

12.4 Unscrew the damper and drain the oil

## 12 SU carburettors – dismantling, overhaul and reassembly

1   Remove the carburettor, as described in the previous Section.

2   Clean off the exterior of the carburettor using paraffin, or a suitable solvent and wipe dry.

3   Unscrew the three retaining screws and lift off the suction chamber and piston assembly (photo).

4   Unscrew the damper from the suction chamber (photo), and drain the oil from the piston rod.

5   Push the piston up to expose the retaining circlip (photo). Extract the circlip and withdraw the piston and spring assembly (photo).

6   Unscrew the needle guide locking screw and withdraw the needle, guide and spring (photo).

7   Mark the relationship of the float chamber cover to the carburettor body. Unscrew the four retaining screws and lift off the cover and O-ring seal (photo).

8   Unscrew the jet adjusting lever retaining screw and withdraw the jet and adjusting lever assembly (photo). Disengage the jet from the lever.

9   Unscrew the float pivot screw (photo) and lift out the float and fuel needle valve (photo). Unscrew the needle valve seat from the base of the float chamber (photo).

10   Unscrew the jet bearing locking nut (photo) and remove the jet bearing (photo).

11   Undo the three retaining screws and remove the fuel shut-off valve and solenoid assembly (photo). Recover the gasket.

12   Dismantling the remaining components is not recommended, as

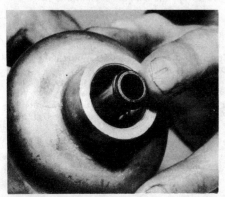

12.5A Extract the retaining circlip ...

12.5B ... to allow removal of the piston and spring

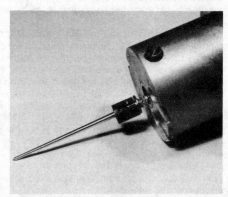

12.6 Slacken the needle guide locking screw and withdraw the needle, guide and spring

12.7 Lift off the float chamber cover after removing the four screws

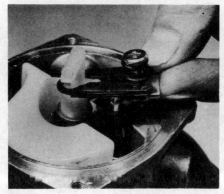

12.8 Removing the jet and adjusting lever assembly

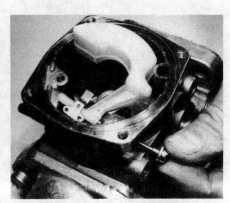

12.9A Unscrew the float pivot screw and remove the float ...

12.9B ... followed by the fuel needle valve ...

12.9C ... then unscrew the needle valve seat

12.10A Unscrew the jet bearing locknut ...

12.10B ... and lift out the jet bearing

12.11 Remove the fuel shut-off valve and solenoid

12.13 Ensure that the filter in the fuel needle valve seat is clean

these parts are not available separately. If the throttle levers, linkage or spindle appear worn, or in any way damaged, it will be necessary to renew the complete carburettor. Do not remove the mixture control stepper motor or vacuum switch. These components are set to each individual carburettor during manufacture and may not operate correctly if disturbed.

13 Check the condition of the float needle valve and seat and renew these components if there is any sign of pitting or wear ridges, particularly on the needle. Ensure that the filter in the seat assembly is clean.

14 Examine the carburettor body for cracks and damage, and ensure that the brass fittings and piston guide are secure.

15 Clean the inside of the suction chamber and the outer circumference of the piston with a petrol-moistened rag and allow to air dry. Insert the piston into the suction chamber without the spring. Hold the assembly in a horizontal position and spin the piston. If there is any tendency for the piston to bind, renew the piston and dashpot assembly.

16 Connect a 12 volt supply to the terminals of the fuel shut-off valve solenoid and ensure that the valve closes. If not, renew the solenoid.

17 Check the piston needle and jet bearing for any signs of ovality, or wear ridges.

18 Shake the float and listen for any trapped fuel which may have entered through a tiny crack or fracture.

19 Check the condition of all gaskets, seals and connecting hoses and renew any that show signs of deterioration.

20 Begin reassembly by refitting the jet bearing and retaining nut to the carburettor body.

21 Refit the fuel needle valve and seat, followed by the float and float pivot screw.

22 Allow the float to close the needle valve under its own weight and measure the distance from the centre of the float to the face of the carburettor body, as shown in Fig. 3.6. If the measured dimension is outside the float height setting given in the Specifications, carefully bend the brass contact pad on the float to achieve the required setting.

23 Engage the jet with the cut-out in the adjusting lever, ensuring that the jet head moves freely. Position the jet in the jet bearing and, at the same time, engage the slot in the adjusting lever with the protruding tip of the mixture adjustment screw. Secure the assembly with the retaining screw.

24 Turn the mixture adjustment screw as necessary to bring the top of the jet flush with the jet bearing upper face when viewed from above. Now turn the adjustment screw two complete turns clockwise to obtain an initial mixture setting.

25 Fit a new O-ring seal to the float chamber cover. Fit the cover with the previously made marks aligned and secure with the four retaining screws.

26 Refit the piston needle, spring and needle guide to the piston (photo), ensuring that the needle guide is flush with the underside of the piston and the triangular etch mark on the guide is between the two transfer holes in the piston (photo). Refit and tighten the locking screw.

27 Temporarily refit the piston and dashpot to the carburettor body without the spring. Engage the piston in its guide, and, with the suction chamber in its correct position relative to the retaining screws, mark the piston-to-suction chamber relationship. Remove the suction chamber and piston.

28 Fit the spring to the piston, align the previously made marks and slide the suction chamber over the piston and spring. Avoid turning the piston in the suction chamber, otherwise the spring will be wound up.

29 Push the piston up and refit the circlip to the piston rod.

30 Refit the piston and suction chamber, and secure with the three retaining screws tightened evenly. Fill the piston damper with engine oil up to the top of the piston rod and refit the damper.

31 Refit the fuel shut-off valve using a new gasket, if necessary, and secure the unit with the three screws.

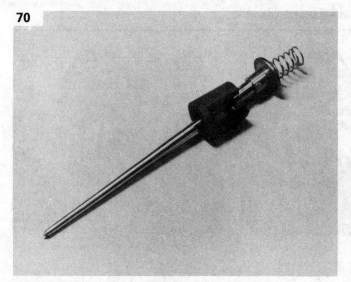

12.26A Refit the needle, spring and guide to the piston ...

12.26B ... with the guide flush with the piston base and the etch mark (arrowed) positioned as shown

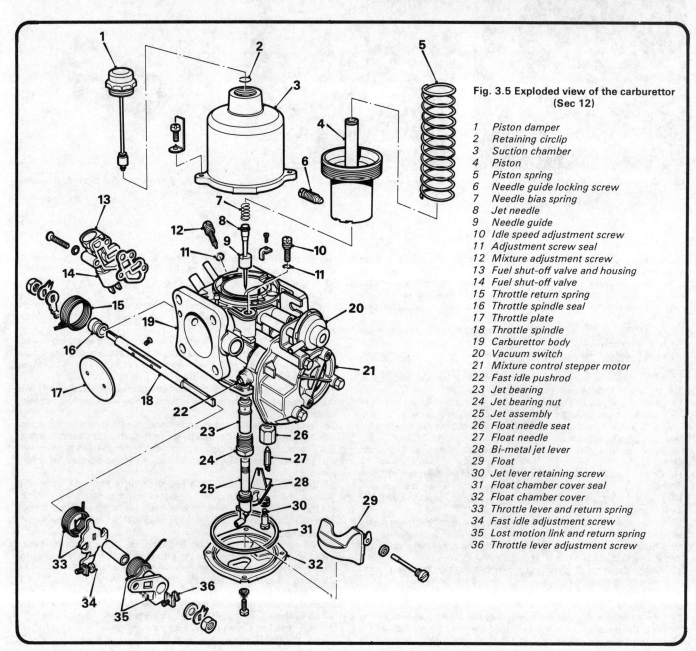

**Fig. 3.5 Exploded view of the carburettor (Sec 12)**

1  Piston damper
2  Retaining circlip
3  Suction chamber
4  Piston
5  Piston spring
6  Needle guide locking screw
7  Needle bias spring
8  Jet needle
9  Needle guide
10  Idle speed adjustment screw
11  Adjustment screw seal
12  Mixture adjustment screw
13  Fuel shut-off valve and housing
14  Fuel shut-off valve
15  Throttle return spring
16  Throttle spindle seal
17  Throttle plate
18  Throttle spindle
19  Carburettor body
20  Vacuum switch
21  Mixture control stepper motor
22  Fast idle pushrod
23  Jet bearing
24  Jet bearing nut
25  Jet assembly
26  Float needle seat
27  Float needle
28  Bi-metal jet lever
29  Float
30  Jet lever retaining screw
31  Float chamber cover seal
32  Float chamber cover
33  Throttle lever and return spring
34  Fast idle adjustment screw
35  Lost motion link and return spring
36  Throttle lever adjustment screw

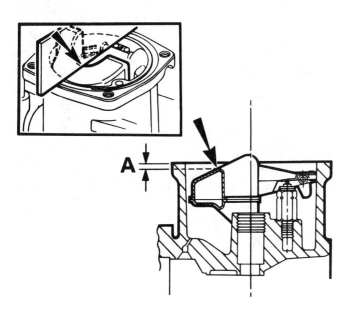

**Fig. 3.6 Carburettor float height adjustment (Sec 12)**

*A = Specified float level height*
*Arrows indicate checking and adjustment points*

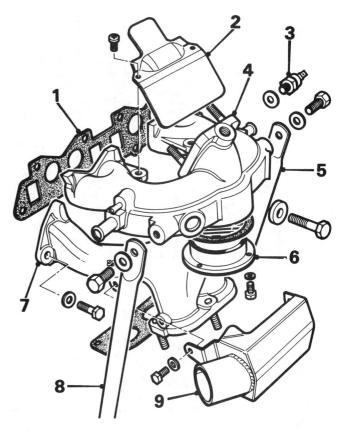

**Fig. 3.7 Inlet and exhaust manifold assemblies (Sec 13)**

| 1 | Manifold gasket | 5 | Steady bar |
|---|---|---|---|
| 2 | Heat shield | 6 | Inlet manifold heater |
| 3 | Manifold temperature sensor | 7 | Exhaust manifold |
| 4 | Inlet manifold | 8 | Steady bar |
| | | 9 | Hot air box |

## 13 Inlet and exhaust manifolds – removal and refitting

1 Disconnect the battery negative terminal.
2 Drain the cooling system as described in Chapter 2.
3 Remove the carburettor as described in Section 11.
4 Undo the banjo union bolt in the centre of the inlet manifold. Recover the two copper washers, one located each side of the banjo union, then place the servo vacuum hose to one side.
5 Slacken the retaining clips and disconnect the heater hoses at the inlet manifold.
6 Slacken the retaining clips securing the hoses to both ends of the water pipes at the rear of the manifold. Undo the bolts securing the pipe support brackets and remove the two water pipes.
7 Jack up the front of the car and support it on axle stands.
8 Undo the nuts securing the exhaust downpipe to the manifold. Separate the flange joint and recover the gasket.
9 Disconnect the electrical lead at the inlet manifold heater and the two wires at the manifold temperature sensor (photo). Disconnect the vacuum hose at the inlet manifold connector.
10 Undo the bolts securing the steady bars to the inlet manifold (photo).
11 Undo the screws securing the heat shield to the centre of the inlet manifold and remove the shield (photo).
12 Undo all the bolts securing the inlet and exhaust manifolds to the cylinder head, noting the location of the float chamber vent hose clip on the end bolt.
13 Withdraw the inlet manifold, undo the remaining bolts and nuts then withdraw the exhaust manifold (photos). Recover the one-piece gasket.
14 Refitting is the reverse sequence to removal, bearing in mind the following points:

  (a) Ensure that the manifold and cylinder head mating faces are perfectly clean and use a new gasket
  (b) Tighten all nuts and bolts to the specified torque where applicable
  (c) Refit the carburettor as described in Section 11
  (d) Refill the cooling system as described in Chapter 2

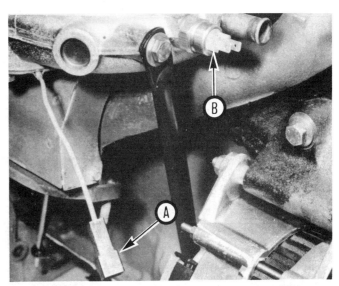

13.9 Inlet manifold heater lead (A) and temperature sensor (B)

13.10 Manifold left-hand steady bar retaining bolt (arrowed)

13.11 Removing the heat shield from the manifold

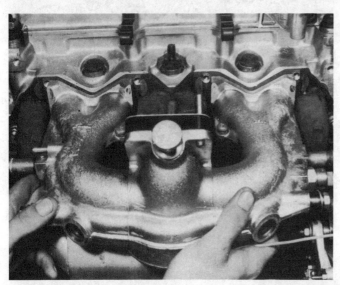

13.13A Withdraw the inlet manifold ...

13.13B ... followed by the exhaust manifold

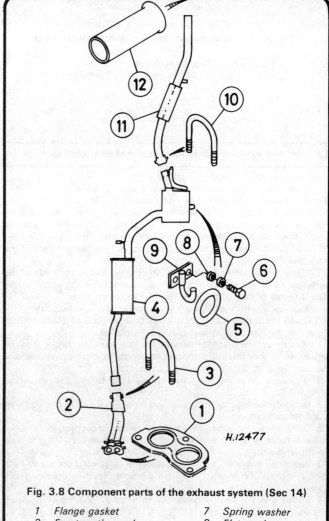

H.12477

Fig. 3.8 Component parts of the exhaust system (Sec 14)

| | | | |
|---|---|---|---|
| 1 | Flange gasket | 7 | Spring washer |
| 2 | Front section and | 8 | Flat washer |
|   | downpipes | 9 | Support bracket |
| 3 | U-clamp | 10 | U-clamp |
| 4 | Intermediate section | 11 | Rear section |
| 5 | Rubber mounting ring | 12 | Tailpipe trim |
| 6 | Bolt | | |

## 14 Exhaust system – checking, removal and refitting

1 The exhaust system should be examined for leaks, damage and security at regular intervals (see Routine Maintenance). To do this, apply the handbrake and in a well ventilated area, allow the engine to idle. Lie down on each side of the car in turn, and check the full length of the exhaust system for leaks while an assistant temporarily places a wad of cloth over the end of the tailpipe. If a leak is evident, stop the engine and use a proprietary repair kit to seal. Holts Flexiwrap and Holts Gun Gum exhaust repair systems can be used for effective repairs to exhaust pipes and silencer boxes, including ends and bends. Holts Flexiwrap is an MOT approved permanent exhaust repair. If the leak is excessive, or damage is evident, renew the section. Check the rubber mountings for deterioration, and renew them if necessary.

2 To remove the system, raise the vehicle by means of axle stands or ramps to provide adequate working clearance underneath.

3 To remove the rear section, undo the nuts of the U-clamp securing the rear section to the intermediate section. Release the rear rubber mounting block (photo) and using a twisting action withdraw the rear

section from the intermediate section. If the joint is stubborn, liberally apply penetrating oil and leave it to soak. Tap the joint with a hammer and it should now be possible to twist it free. If necessary, carefully heat the joint with a blowlamp to assist removal, *but shield the fuel tank, fuel lines and underbody adequately from heat.*

4 To remove the intermediate section first remove the rear section. The intermediate section can now be removed using the procedure described above.

5 To remove the front section first remove the rear and intermediate sections. Undo the bolts securing the anti-roll bar clamps to the front crossmember, lever down the anti-roll bar as far as possible and wedge it in this position using blocks of wood.

6 Undo the nuts securing the downpipe flange to the exhaust manifold (photo), separate the joint and recover the gasket. Lower the front section and remove it from under the car.

7 Refitting is the reverse sequence to removal. Position the joints so that there is adequate clearance between all parts of the system and the underbody, and ensure that there is equal load on all mounting blocks. Always use a new gasket on the downpipe-to-manifold joint face.

14.3 Typical exhaust section rubber mounting block

14.6 Exhaust downpipe flange and flexible joints

## 15 Fault diagnosis – fuel and exhaust systems (conventional carburettor induction)

*Unsatisfactory engine performance, bad starting and excessive fuel consumption are not necessarily the fault of the fuel system or carburettor. In fact they more commonly occur as a result of ignition and timing faults. Before acting on the following, it is necessary to check the ignition system first. Even though a fault may lie in the fuel system, it will be difficult to trace unless the ignition system is correct. The faults below, therefore, assume that, where applicable, this has been attended to first. If during the fault diagnosis procedure it is suspected that the carburettor electronic mixture control or any of its related systems may be at fault, it is recommended that the help of a reputable BL dealer is sought. Accurate testing of the system and its components entails the use of a step-by-step systematic checking procedure using specialist equipment and this is considered beyond the scope of the average home mechanic.*

| Symptom | Reason(s) |
|---|---|
| Difficult starting when cold | Faulty electronic mixture control system or related component<br>Carburettor piston sticking<br>Fuel tank empty or pump defective<br>Incorrect float chamber fuel level |
| Difficult starting when hot | Faulty electronic mixture control system or related component<br>Air cleaner choked<br>Carburettor piston sticking<br>Float chamber flooding or incorrect fuel level<br>Fuel tank empty or pump defective<br>Carburettor idle mixture adjustment incorrect |

| Symptom | Reason(s) |
|---|---|
| Excessive fuel consumption | Leakage from tank, pipes, pump or carburettor<br>Air cleaner choked<br>Carburettor idle mixture adjustment incorrect<br>Carburettor float chamber flooding<br>Faulty fuel shut-off solenoid or control system<br>Carburettor worn<br>Excessive engine wear or other internal fault<br>Tyres underinflated<br>Brakes binding |
| Fuel starvation | Fuel level low<br>Leak on suction side of pump<br>Fuel pump faulty<br>Float chamber fuel level incorrect<br>Fuel tank breather restricted<br>Fuel tank inlet or carburettor inlet filter blocked |
| Poor performance, hesitation or erratic running | Carburettor idle mixture adjustment incorrect<br>Accelerator cable out of adjustment<br>Carburettor piston damper oil level low<br>Leaking manifold gasket<br>Fuel starvation<br>Carburettor worn<br>Excessive engine wear or other internal fault |
| Excessive exhaust system noise | Leaking, corroded or damaged silencer or pipe<br>Leaking pipe-to-pipe or pipe-to-manifold joints<br>System in contact with vehicle underbody |

## Part B: Electronic fuel injection system

### 16 General description

The fuel system consists of a centrally mounted fuel tank, electric fuel pump and Lucas indirect multi-point electronic fuel injection (EFI) together with its related electrical and mechanical components.

The exhaust system is identical to that used on models with carburettor induction, consisting of a front, intermediate and rear section suspended from the underbody on rubber mountings and bolted to a cast iron manifold at the front.

**Warning:** *Many of the procedures in this Chapter entail the removal of fuel pipes and connections which may result in some fuel spillage. Before carrying out any operation on the fuel system refer to the precautions given in Safety First! at the beginning of this Manual and follow them implicitly. Petrol is a highly dangerous and volatile liquid and the precautions necessary when handling it cannot be overstressed.*

### 17 Maintenance and inspection

1    At the intervals given in Routine Maintenance at the beginning of this Manual, carry out the following service operations to the fuel system components.

2    With the car over a pit, raised on a vehicle lift, or securely supported on axle stands, carefully inspect the fuel pipes, hoses and unions for chafing, leaks and corrosion. Renew any pipes that are severely pitted with corrosion or in any way damaged. Renew any hoses that show signs of cracking or other deterioration.

3    Examine the fuel tank for leaks, particularly around the fuel gauge sender unit, and for signs of corrosion or damage.

4    Check condition of exhaust system, as described in Section 39.

5    From within the engine compartment, check the security of all fuel hose attachments and inspect the fuel hoses and vacuum hoses for kinks, chafing or deterioration.

6    Renew the air cleaner element and clean the air cleaner body and cover.

7    Check the operation of the accelerator linkage and lubricate the linkage, cable and pedal pivot with a few drops of engine oil.

8    Check the fuel injection system idle speed and mixture settings, as described in Section 27, where necessary.

9    At the less frequent intervals given in Routine Maintenance, renew the fuel filter as described in Section 20.

### 18 Air cleaner and element – removal and refitting

1    To renew the air cleaner element, spring back the retaining clips and lift off the air cleaner cover.

2    Withdraw the element then wipe clean the inside of the air cleaner body and cover (photo).

3    Fit a new element, locate the cover on the air cleaner body and secure with the retaining clips.

4    To remove the air cleaner and bracket first disconnect the battery negative terminal.

5    Slacken the retaining clip and detach the cold air intake hose from the air cleaner body (photo).

18.2 Withdraw the paper element from the air cleaner body

6   Disconnect the small air hose from the centre of the air cleaner cover (photo).

7   Disconnect the air flow meter wiring plug, undo the two bolts and withdraw the air flow meter from the air cleaner (photos). Move the air flow meter with its hose still attached, to one side.

8   Remove the two relays, with their sockets, from the air cleaner bracket. Place the relays, with wires still attached, to one side (photo).

9   Disconnect the HT lead from the centre of the coil.

10   Undo the two coil bracket mountings bolts and move the coil, with LT wiring still connected, to one side (photo).

11   Mark their locations then disconnect the two wires at the ballast resistor located below the two relays (photo).

12   Undo the two bolts (one long and one short – photo) securing the air cleaner bracket to the inner wing valance. Withdraw the air cleaner and bracket assembly from the engine compartment (photo).

13   Refitting is a direct reversal of the removal sequence.

## 19  Fuel system – depressurising

1   The fuel system must be depressurised as follows before any of the pipes or hoses between the fuel pump and the fuel pressure regulator are disconnected.

2   Detach the main and fuel pump relays from the air cleaner bracket by pulling them upwards, with their wiring plugs still attached (photo 18.8).

3   Identify the fuel pump relay by noting the wiring colours in the relay plug as follows:

*White/pink*
*Brown*
*Brown/pink*
*Blue/red*

18.5 Slacken the clip and detach the cold air intake hose

18.6 Disconnect the air hose from the outlet (arrrowed) on the air cleaner cover

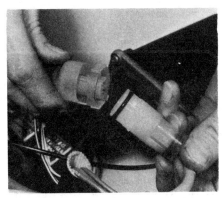

18.7A Disconnect the air flow meter wiring plug ...

18.7B ... undo the bolts and withdraw the air flow meter

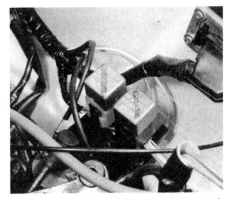

18.8 Lift the main and fuel pump relays out of their air cleaner bracket attachments

18.10 Undo the coil bracket mounting bolts (arrowed)

18.11 Disconnect the two wires at the ballast resistor

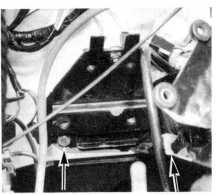

18.12A Undo the two air cleaner bracket retaining bolts (arrowed) ...

18.12B ... then remove the air cleaner and bracket assembly

4   Having ascertained which is in the fuel pump relay, disconnect it from the wiring plug. Refit the other relay back into its location in the air cleaner bracket.

5   Operate the starter and allow the engine to run until it stops. If it is unsafe for the engine to run, disconnect the white wire at the coil '+' terminal and crank the engine on the starter motor for at least 10 seconds.

6   On completion of the work for which the system was depressurised, reconnect the fuel pump and coil leads and refit the relay.

## 20  Fuel filter – removal and refitting

Note: Refer to the warning in Section 16 before proceeding.

1   Depressurise the fuel system as described in Section 19 then disconnect the battery negative terminal.

2   Place some absorbent rag beneath the fuel filter and disconnect the outlet hose from the filter nozzle (photo). Plug the hose after removal.

3   Slacken the retaining clamp pinch-bolt and slide the filter forward out of its clamp.

4   Disconnect the inlet hose at the filter nozzle and remove the filter. Plug the inlet hose if it is to remain disconnected for some time.

5   Refit the filter using the reverse sequence of removal, ensuring that the arrow on the filter points towards the outlet hose. Reconnect the fuel relay circuit as described in Section 19 then reconnect the battery.

## 21  Fuel pump – removal and refitting

Note: Refer to the warning note in Section 16 before proceeding.

1   Depressurise the fuel system as described in Section 19 then disconnect the battery negative terminal.

2   Syphon, or hand pump all the fuel from the fuel tank.

3   Jack up the rear of the car and support it on axle stands.

4   Disconnect the wiring connectors at the fuel pump which is located to the rear of the fuel tank, just above the rear axle transverse member (photo).

5   Undo the two nuts and washers securing the pump retaining strap to the mounting bracket.

6   Ease the strap studs out of the mounting bracket, then manoeuvre the pump down between the rear axle member and the spare wheel well.

7   Position a suitable bowl beneath the pump and disconnect the inlet and outlet hoses from the pump nozzles. Plug the hoses after removal. Note: a quantity of fuel will be released during this operation.

8   Note the fitted relationship of the pump terminals to the strap studs and push the pump out of the strap support rubbers. Free the pump from the rubbers if adhesive has been used for retention.

9   To refit the pump apply adhesive to the strap rubbers and insert the pump into the strap. Position the pump in the strap as noted during

20.2 Disconnect the fuel filter outlet hose (arrowed)

removal so that the terminals will be vertical when the pump is installed.

10  Reconnect the inlet and outlet hoses to the pump nozzles.

11  Fit the plain washers to the strap mounting studs, manoeuvre the pump into position and secure with the plain washers and nuts.

12  Reconnect the pump wiring and lower the car to the ground.

13  Reconnect the fuel relay circuit as described in Section 19, reconnect the battery and fill the fuel tank.

## 22  Fuel tank – removal, servicing and refitting

Refer to Part A, Section 5, but note that there is a fuel return hose as well as a feed hose at the fuel gauge sender unit and both must be disconnected.

## 23  Fuel gauge sender unit – removal and refitting

Refer to Part A, Section 6.

## 24  Accelerator cable – removal and refitting

1   Slacken the screw to release the inner cable from the connector on the throttle housing linkage (photo).

2   Release the outer cable at the support bracket and withdraw the cable from the engine (photo).

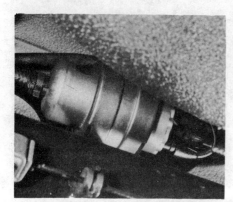

21.4 Fuel pump location above the rear axle member

24.1 Slacken the screw (arrowed) to release the accelerator cable from the linkage connector

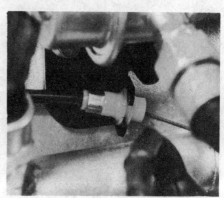

24.2 Release the accelerator outer cable from the support bracket

3  Working inside the car, prise the retaining clip from the top of the accelerator pedal and disconnect the inner cable.

4  Release the accelerator cable from the engine compartment bulkhead and withdraw the complete cable from the car.

5  Refitting is the reverse sequence of removal, but adjust the position of the inner cable in its connector to give a small amount of free play with the throttle closed. Check that with the accelerator pedal fully depressed the linkage is fully open, and with the pedal released the linkage closes fully.

## 25  Accelerator pedal – removal and refitting

Refer to Part A, Section 8.

## 26  Electronic fuel injection system – description and operation

The electronic fuel injection (EFI) system is a microprocessor controlled fuel management system designed to overcome the limitations associated with conventional carburettor induction. This is achieved by continuously monitoring the engine using various sensors whose data is input to the fuel system electronic control unit (ECU). Based on this information, the ECU program and memory then determine the exact amount of fuel necessary, which is injected directly into the inlet manifold, for all actual and anticipated driving conditions.

The main components of the system are shown in Fig. 3.10 and their individual operation is as follows.

*Fuel ECU:* The fuel ECU is a microprocessor which controls the entire operation of the fuel injection system. Contained in the ECU memory is a program which controls the fuel supply to the injectors and their opening duration. The program enters sub-routines to alter these parameters according to inputs from the other components of the system.

*Injectors:* Each fuel injector consists of a solenoid operated needle valve which opens under commands from the fuel ECU. Fuel from the fuel rail is then delivered through the injector nozzle into the inlet manifold.

*Coolant thermister:* The thermister senses engine temperature through an element whose resistance alters with changes in temperature. This information is then supplied to the fuel ECU.

*Air flow meter:* The air flow meter contains two resistive elements

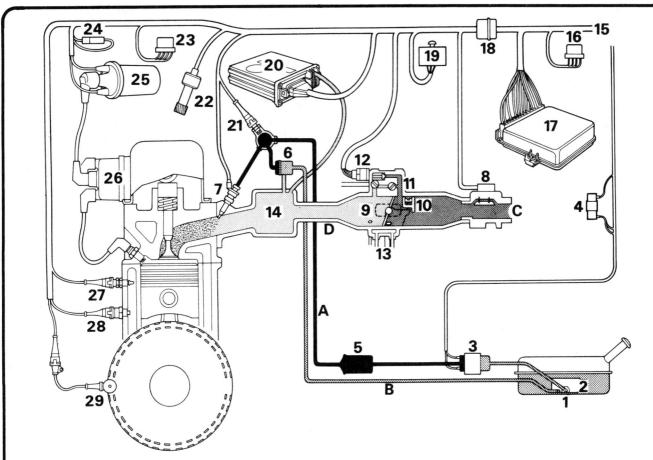

**Fig. 3.9 Schematic layout of the electronic fuel injection system (Sec 26)**

| | |
|---|---|
| 1 | Fuel tank pick-up strainer |
| 2 | Fuel spill return swirl pot |
| 3 | Fuel pump |
| 4 | Fuel pump ballast resistor |
| 5 | Fuel filter |
| 6 | Fuel pressure regulator |
| 7 | Injector |
| 8 | Air flow meter |
| 9 | Throttle potentiometer |
| 10 | Throttle stop screw |
| 11 | Idle speed and mixture adjustment screws |
| 12 | Air valve stepper motor |
| 13 | Throttle housing – coolant and crankcase ventilation ports |
| 14 | Inlet manifold |
| 15 | To main harness multi-plug |
| 16 | Main relay |
| 17 | Fuel ECU |
| 18 | Engine harness multi-plug |
| 19 | Inertia switch |
| 20 | Ignition ECU |
| 21 | Fuel temperature switch |
| 22 | Speed sensor |
| 23 | Fuel relay |
| 24 | Ignition coil ballast resistor |
| 25 | Ignition coil |
| 26 | Distributor cap and spark plug leads |
| 27 | Coolant temperature thermister |
| 28 | Knock sensor |
| 29 | Crankshaft sensor |
| A | Regulated fuel pressure |
| B | Spill return low pressure fuel |
| C | Inlet air flow |
| D | Manifold depression |

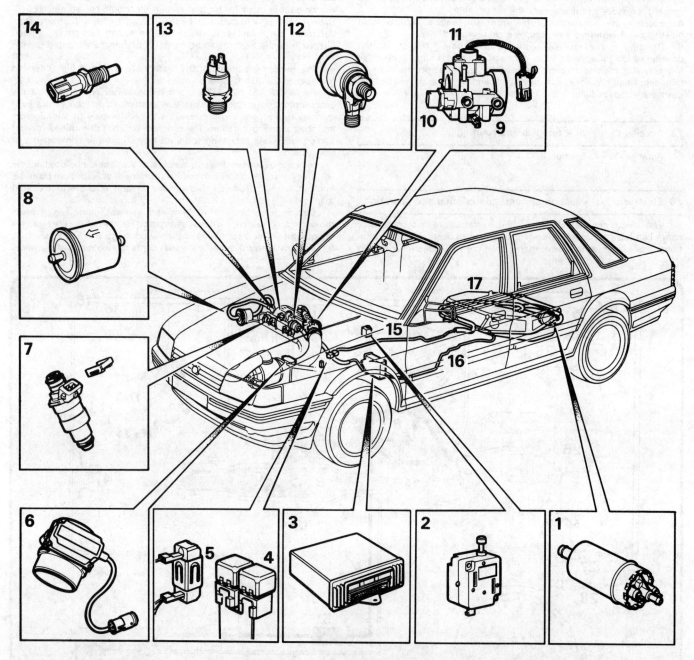

**Fig. 3.10 Electronic fuel injection system component location (Sec 26)**

| | |
|---|---|
| 1 Fuel pump | 6 Air flow meter |
| 2 Inertia switch | 7 Fuel injector and |
| 3 Fuel ECU | retaining clip |
| 4 Main relay | 8 Fuel filter |
| 5 Fuel relay and pump | 9 Throttle housing |
| ballast resistor | 10 Air valve stepper motor |

| | |
|---|---|
| 11 Throttle potentiometer | 15 Fuel feed pipe |
| 12 Fuel pressure regulator | 16 Fuel spill return pipe |
| 13 Fuel temperature switch | 17 Fuel tank with return |
| 14 Coolant temperature | swirl pot |
| thermister | |

mounted in the intake air stream, one of which is heated by a current passing through it. Air passing over the heated wire alters its resistance by cooling it, while the temperature of the air is sensed by the other element. An electronic module within the air flow meter monitors the reaction of the elements to the air flow and provides a proportional signal to the fuel ECU.

*Throttle potentiometer:* The potentiometer is a variable resistor attached to the throttle shaft in the throttle housing. The unit is supplied with a constant input voltage and as the resistance of the potentiometer varies with throttle shaft movement, the output voltage is proportionately affected. This allows the fuel ECU to determine

throttle valve position and rate of change.

*Air valve stepper motor:* The stepper motor controls idling speed in response to signals received from the fuel ECU. The motor has two control windings enabling it to rotate in either direction. Thus the air valve in the throttle housing can be opened or closed to bypass the throttle valve and maintain a stabilised idling speed.

*Fuel pump:* The electric fuel pump is a self priming centrifugal type of the 'wet' variety whereby the pump and motor are both immersed in fuel. The pump supplies fuel under pressure to the fuel rail and pressure regulator, via the fuel filter.

*Fuel pressure regulator:* The regulator is a vacuum operated mechanical device which ensures that the pressure differential be-tween fuel in the fuel rail and the inlet manifold is maintained at a constant value. As manifold depression increases, the regulated fuel pressure is reduced in direct proportion. When fuel pressure in the fuel rail exceeds the regulator setting, the regulator opens to allow fuel to return to the tank.

*Relays:* Two relays are used in the system and are mounted on the air cleaner bracket. The main relay is energised when the ignition is switched on and provides the fuel ECU supply voltage. The fuel relay provides a supply voltage to the fuel pump. The relay is energised for three seconds after the ignition is initially switched on and then continuously when the engine is running.

*Ballast resistor:* The ballast resistor is located beneath the relays and is connected in the electrical circuit between fuel relay and fuel pump. When the ignition is initially switched on and when the engine is running the ballast resistor reduces the voltage supplied to the fuel pump to approximately 7 volts. When the engine is cranking on the starter the ballast resistor is bypassed and full battery voltage is supplied to the fuel pump.

*Fuel temperature switch:* The fuel temperature switch contacts remain open during normal engine operation and only close when the fuel rail temperature exceeds 194°F (90°C). When the contacts close, a signal is sent to the fuel ECU overriding the coolant thermister signal. The ECU then alters the opening duration of the injectors accordingly to minimise the effects of fuel vaporisation.

*Inertia switch:* The switch is a mechanically controlled ac-celerometer connected in the electrical circuit between the ignition switch and the fuel ECU and fuel relay. Under violent deceleration or impact, the switch trips out and cuts off the supply voltage. Depressing a button resets the switch.

The electronic fuel injection system operates in the following way. When the ignition is switched on a voltage is supplied via the inertia switch to the main relay, fuel ECU and ignition ECU. The fuel ECU energises the fuel pump relay and a voltage is supplied to the pump via the ballast resistor. The pump is allowed to run for three seconds to pressurise the system. Excess fuel is returned to the tank by the action of the fuel pressure regulator.

When the starter is operated the ballast resistor is bypassed and battery voltage is supplied directly to the pump which runs at maximum speed. Inputs to the fuel ECU from the road speed transducer, coolant thermister, throttle potentiometer, air flow meter and ignition ECU enable the fuel ECU to establish the amount of fuel required, and the injector opening duration, to allow the engine to start and run. During starting the injectors operate simultaneously, and at each ignition pulse so that fuel is sprayed into the inlet manifold at twice the normal rate, giving the necessary enrichment for starting.

When the engine fires and runs the voltage to the fuel pump is diverted back through the ballast resistor thus reducing the pump running speed. During engine idling the fuel ECU modifies the injector opening duration and fuel supply rate according to data received from the various sensors. The ignition ECU is also signalled to hold the ignition timing at a preset value and the air valve stepper motor is activated to control the intake air supply as necessary.

During normal driving any changes in the information from the sensors cause the ECU program to enter a sub-routine and determine the new fuel supply and injector duration requirements accordingly.

During full throttle acceleration the injectors are held open for a longer duration thus providing the necessary enrichment to avoid hesitation. Under overrun conditions the fuel supply is cut off by the fuel ECU providing the engine has reached a predetermined temperature and the accelerator pedal is released. When the engine speed decreases or the accelerator pedal is activated, the fuel supply is gradually reinstated to eliminate hesitation.

During hot start conditions inputs to the fuel ECU from the coolant thermister and fuel temperature switch cause the fuel ECU to alter the injector opening duration accordingly to counteract vaporisation.

Under conditions of abrupt deceleration or impact the inertia switch opens and breaks the system supply voltage. This shuts down the fuel system, stops the engine and reduces the fire hazard.

## 27 Electronic fuel injection system – adjustments

**Warning:** *Austin Rover consider it essential to reset the position of the throttle potentiometer on the throttle housing whenever the position of the throttle is altered via the throttle stop screw. The potentiometer*

position can only be accurately set using Austin Rover electronic test equipment, so this work should ideally be carried out by your dealer. However, the following procedure may be used as a temporary measure.

**Note:** *Before making any changes to the settings of the fuel injection system, ensure that the spark plug gaps and valve clearances are correctly set and that the ignition system is operating correctly. The function of the system is such that the engine idling speed is controlled by the fuel system ECU which, in conjunction with the ignition ECU also maintains a constant ignition timing setting at the idling speed. Therefore unless a new component has been fitted, the idle speed or mixture screws have been tampered with, or it is known that the system requires attention, no adjustment should normally be necessary. If however the settings are to be altered an accurate tachometer and an exhaust gas analyser (CO meter) will be required.*

1   Undo the clip and disconnect the intake air hose from the throttle housing.

2   Using feeler gauges measure the clearance between the base of the throttle valve and the housing. A 0.0015 in (0.038 mm) feeler blade should be a sliding fit. If adjustment is necessary, hook off the tamperproof cap over the throttle stop screw using a screwdriver (photo) then turn the screw as required to give the correct clearance. Refit the intake air hose after this adjustment.

27.2 Hook out the throttle stop screw tamperproof cap using a screwdriver

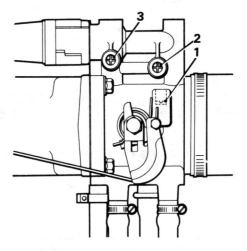

Fig. 3.11 Throttle housing adjustment points (Sec 27)

1   Throttle stop screw
2   Idle speed adjustment screw (early models)
3   Idle mixture adjustment screw

3    Start the engine and run it until normal operating temperature is obtained. The electric cooling fan should cut-in at least once.

4    Switch off the engine and connect a tachometer to the coil '–' terminal.

5    Switch on the ignition and disconnect the wiring multi-plug from the stepper motor on top of the throttle housing. Switch off the ignition, wait five seconds then reconnect the multi-plug. Switch on the ignition, wait five seconds and disconnect the multi-plug once more. This will ensure that the stepper motor air valve is closed and will remain closed during adjustment.

6    Start the engine and allow it to idle. On early models with an idle speed adjustment screw, turn the screw fully clockwise to close the throttle bypass. Turn the throttle stop adjustment screw as necessary until the engine is idling at 670 rpm (650 rpm if the ambient temperature is above 77°F/25°C). Turning the screw clockwise will decrease the speed and turning it anti-clockwise will increase it.

7    Reconnect the multi-plug to the stepper motor. Automatic idling will now take over and the speed will be controlled by the fuel system ECU.

8    Connect an exhaust gas analyser in accordance with the manufacturer's instructions.

9    With the engine idling take a reading of the exhaust gas CO content. If this is not as given in the Specifications, turn the idle mixture adjustment screw by the smallest amount necessary to obtain the specified setting. Turning the screw clockwise increases the CO content and turning the screw anti-clockwise decreases it.

10    After completing the adjustments, switch off the ignition and disconnect the instruments. Turn the idle speed adjustment screw (early models) back to its original setting (fully anti-clockwise).

11    Failure to rectify a weak mixture (CO level) may be due to partial clogging of the fuel injectors. Check their output as described in Section 34.

## 28 Air-flow meter – removal and refitting

1    Disconnect the air flow meter wiring plug.

2    Release the retaining clip and detach the intake air hose from the air flow meter (photo).

3    Undo the two bolts securing the meter to the air cleaner bracket and withdraw the unit from the bracket and air cleaner body.

4    Refitting is the reverse sequence to removal.

28.2 Intake air hose attachment at the air flow meter

## 29 Air valve stepper motor – removal and refitting

1    Slide back the rubber cover and disconnect the stepper motor multi-plug (photo).

2    Using a 32 mm spanner, unscrew the stepper motor from the throttle housing.

29.1 Disconnect the multi-plug from the stepper motor (arrowed)

3    To refit the stepper motor, fit a new sealing washer and apply a thread sealant to the stepper motor threads.

4    Screw the unit in finger tight then carefully tighten with the spanner.

5    Refit the multi-plug and the rubber cover.

## 30 Fuel temperature switch – removal and refitting

1    Disconnect the wiring multi-plug and unscrew the switch from the top of the fuel rail.

2    Refit the switch using the reverse sequence of removal.

## 31 Fuel pressure regulator – removal and refitting

**Note:** *Refer to the warning note in Section 16 before proceeding.*

1    Depressurise the fuel system as described in Section 19 then disconnect the battery negative terminal.

2    Place absorbent rags beneath the regulator, unscrew the union nuts and remove the two fuel hoses (photo).

3    Disconnect the vacuum hose from the rear of the regulator.

31.2 Fuel hose union nuts (A) vacuum hose (B) and retaining nut (C) at the fuel pressure regulator

4  Undo the retaining nut and washer then withdraw the unit from its support bracket.
5  Refitting is the reverse sequence of removal. After fitting reconnect the fuel relay circuit as described in Section 19 and reconnect the battery.

## 32 Fuel rail – removal and refitting

**Note:** *Refer to the warning note in Section 16 before proceeding.*
1  Depressurise the fuel system as described in Section 19 then disconnect the battery negative terminal.
2  Place absorbent rag beneath the fuel rail.
3  Unscrew the fuel pressure regulator to fuel rail hose union at the regulator. Plug the hose and regulator after removal.
4  Slacken the fuel pressure regulator retaining nut, slide the regulator out of its support bracket and place it to one side.
5  Unscrew the fuel inlet hose union at the fuel rail (photo), remove the hose and plug its end.
6  Disconnect the wiring multi-plug at the fuel temperature switch and the multi-plug at each fuel injector (photo).
7  Extract the clips securing the fuel injectors to the fuel rail.
8  Undo the two bolts securing the fuel rail to the manifold (photo),

32.8 Undo the fuel rail to manifold retaining bolts (arrowed)

hold the injectors to prevent displacement and remove the fuel rail.
9  Before refitting the fuel rail, renew the O-ring seals on each fuel injector.
10  Refitting the fuel rail is the reverse sequence of removal. After fitting reconnect the fuel relay circuit as described in Section 19 and reconnect the battery.

## 33 Fuel injectors – removal and refitting

1  Depressurise the fuel system as described in Section 19 then disconnect the battery negative terminal.
2  Slacken the fuel pressure regulator retaining nut, slide the

32.5 Fuel hose to fuel rail union nut (arrowed)

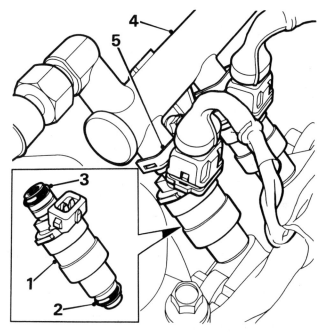

Fig. 3.12 Fuel injector details (Sec 33)

1  Fuel injector
2  Inlet manifold seating O-ring seal
3  Fuel rail seating O-ring seal
4  Fuel rail
5  Injector retaining clip

32.6 Disconnect the wiring multi-plugs at the fuel injectors

regulator out of its support bracket and lay it on the manifold with hoses still attached.

3   Disconnect the wiring multi-plug at the fuel temperature switch and the multi-plug at each fuel injector.

4   Undo the two bolts securing the fuel rail to the manifold.

5   Lift the fuel rail up and at the same time ease the injectors out of their locations in the manifold. Extract the retaining clip(s) and withdraw the injector(s) from the fuel rail.

6   To refit the injector(s), wipe clean the injector and manifold seating areas and fit a new O-ring seal to the base of each injector. Fit a new O-ring seal to the top of any injector that was removed from the fuel rail.

7   Fit the injectors to the fuel rail and secure with the retaining clips.

8   Align the injectors with their seatings in the manifold and press them fully into position.

9   Refit the two fuel rail retaining bolts and tighten securely.

10  Reconnect the fuel temperature switch and fuel injector multi-plugs.

11  Slide the fuel pressure regulator back into its support bracket and tighten the retaining nut.

12  Reconnect the fuel relay circuit as described in Section 19 then reconnect the battery.

### 34  Fuel injectors – testing

**Note:** *Refer to the warning note in Section 16 before proceeding.*

1   Slacken the fuel pressure regulator retaining nut, slide the regulator out of its support bracket and lay it on the manifold with hoses still attached.

2   Undo the two bolts securing the fuel rail to the manifold. Lift up the fuel rail and at the same time ease the injectors out of their seatings in the manifold. Lay the injectors over the engine wiring harness.

3   Switch on the ignition, wait ten seconds, then switch off. The fuel pump will operate for approximately three seconds while the ignition is switched on and will pressurise the injectors.

4   Observe each injector and renew any that leak more than two drops of fuel per minute.

5   To check the injector spray pattern, disconnect all the injector multi-plugs except the one on the injector to be tested.

6   Place a glass jar under the connected injector so that the sprayed fuel can be contained, but the pattern still observed.

7   Ensure that the ignition is switched off then disconnect the wiring multi-plug from the fuel system ECU. The ECU is located under the carpet above the passenger's footwell.

8   Using suitable connectors, connect pin No 1 and pin No 16 in the ECU multi-plug connector to earth. Ensure that the connectors do not touch any of the pins in the plug. Observe the fuel spray which will now be emitting from the injector. The spray pattern should be regular and at a rate of 185cc/min. If this is not the case the injector should be renewed.

9   After testing, remove the earth connector from the ECU multi-plug. Reconnect the multi-plug to the next injector to be tested and position it in the jar, Remove the multi-plug from the previously tested injector then repeat the test as described in paragraph 8.

10  After all the injectors have been tested, renew any that are faulty with reference to Section 33.

11  Fit new O-ring seals to the base of all the injectors and wipe clean their seats.

12  Locate the injectors into their seats in the manifold, push them down securely and refit the two fuel rail retaining bolts. Reconnect the injector multi-plugs.

13  Slide the fuel pressure regulator into its support bracket and tighten the retaining nut.

### 35  Throttle housing – removal and refitting

1   Disconnect the battery negative terminal.

2   Disconnect the wiring multi-plug at the end of the throttle potentiometer lead.

3   Slide back the rubber cover and disconnect the wiring multi-plug at the air valve stepper motor.

4   Slacken the retaining clip and detach the intake air hose from the throttle housing.

5   Slacken the retaining screw and release the accelerator cable from the connector on the throttle housing linkage.

6   Unscrew the cooling system expansion tank filler cap to release all pressure in the cooling system. Remove the cap slowly if the system is hot.

7   Disconnect the air and vacuum hoses at the rear of the throttle housing, noting their locations (photo).

8   Undo the bolts securing the throttle housing to the manifold and ease the unit off the flange.

9   Disconnect the coolant and crankcase breather hoses from beneath the housing and remove the housing from the engine. Plug the coolant hose to prevent water loss.

10  The air valve stepper motor may be removed from the throttle housing by simply unscrewing it using a 32 mm spanner. It is advisable to leave the throttle potentiometer undisturbed so as not to lose its set fitted position. If the potentiometer is disturbed, recalibration by a BL dealer will be necessary.

11  Refitting the throttle housing is the reverse sequence of removal bearing in mind the following points:

(a)   *Ensure that the throttle housing and manifold mating faces are perfectly clean and if necessary use a new gasket*

(b)   *Reconnect the accelerator cable with reference to Section 24*

(c)   *Top-up the cooling system if necessary to the level indicated on the expansion tank*

### 36  Inertia switch – removal and refitting

1   The switch is located behind the radio cassette player on the right-hand side of the centre console. Access to the unit can be gained by working through the driver's footwell.

2   Undo the bolt securing the switch mounting bracket to the base of the heater (photo).

3   Withdraw the switch, disconnect the wiring multi-plug and remove the switch from the car.

4   Undo the two through bolts and remove the switch from its mounting bracket.

5   To test the switch operation, depress the red button and firmly strike the rear of the unit against the palm of your hand. The button should spring out and lock in the raised position. If this is not the case the unit should be renewed.

6   Refitting the switch is the reverse sequence of removal.

### 37  Fuel system ECU – removal and refitting

1   The fuel system ECU is located behind the carpet above the front passenger's footwell.

2   To remove the unit, disconnect the battery negative terminal then disconnect the ECU wiring multi-plug.

3   Undo the retaining screws and remove the unit from its mounting bracket.

4   Refitting is the reverse sequence of removal.

### 38  Inlet and exhaust manifolds – removal and refitting

**Note:** *Refer to the warning note in Section 16 before proceeding.*

1   Depressurise the fuel system as described in Section 19 then disconnect the battery negative terminal.

2   Undo the bolt securing the servo vacuum hose banjo union to the inlet manifold (photo). Recover the copper washers located each side of the union and move the hose clear of the manifold.

3   Disconnect the air valve stepper motor and throttle potentiometer wiring multi-plugs.

4   Note the location of the vacuum and air hoses at the rear of the throttle housing and disconnect them.

5   Disconnect the wiring multi-plugs from the injectors and from the fuel temperature switch on the fuel rail.

6   Disconnect the fuel pressure regulator vacuum hose from the inlet manifold.

7   Place absorbent rags beneath the fuel rail and unscrew the fuel rail to fuel pressure regulator hose at the regulator. Plug the hose to prevent further loss of fuel.

35.7 Air and vacuum hose attachments at the throttle housing

36.2 Inertia switch mounting bracket retaining bolt (arrowed)

38.2 Servo vacuum hose banjo union and retaining bolt

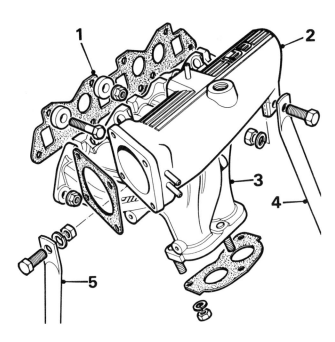

**Fig. 3.13 Exploded view of the inlet and exhaust manifolds (Sec 38)**

1  *Manifold gasket*
2  *Inlet manifold*
3  *Exhaust manifold*
4  *Right-hand steady bar*
5  *Left-hand steady bar*

8  Slacken the fuel pressure regulator retaining nut and slide the regulator out of its support bracket.
9  Undo the two fuel rail retaining bolts, lift up the fuel rail and at the same time ease the injectors out of their seatings in the manifold. Position the fuel rail and injectors to one side. Undo the two heater pipe retaining bolts.
10  Slacken the screw securing the accelerator cable to its connector on the throttle housing linkage. Remove the cable from the connector and mounting bracket and move it clear of the engine.

11  Undo the four bolts securing the throttle housing to the inlet manifold and recover the gasket.
12  Undo the manifold nuts and bolts which are accessible from above including the exhaust manifold nut located under the thermostat housing. Slacken the two bolts and nuts securing the steady bars to the manifold.
13  Jack up the front of the car and support it on axle stands.
14  From under the car undo the nuts securing the exhaust downpipe flange to the manifold. Separate the joint and recover the gasket.
15  Undo the bolts securing the manifold steady bars to the engine and gearbox. Undo the remaining bolt and lift off the alternator heat shield.
16  Undo the remaining manifold nuts and bolts which are accessible from below.
17  From above, carefully ease the two manifolds off the cylinder head and remove them from the engine. Recover the manifold to cylinder head gasket.
18  Refitting the manifolds is the reverse sequence of removal bearing in mind the following points:

(a)  *Ensure that all the mating faces are clean and use new gaskets throughout. Locate the exhaust manifold in place first and secure with the outer nuts finger tight. Fit the inlet manifold and all the remaining nuts and bolts then tighten them progressively in a diagonal sequence*
(b)  *Fit new O-ring seals to the injectors and ensure that their seatings are clean*
(c)  *Connect the accelerator cable to the throttle housing linkage with reference to Section 24*
(d)  *Reconnect the fuel relay circuit as described in Section 19 and reconnect the battery*

## 39 Exhaust system – checking, removal and refitting

Refer to Part A, Section 14.

## 40 Fault diagnosis – fuel and exhaust systems (electronic fuel injection system)

Owing to the complexity of the electronic circuitry and the nature of the computer controlled operation, special test equipment has been developed for fault diagnosis on the fuel injection system. Therefore any suspected faults on the system or its related components should be referred to a suitably equipped BL dealer.

# Chapter 4 Ignition system

*For modifications, and information applicable to later models, see Supplement at end of manual*

## Contents

## Specifications

**System type** ................................................ Programmed electronic ignition

**Ignition coil**
Type ................................................ Lucas 45328 or Bosch 0.22 122 360
Current consumption – engine idling ................................................ 2.3 to 2.7 Amps
Primary resistance at 20°C (68°F) ................................................ 0.82 ohms ± 5%

**Ignition electronic control unit**
Type:
    All models except MG EFi ................................................ Lucas AB17 – 84187
    MG EFi models ................................................ Lucas AB17 – 84381

**Crankshaft sensor**
Type ................................................ Lucas 84229

**Knock sensor**
Type ................................................ Lucas or Lamerholm type VP50/1-M12

**Firing order** ................................................ 1-3-4-2

**Direction of rotor arm rotation** ................................................ Anti-clockwise

**Ignition timing***
Vacuum connected:
    All models except MG EFi ................................................ 37° to 42° BTDC at 1500 rpm
    MG EFi models ................................................ 40° BTDC at 1500 rpm
Vacuum disconnected:
    All models except MG EFi ................................................ 20° to 26° BTDC at 1500 rpm
    MG EFi models ................................................ 15° BTDC at 1500 rpm
*Non-adjustable, for checking purposes only

**Spark plugs**
Type:
    Carburettor models (1984 to 1988) ................................................ Champion RN9YCC or RN9YC
    EFi models ................................................ Champion RN7YCC or RN7YC
Electrode gap ................................................ 0.8 mm (0.032 in)

**HT leads** ................................................ Champion LS-05 boxed set

**Torque wrench settings**

| | lbf ft | Nm |
|---|---|---|
| Spark plugs | 13 | 18 |

## 1 General description

Montego models covered by this manual are equipped with a programmed electronic ignition system which utilises computer technology and electro-magnetic circuitry to simulate the main functions of a conventional ignition distributor.

A reluctor ring on the periphery of the engine flywheel, and a crankshaft sensor whose inductive head runs between the reluctor ring teeth, replace the operation of the contact breaker points in a conventional system. The reluctor ring utilizes 34 teeth spaced at 10° intervals with two spaces 100° apart corresponding to TDC for Nos 1 and 4 pistons and Nos 2 and 3 pistons respectively. As the crankshaft rotates, the reluctor ring teeth pass over the crankshaft sensor which transmits a pulse to the ignition electronic control unit (ECU) every

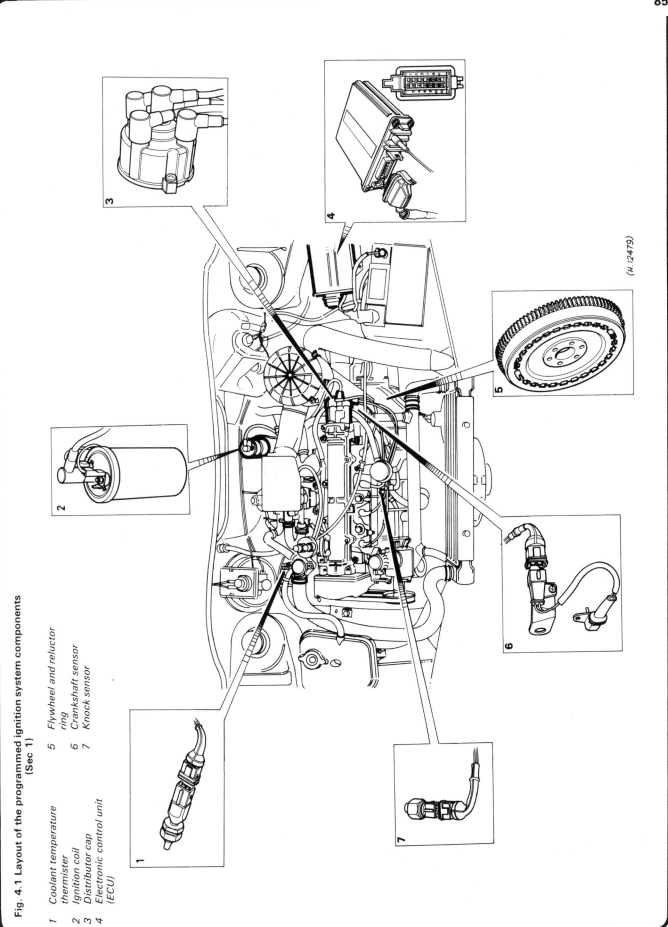

Fig. 4.1 Layout of the programmed ignition system components
(Sec 1)

1 Coolant temperature
  thermister
2 Ignition coil
3 Distributor cap
4 Electronic control unit
  (ECU)

5 Flywheel and reluctor
  ring
6 Crankshaft sensor
7 Knock sensor

(H. 12479)

time a tooth passes over it. The ECU recognises the absence of a pulse every 180° and consequently establishes the TDC position. Each subsequent pulse then represents 10° of crankshaft rotation. This, and the time interval between pulses allows the ECU to accurately determine crankshaft position and speed.

A small bore pipe connecting the inlet manifold to a pressure transducer within the ECU supplies the unit with information on engine load. From this constantly changing data, the ECU selects a particular advance setting from a range of ignition characteristics stored in its memory. This basic setting can be further advanced or retarded according to information sent to the ECU from the coolant temperature thermister and knock sensor.

With the firing point established, the ECU triggers the ignition coil which delivers HT voltage to the spark plugs in the conventional manner. The cycle is then repeated, many times a second for each cylinder in turn.

On MG models with electronic fuel injection, many of the ignition system components have a second function in the control and operation of the fuel injection system. Further details will be found in Chapter 3.

**Warning:** *The voltages produced by the electronic ignition system are considerably higher than those produced by a conventional system. Extreme care must be used when working on the system with the ignition switched on, particularly by persons fitted with a cardiac pacemaker.*

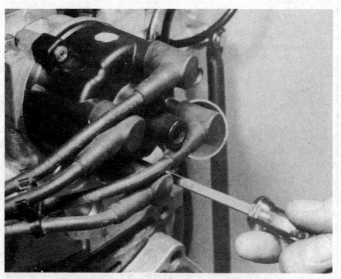

3.1 Undo the two distributor cap retaining screws

## 2 Maintenance and inspection

1 At the service intervals given in Routine Maintenance at the beginning of this Manual, remove the distributor cap and thoroughly clean it inside and out with a dry lint-free rag. Examine the four HT lead segments inside the cap. If the segments appear badly burnt or pitted, renew the cap. Check the carbon brush in the centre of the cap ensuring that it is free to move and that it stands proud of its holder.
2 Check all ignition wiring, cables and HT leads for security and cleanliness and wipe them over with a clean rag if necessary. Pay particular attention to the wiring and HT lead connections at the coil tower. Dirt or moisture in this area can increase the likelihood of HT leakage due to arcing.
3 Renew the spark plugs at the recommended interval (see *Routine maintenance*).

## 3 Distributor cap and rotor arm – removal and refitting

1 Undo the two screws and lift the cap off the cylinder head (photo). Thoroughly clean the cap inside and out with a dry lint-free rag. Examine the four HT lead segments and the carbon brush inside the cap. Renew the cap if the segments are badly burnt or pitted or if the carbon brush does not stand proud of its holder.
2 If renewal of the cap is necessary, record the position of the HT leads in relation to the cap then pull them off. Transfer the leads to a new cap, refitting them in the same position.
3 To remove the rotor arm, undo the retaining screw using a suitable Allen key and withdraw the rotor arm from the end of the camshaft (photo). Recover the rotor arm shield.
4 Refitting the shield, rotor arm and distributor cap is the reverse sequence of removal. Before refitting the rotor arm, clean out its retaining screw thread with an M6 tap to remove all traces of any remaining locking solution left by the screw.

## 4 Crankshaft sensor – removal and refitting

1 Disconnect the battery negative terminal.
2 Disconnect the wiring multi-plug and undo the wiring plug screw. Undo the two bolts and withdraw the unit from the gearbox adaptor plate.
3 To refit the sensor ensure that the correct spacer is fitted to the sensor then position the unit on the adaptor plate and secure with the two retaining bolts and one screw.
4 Reconnect the wiring multi-plug and the battery negative terminal.

3.3 Undo the rotor arm retaining screw using a suitable Allen key

## 5 Knock sensor – removal and refitting

1 The knock sensor is located on the front facing side of the cylinder block in the centre. To remove the unit disconnect the battery negative terminal and the wiring multi-plug then unscrew the sensor from its location.
2 Refitting is the reverse sequence of removal, but ensure that the sensor and cylinder block mating faces are clean.

## 6 Coolant thermister – removal and refitting

1 Removal and refitting procedures for this component are contained in Chapter 2.

## 7 Electronic control unit – removal and refitting

1 Disconnect the battery negative terminal.
2 Release the catch and disconnect the wiring multi-plug from the front of the unit (photo).

7.2 ECU multi-plug connector (A) and retaining screw (B)

3  Disconnect the ignition vacuum supply hose.
4  Undo the retaining screw, slip the unit out of the retaining tags and remove it from the engine compartment.
5  Refitting is the reverse sequence of removal.

## 8  Ignition coil – description and testing

1  The coil is mounted in the centre of the engine compartment bulkhead and it should be kept clean to prevent HT voltage loss through arcing.
2  To ensure correct HT polarity at the spark plugs, the coil LT leads must always be connected correctly (ie white lead to the coil positive terminal and white/black lead to the coil negative terminal). Incorrect connections can cause bad starting, misfiring and short spark plug life.
3  Apart from the tests of the low tension circuit contained in the fault diagnosis test procedure (Section 10), accurate checking of the coil output requires special equipment and for the home mechanic the easiest test is by substitution of a new unit.
4  If a new coil is to be fitted, ensure that it is of the correct type specifically for use on the programmed electronic ignition system. Failure to do so could cause irreparable damage to the electronic control unit.
5  To remove the coil, disconnect the HT and LT wiring, undo the two retaining bolts and lift away the coil. Refitting is the reverse sequence of removal.

## 9  Spark plugs and HT leads – general

1  The correct functioning of the spark plugs is vital for the correct running and efficiency of the engine. It is essential that the plugs fitted are appropriate for the engine, and the suitable type is specified at the beginning of this chapter. If this type is used and the engine is in good condition, the spark plugs should not need attention between scheduled replacement intervals. Spark plug cleaning is rarely necessary and should not be attempted unless specialised equipment is available as damage can easily be caused to the firing ends.
2  To remove the plugs, first mark the HT leads to ensure correct refitment, and then pull them off the plugs. Using a spark plug spanner, or suitable deep socket and extension bar, unscrew the plugs and remove them from the engine.
3  The condition of the spark plugs will also tell much about the overall condition of the engine.
4  If the insulator nose of the spark plug is clean and white, with no deposits, this is indicative of a weak mixture, or too hot a plug. (A hot plug transfers heat away from the electrode slowly – a cold plug

transfers it away quickly).
5  If the tip and insulator nose are covered with hard black-looking deposits, then this is indicative that the mixture is too rich. Should the plug be black and oily, then it is likely that the engine is fairly worn, as well as the mixture being too rich.
6  If the insulator nose is covered with light tan to greyish brown deposits, then the mixture is correct and it is likely that the engine is in good condition.
7  The spark plug gap is of considerable importance, as if it is too large or too small, the size of the spark and its efficiency will be seriously impaired. The spark plug gap should be set to the figure given in the Specifications at the beginning of this Chapter.
8  To set it, measure the gap with a feeler gauge, and then bend open, or close, the *outer* plug electrode until the correct gap is achieved. The centre electrode should *never* be bent as this may crack the insulation and cause plug failure, if nothing worse.
9  To refit the plugs, screw them in by hand initially and then fully tighten to the specified torque. If a torque wrench is not available, tighten the plugs until initial resistance is felt as the sealing washer contacts its seat and then tighten by a further eighth of a turn. Refit the HT leads in the correct order, ensuring that they are a tight fit over the plug ends. Periodically wipe the leads clean to reduce the risk of HT leakage by arcing.

## 10  Fault diagnosis – ignition system

Problems associated with the programmed electronic ignition system can usually be grouped into one of two areas, those caused by the more conventional HT side of the system such as spark plugs, HT leads, rotor arm and distributor cap, and those caused by the LT circuitry including the electronic control unit and its related components.

It is recommended that the checks described in Part 1 under the headings 'Engine fails to start' or 'Engine misfires' should be carried out first, according to the symptoms. If the fault still exists, the test procedure contained in Part 2 should be used. For these tests a good quality 0 to 12 volt voltmeter and an ohmmeter will be required.

Before carrying out any of the following tests ensure that the battery terminals are clean and secure and that the battery is fully charged and capable of cranking the engine on the starter motor. If this is not the case refer to Chapter 9.

## Part 1
### Engine fails to start

1  One of the most common reasons for bad starting is wet or damp spark plug leads and distributor cap. Remove the distributor cap. If condensation is visible internally dry the cap with a rag and wipe over the leads. Holts Wet Start can be very effective. To prevent the problem recurring, Holts Damp Start can be used to provide a sealing coat, so excluding any further moisture from the ignition system. In extreme difficulty, Holts Cold Start will help to start a car when only a very poor spark occurs. Refit the cap.
2  If the engine still fails to start, check that current is reaching the plugs, by disconnecting each plug lead in turn at the spark plug end, and holding the end of the cable about $\frac{3}{16}$ inch (5 mm) away from the cylinder block. Spin the engine on the starter motor.
3  Sparking between the end of the cable and the block should be fairly strong with a regular blue spark. (Hold the lead with rubber to avoid electric shocks). If current is reaching the plugs, then remove and regap them. The engine should now start.
4  If there is no spark at the plug leads, take off the HT lead from the centre of the distributor cap and hold it to the block as before. Spin the engine on the starter once more. A rapid succession of blue sparks between the end of the lead and the block indicate that the coil is in order and that the distributor cap is cracked, the rotor arm faulty or the carbon brush in the top of the distributor cap is not making good contact with the rotor arm.
5  If there are no sparks from the end of the coil lead check the connections at the coil for security. If these are in order carry out the checks contained in Part 2 of this Section.

## Engine misfires

6   If the misfire is regular and even, run the engine at a fast idle and pull off each of the plug HT leads in turn while listening to the note of the engine. Hold the lead with a dry cloth or rubber glove as additional protection against shock from the HT supply.

7   No difference in engine running will be noticed when the lead from the defective circuit is removed. Removing the lead from one of the good cylinders will accentuate the misfire.

8   Remove the plug lead from the end of the defective plug and hold it about $\frac{3}{16}$ inch (5 mm) away from the block. Restart the engine. If the sparking is fairly strong and regular, the fault must lie in the spark plug.

9   The plug may be loose, the insulation may be cracked, or the electrodes may have burnt away, giving too wide a gap for the spark to jump. Worse still, one of the electrodes may have broken off. Either renew the plug, or reset the gap, and then test it.

10  If there is no spark at the end of the plug lead, or if it is weak and intermittent, check the HT lead from the distributor cap to the plug. If the insulation is cracked or perished, renew the lead. Check the connections at the distributor cap.

11  If there is still no spark, examine the distributor cap carefully for tracking. This can be recognised by a very thin black line running between two or more electrodes, or between an electrode and some other part of the cap. These lines are paths which now conduct electricity across the cap, thus letting it run to earth. The only answer in this case is a new distributor cap.

12  Other causes of misfiring have already been described under the section dealing with the failure of the engine to start. To recap, these are:

(a)   The coil may be faulty giving an intermittent misfire
(b)   There may be a damaged wire or loose connection in the low tension circuit
(c)   There may be a fault in the electronic ignition system

13  If all these areas appear satisfactory then the fault may lie with the fuel system or there may be an internal engine fault. Further information will be found in Chapters 3 and 1 respectively.

## Part 2 – Programmed ignition system test procedure
### Engine fails to start

| Test | Remedy |
| --- | --- |
| 1   Connect a voltmeter across pins 9(+) and 12(−) of the electronic control unit (ECU) wiring connector. Does the voltmeter indicate battery voltage 10 seconds after switching on the ignition? | Yes: Proceed to test 2<br>No: Check the wiring between the ignition switch and pin 9, and between pin 12 and earth. Rectify as required |
| 2   Connect a voltmeter across pins 10(+) and 12(−) of the ECU wiring connector. Does the voltmeter indicate battery voltage 10 seconds after switching on the ignition? | Yes: Proceed to test 3<br>No: Check the wiring between the ignition switch and coil (+) terminal and between pin 10 and the coil (−) terminal. Rectify as required |
| 3   Connect an ohmmeter across the coil terminals. Is the coil primary winding resistance between 0.73 and 0.83 ohms? | Yes: Proceed to test 4<br>No: Renew the coil |
| 4   Connect a voltmeter between the battery (+) terminal and the coil (−) terminal. Does the reading on the voltmeter increase when the engine is cranking? | Yes: Engine should start. If not check ignition HT components, fuel system and engine internal components<br>No: Proceed to test 5 |
| 5   Switch ignition off and connect an ohmmeter across terminals 4 and 6 of the ECU. Does the ohmmeter register 1.5 k ohms approximately? | Yes: Probable ECU fault<br>No: Check crankshaft sensor wiring and connections. If satisfactory, sensor is suspect |

### Engine misfires and performance is unsatisfactory

| Test | Remedy |
| --- | --- |
| 1   Highlight ignition timing marks (notch on crankshaft pulley, corresponding notch on water pump bracket) with white chalk, connect a stroboscopic timing light, disconnect ECU vacuum pipe at manifold and start engine. Does the pulley mark advance as engine speed is increased? | Yes: Proceed to test 2<br>No: Probable ECU fault |
| 2   With the engine operating as in test 1 above, apply suction to the end of the disconnected ECU vacuum pipe. Does the pulley mark advance as suction is applied? | Yes: Ignition system is satisfactory, fault lies elsewhere<br>No: Check for leaks in vacuum pipe and connections. If satisfactory, ECU is faulty. |

# Chapter 5  Clutch

*For modifications, and information applicable to later models, see Supplement at end of manual*

## Contents

## Specifications

**Type** ............................................................................................. Single dry plate operated by self-adjusting cable

**Clutch disc diameter** ............................................................... 8.46 in (215 mm)

**Clutch pedal free travel (all models)** ................................. 0.47 to 1.10 in (12.0 to 28.0 mm)

| **Torque wrench settings** | lbf ft | Nm |
| --- | --- | --- |
| Clutch cover to flywheel ................................................. | 17 | 23 |

## 1  General description

Montego models covered by this manual are equipped with a conventional diaphragm spring clutch, operated mechanically by a self-adjusting cable.

The clutch components comprise a steel cover assembly, clutch disc or driven plate, release bearing and release mechanism. The cover assembly which is bolted and dowelled to the rear face of the flywheel contains the pressure plate and diaphragm spring.

The clutch disc is free to slide along the gearbox input shaft splines and is held in position between the flywheel and pressure plate by the pressure of the diaphragm spring.

Friction material is riveted to the clutch disc, which has a spring-cushioned hub to absorb transmission shocks and to help ensure a smooth take-up of the drive.

Depressing the clutch pedal moves the clutch release lever on the gearbox by means of the clutch cable. This movement is transmitted to the release bearing which moves inwards against the fingers of the diaphragm spring. The spring is sandwiched between two annular rings which act as fulcrum points. As the release bearing pushes the spring fingers in, the outer circumference pivots out, so moving the pressure plate away from the flywheel and releasing its grip on the clutch disc.

When the pedal is released, the diaphragm spring forces the pressure plate into contact with the friction linings of the clutch disc. The disc is now firmly sandwiched between the pressure plate and the flywheel, thus transmitting engine power to the gearbox.

Adjustment of the clutch to compensate for wear of the clutch disc friction linings is automatically taken up by the self-adjusting mechanism incorporated in the cable.

## 2  Clutch cable – removal and refitting

1  Working in the engine compartment, release the clutch cable from its retaining clips and cable ties.

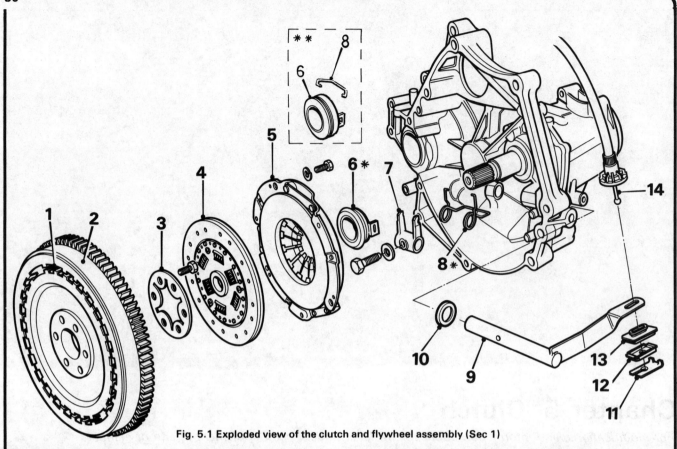

**Fig. 5.1 Exploded view of the clutch and flywheel assembly (Sec 1)**

1  Reluctor ring teeth
2  Flywheel
3  Locking plate
4  Clutch disc
5  Clutch cover assembly

6  Release bearing
7  Release fork
8  Release bearing retaining
   spring wire

9  Operating lever
10 Operating lever seal
11 Cable retaining clip
12 Cable seating plate

13 Rubber pad
14 Clutch cable
 *  Early models
 ** Later models

**Fig. 5.2 Clutch cable and self-adjusting mechanism (Sec 2)**

1  Operating lever
2  Cable retaining clip

3  Cable seating plate
4  Rubber pad

5  Outer cable
6  Spring retaining clip
7  Self-adjusting spring

8  Self-adjusting mechanism
   housing
9  Clutch pedal

2.2 Withdraw the retaining clip from the cable at the end of the self-adjusting spring

2.3 Slide out the retaining clip (arrowed) and cable seating plate

2.4 Withdraw the cable and rubber pad from the mounting bracket

2    Using pliers withdraw the retaining clip from the cable at the end of the self-adjusting spring (photo).
3    Release the inner cable from the clutch operating lever by sliding out the retaining clip and cable seating plate located on the underside of the lever (photo).
4    Withdraw the inner cable end from the operating lever rubber pad, release the rubber retainer and withdrawn the cable assembly from the engine mounting bracket (photo).
5    From inside the car unhook the cable end from the clutch pedal and withdraw the cable into the engine compartment. Remove the cable assembly from the car.
6    To refit the cable, feed the hooked end through the engine compartment bulkhead and connect it to the pedal. Ensure that the outer cable is located correctly in the bulkhead tube.
7    Route the cable through the engine compartment locating it in its retaining clips and cable ties.
8    Feed the cable through the mounting bracket until the guide sleeve is seated squarely in the bracket.
9    Feed the inner cable through the rubber pad of the operating lever and slide on the cable seating plate and retaining clip.
10  Refit the retaining clip to the cable at the end of the self-adjusting spring. Press down on the clutch operating lever and at the same time pull up on the cable to operate the self-adjusting mechanism.

### 3    Clutch pedal – removal and refitting

The clutch pedal and brake pedal are hinged on a common pivot and reference should be made to Chapter 8, Section 21, for the removal and refitting procedures.

### 4    Clutch assembly – removal, inspection and refitting

1    Remove the gearbox as described in Chapter 6.
2    In a diagonal sequence, half a turn at a time, slacken the bolts securing the clutch cover assembly to the flywheel.
3    When all the bolts are slack, remove them and then ease the cover assembly off the locating dowels. Collect the clutch disc which will drop out when the cover assembly is removed.
4    With the clutch assembly removed, clean off all traces of asbestos dust using a dry cloth. This is best done outside or in a well ventilated area; *asbestos dust is harmful, and must not be inhaled.*
5    Examine the linings of the clutch disc for wear and loose rivets, and the disc for rim distortion, cracks, broken torsion springs and worn splines. The surface of the friction linings may be highly glazed, but, as long as the friction material pattern can be clearly seen, this is satisfactory. If there is any sign of oil contamination, indicated by a continuous, or patchy, shiny black discolouration, the disc must be renewed and the source of the contamination traced and rectified. This will be either a leaking crankshaft oil seal or gearbox mainshaft oil seal – or both. Renewal procedures are given in Chapter 1 and Chapter 6

respectively. The disc must also be renewed if the lining thickness has worn down to, or just above, the level of the rivet heads.
6    Check the machined faces of the flywheel and pressure plate. If either is grooved, or heavily scored, renewal is necessary. The pressure plate must also be renewed if any cracks are apparent, or if the diaphragm spring is damaged or its pressure suspect.
7    With the gearbox removed it is advisable to check the condition of the release bearing, as described in the following Section.
8    To refit the clutch assembly, place the clutch disc in position with the raised portion of the spring housing facing away from the flywheel. The words FLYWHEEL SIDE will also usually be found on the other side of the disc that faces the flywheel.
9    Hold the disc in place and refit the cover assembly loosely on the dowels (photo). Refit the retaining bolts and tighten them finger tight so that the clutch disc is gripped, but can still be moved.
10  The clutch disc must now be centralised so that when the engine and gearbox are mated, the gearbox input shaft splines will pass through the splines in the centre of the hub.
11  Centralisation can be carried out quite easily by inserting a round bar or long screwdriver through the hole in the centre of the clutch disc, so that the end of the bar rests in the hole in the centre of the crankshaft. Moving the bar sideways or up and down will move the clutch disc in whichever direction is necessary to achieve centralisation. With the bar removed, view the clutch disc hub in relation to the hole in the end of the crankshaft and the circle created

4.9 Refitting the clutch disc and cover

4.11 Centralise the disc so that the clutch disc hub is in line with the hole in the crankshaft and the circle formed by the diaphragm spring fingers

by the ends of the diaphragm spring fingers (photo). When the hub appears exactly in the centre, all is correct. Alternatively, if a clutch aligning tool can be obtained this will eliminate all the guesswork, obviating the need for visual alignment.
12  Tighten the cover retaining bolts gradually, in a diagonal sequence to the specified torque wrench setting.
13  The gearbox can now be refitted as described in Chapter 6.

## 5    Clutch release bearing – removal, inspection and refitting

1    Remove the gearbox as described in Chapter 6.
2    Slip the retaining spring wire ends out of the slots on the bearing body then slide the bearing off the gearbox input shaft (photo).
3    Check the bearing for smoothness of operation and renew it if there is any roughness or harshness as the bearing is spun.
4    Refitting is the reverse sequence of removal, but ensure that the retaining spring wire ends are located behind the release fork.

5.2 Slip off the retaining wire ends and withdraw the release bearing

## 6  Fault diagnosis – clutch

| Symptom | Reason(s) |
| --- | --- |
| Judder when taking up drive | Loose or worn engine/gearbox mountings<br>Clutch disc linings contaminated with oil or worn<br>Clutch cable sticking or defective<br>Clutch disc hub sticking on input shaft splines |
| Clutch fails to disengage | Clutch cable sticking or defective<br>Excessive free play in the cable or release mechanism<br>Clutch disc linings contaminated with oil<br>Clutch disc hub sticking on input shaft splines |
| Clutch slips | Clutch cable sticking or defective<br>Clutch release mechanism sticking or partially seized<br>Faulty pressure plate or diaphragm assembly<br>Clutch disc linings worn or contaminated with oil |
| Noise when depressing clutch pedal | Worn release bearing<br>Defective release mechanism<br>Faulty pressure plate or diaphragm assembly |
| Noise when releasing clutch pedal | Faulty pressure plate assembly<br>Broken clutch disc torsion springs<br>Gearbox internal wear |
| Self-adjusting mechanism not working | Adjuster body not fully home on bulkhead<br>Incorrect free play in clutch pedal<br>Cable end fitting at pedal not in contact with trigger tube of adjusting mechanism when clutch engaged |

# Chapter 6 Gearbox

*For modifications, and information applicable to later models, see Supplement at end of manual*

## Contents

## Specifications

**Type** ............................................................................................. Five forward speeds (all synchromesh) and reverse. Final drive integral with main gearbox

## Lubrication

Oil type/specification* ..................................................................... Multigrade engine oil, viscosity SAE 10W/40 (or equivalent multigrade engine oil having a viscosity rating compatible with temperatures in the vehicle operating area – see manufacturer's handbook) (Duckhams QXR, Hypergrade, or 10W/40 Motor Oil)

*Note: Austin Rover specify a 10W/40 oil to meet warranty requirements for models produced after August 1983. Duckhams QXR and 10W/40 Motor Oil are available to meet these requirements.*

## Gearbox ratios

All models except MG EFi:

| | |
|---|---|
| 1st ..................................................................................... | 2.92 : 1 |
| 2nd .................................................................................... | 1.75 : 1 |
| 3rd ..................................................................................... | 1.15 : 1 |
| 4th ..................................................................................... | 0.87 : 1 |
| 5th ..................................................................................... | 0.66 : 1 |
| Reverse ............................................................................. | 3.00 : 1 |

MG EFi models:

| | |
|---|---|
| 1st ..................................................................................... | 3.25 : 1 |
| 2nd .................................................................................... | 1.89 : 1 |
| 3rd ..................................................................................... | 1.33 : 1 |
| 4th ..................................................................................... | 1.04 : 1 |
| 5th ..................................................................................... | 0.85 : 1 |
| Reverse ............................................................................. | 3.00 : 1 |

Final drive ratios:

| | |
|---|---|
| All models except MG EFi ................................................. | 3.87 : 1 |
| MG EFi models .................................................................. | 3.875 : 1 |

## Gearbox overhaul data

| | |
|---|---|
| Mainshaft endfloat .......................................................... | 0.005 to 0.008 in (0.14 to 0.21 mm) |
| Endfloat adjustment ......................................................... | Selective circlips |
| Countershaft endfloat ...................................................... | 0.004 to 0.013 in (0.1 to 0.35 mm) |
| Endfloat adjustment ......................................................... | Selective distance collars and thrustwashers |
| Minimum baulk ring to gear clearance ........................... | 0.016 in (0.40 mm) |
| Maximum selector fork to synchro sleeve clearance ..... | 0.040 in (1.0 mm) |
| Maximum reverse gear fork to gear clearance ............... | 0.028 in (0.7 mm) |
| Differential endfloat ........................................................ | 0.006 in (0.15 mm) |
| Endfloat adjustment ......................................................... | Selective circlips |

## Torque wrench settings

| | lbf ft | Nm |
|---|---|---|
| Bellhousing to engine adaptor plate | 66 | 90 |
| Gearbox steady rod bolt | 6 | 8 |
| Gear lever to gearchange rod | 18 | 25 |
| Remote control housing to underbody | 18 | 25 |
| Oil filler plug | 33 | 45 |
| Oil drain plug | 30 | 40 |
| Countershaft access plug | 52 | 70 |
| Reversing lamp switch | 18 | 25 |
| Reverse idler shaft retaining bolt | 49 | 67 |
| Gearcase to bellhousing bolts | 20 | 27 |
| Speedometer pinion retaining bolt | 8 | 11 |
| Clutch release fork to operating lever bolt | 21 | 29 |
| Countershaft retaining nut | 81 | 110 |
| Reverse gear fork bracket bolts | 10 | 14 |
| Gearchange holder and interlock retaining bolts: | | |
| M8 bolts | 21 | 29 |
| M6 bolts | 9 | 13 |
| Gearchange shaft detent plug | 16 | 22 |

## 1 General description

The gearbox is of Honda manufacture and is equipped with five forward and one reverse gear. Synchromesh gear engagement is used on all forward gears.

The mainshaft and countershaft carry the constant mesh gear cluster assemblies and are supported on ball and roller bearings. The short input end of the mainshaft eliminates the need for additional support from a crankshaft spigot bearing. The synchromesh gear engagement is by spring rings which act against baulk rings under the movement of the synchroniser sleeves. Gear selection is by means of a floor mounted lever connected by a remote control housing and gear change rod to the gear change shaft in the gearbox. Gear change shaft movement is transmitted to the selector forks via the gear change holder and interlock assembly.

The final drive (differential) unit is integral with the main gearbox and is located between the bellhousing and gearcase. The gearbox and final drive components both share the same lubricating oil.

## 2 Maintenance and inspection

1 At the intervals given in Routine Maintenance at the beginning of this manual carry out the following service operations on the gearbox.
2 Carefully inspect the gearbox joint faces and oil seals for signs of damage, deterioration or oil leakage.
3 At the same service interval check, and if necessary, top up the gearbox oil. The filler plug is located on the left-hand side of the gearcase and can be reached from above, through the engine compartment (photo). Wipe the area around the filler plug with a rag before unscrewing the plug. Top up if necessary using the specified gear oil to bring the level up to the filler plug orifice.
4 The gearbox oil must be renewed at the interval specified in *Routine maintenance* at the front of the manual. To drain the old oil, position a suitable container beneath the drain plug, located on the gearbox filler plug side, below the driveshaft inner joint (photo). A

square key is required to remove and refit the drain plug, but the ⅜ in square drive end of a socket bar will suffice. When the oil is fully drained, refit the drain plug, tighten it to the specified torque wrench setting, then refill the gearbox with the specified quantity and grade of oil through the filler hole.
5 At less frequent intervals check for excess free play or wear in the gear linkage and gear lever joints and pivots.

## 3 Gearbox – removal and refitting

1 Disconnect the battery negative terminal then refer to Chapter 7 and remove the left-hand driveshaft.
2 Remove the air cleaner as described in Chapter 3 and the starter motor as described in Chapter 9.
3 Undo the small retaining bolt and withdraw the speedometer cable and pinion or speed transducer assembly from the top of the gearbox.
4 Undo the retaining nut and bolt and move the crankshaft sensor multi-plug bracket to one side (photo).
5 Disconnect the reversing lamp switch wires at the two cable connectors (photo).
6 Where applicable, undo the bolt securing the inlet manifold support strut bolt to the bellhousing.
7 Undo the bolt securing the gearbox breather bracket and earth strap to the gearcase. Remove the earth strap then refit the bolt (photo).
8 Using the procedure described in Chapter 5, disconnect the clutch cable from its support and operating lever.
9 Remove the right-hand road wheel and the two bolts and nuts securing the right-hand suspension strut to the swivel hub.
10 Undo the nut securing the tie-rod outer balljoint to the swivel hub and release the joint using a suitable balljoint separator tool.
11 Tip the swivel hub outwards at the top as far as possible without straining the brake hose. Release the driveshaft inner constant velocity joint and withdraw the joint from the gearbox.
12 From under the car undo and remove the small bolt in the centre

2.3 Gearbox filler plug location (arrowed)

2.4 Gearbox drain plug location (arrowed)

3.4 Undo the crankshaft sensor multi-plug bracket bolt (arrowed)

3.5 Reversing lamp switch wire connectors (arrowed)

3.7 Gearbox breather bracket and earth strap retaining bolt

3.12A Remove the dished washer and steady rod ...

3.12B ... followed by the flat washer

3.13 Gearchange rod retaining roll pin (arrowed)

3.21 Gearbox left-hand mounting components

of the gearbox steady rod. Remove the dished washer, slide off the steady rod and remove the flat washer. Note that the lip on the inner washer faces the steady rod rubber bush (photos).
13 Remove the spring clip to expose the gearchange rod to gearchange shaft retaining roll pin (photo).
14 Using a suitable punch, drive out the roll pin and slide the gearchange rod rearwards, off the shaft.
15 Place a jack and interposed block of wood beneath the engine sump and just take the weight of the engine and gearbox assembly.
16 Undo the four bolts securing the rear engine mounting to the gearbox and the two nuts securing the mounting to its support bracket.
17 Undo the two nuts and bolts securing the front snubber bracket to the gearbox. Withdraw the snubber from the gearbox and snubber cup.
18 Undo all the bolts securing the gearbox bellhousing to the engine adaptor plate.
19 From under the left-hand wheel arch undo the retaining screws and lift away the access panel.
20 Position a second jack beneath the gearbox and just take its weight.
21 Undo all the bolts securing the gearbox left-hand mounting to the gearcase and body, and the through bolt securing the two parts of the mounting together. Separate the mounting, recover the two large flat washers and remove the mounting assembly from the car (photo).
22 Make a final check that everything securing the gearbox to the engine has been disconnected and moved well clear.
23 With the help of an assistant, raise both the jacks until there is sufficient clearance, then lift out the rear mounting. Lower the two jacks slowly, ease the gearbox bellhousing off the locating dowels, and withdraw the unit from under the wheel arch.
24 Refitting the gearbox is the reverse sequence to removal, bearing in mind the following points:

(a) *Tighten all retaining nuts and bolts to the specified torque where applicable*
(b) *Refit the clutch cable with reference to Chapter 5*
(c) *With the gearbox installed and all mountings tightened,*

*slacken the front snubber cup retaining bolts, centralize the cup around the snubber rubber, then tighten the bolts*
(d) *Fill the gearbox with the specified lubricant as described in Section 2*

## 4 Gearbox overhaul – general

Dismantling, overhaul and reassembly of the gearbox is reasonably straightforward and can be carried out without recourse to the manufacturer's special tools. It should be noted however that any repair or overhaul work on the final drive differential must be limited to the renewal of the carrier support bearings. Owing to the complicated nature of this unit and the costs involved, the advice of a BL dealer should be sought if further repair is necessary.
Before starting any work on the gearbox, clean the exterior of the casings using paraffin as a suitable solvent. Dry the unit with a lint free rag. Make sure that an uncluttered working area is available with some small containers and trays handy to store the various parts. Label everything as it is removed.
Before starting reassembly, all the components must be spotlessly clean and should be lubricated with the recommended grade of gear oil during reassembly.

## 5 Gearbox – dismantling

1 Stand the gearbox on its bellhousing face on the bench, and begin dismantling by removing the reversing lamp switch (photo).
2 Undo the retaining bolt and plate and lift out the speedometer pinion assembly if this was not done during gearbox removal.
3 Undo all the gearcase to bellhousing retaining bolts, noting the location of the breather pipe and bracket, which are also retained by one of the case bolts. Remove the breather pipe and bracket (photo).
4 Undo the reverse idler shaft retaining bolt located on the side of the gearcase (photo).

5.1 Unscrew the reversing lamp switch

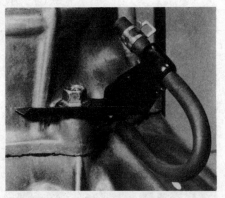

5.3 Gearbox breather, bracket and case retaining bolt

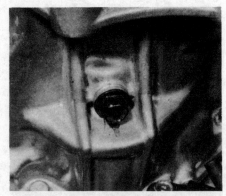

5.4 Reverse idler shaft retaining bolt

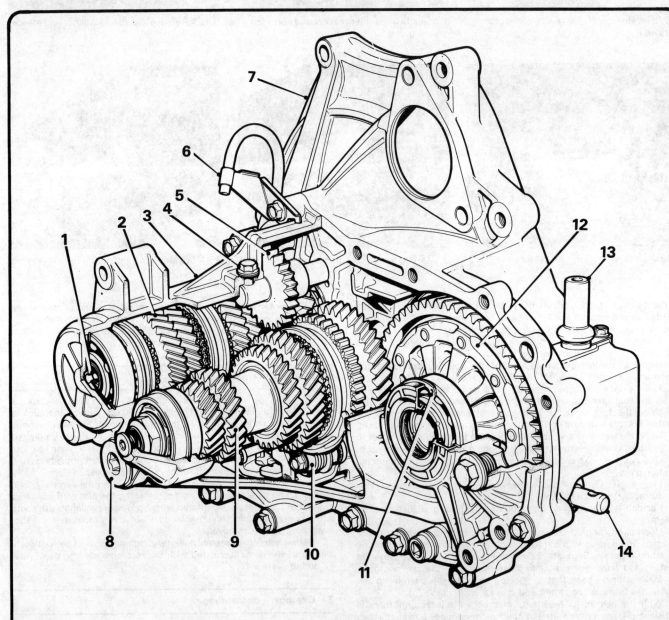

**Fig. 6.1 Cut-away view of the gearbox (Sec 5)**

| | | | |
|---|---|---|---|
| 1 | Oil guide plate | 5 | Reverse idler gear |
| 2 | Mainshaft assembly | 6 | Gearbox breather and bracket |
| 3 | Gearcase | 7 | Bellhousing |
| 4 | Reverse idler shaft retaining bolt | 8 | Countershaft access plug |
| | | 9 | Countershaft assembly |
| 10 | Gearchange holder and interlock assembly | 12 | Final drive differential |
| 11 | Differential endfloat circlip shim | 13 | Speedometer pinion |
| | | 14 | Gearchange shaft |

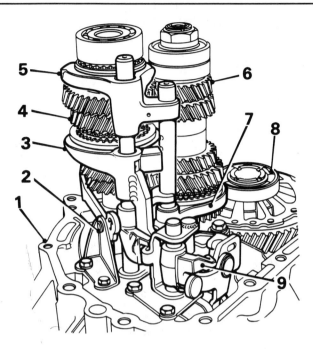

**Fig. 6.2 Geartrain and gear selector components (Sec 5)**

| | |
|---|---|
| 1  Bellhousing | 6  Countershaft assembly |
| 2  Reverse gear fork and bracket | 7  1st/2nd gear selector fork |
| 3  3rd/4th gear selector fork | 8  Final drive differential |
| 4  Mainshaft assembly | 9  Gearchange holder and |
| 5  5th gear selector fork | interlock assembly |

5  Using a large Allen key, hexagonal bar, or suitable bolt with two nuts locked together, undo the countershaft access plug on the end of the gearcase (photo).

6  Using circlip pliers inserted through the access plug aperture, spread the countershaft retaining circlip while at the same time lifting upwards on the gearcase. Tap the case with a soft mallet if necessary. When the circlip is clear of its groove, lift the case up and off the bellhousing and gear clusters (photos).

7  Undo the two retaining bolts and remove the reverse gear fork and bracket.

8  Lift out the reverse gear idler shaft with reverse gear.

9  Undo the three bolts and remove the gearchange holder and interlock assembly. Note that the holder locates in a slot in the 1st/2nd selector shaft.

10  With the help of an assistant lift up the mainshaft and countershaft as an assembly approximately 0.5 in (12.0 mm) and withdraw the selector shafts and forks from the bellhousing and gear clusters (photo).

11  Lift the mainshaft and countershaft out of their respective bearings in the bellhousing (photo).

12  Finally, remove the differential from the bellhousing (photo).

## 6  Mainshaft – dismantling and reassembly

1  Remove the mainshaft bearing, using a two or three-legged puller if necessary, unless the bearing remained in the gearcase during removal.

2  Withdraw the 5th gear synchroniser hub and sleeve assembly from the mainshaft using a two or three-legged puller if it is tight. Recover the 5th gear baulk ring from the cone face of 5th gear and place it, together with the spring ring, on the synchroniser unit.

3  Slide off 5th gear followed by the 5th gear needle roller bearing.

5.5 Countershaft access plug (arrowed)

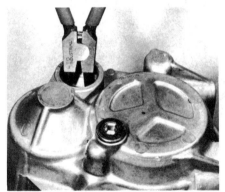

5.6A Release the countershaft bearing retaining circlip ...

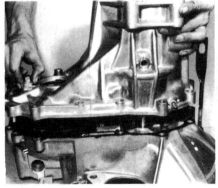

5.6B ... then withdraw the gearcase from the bellhousing and gear clusters

5.10 Lift the mainshaft and countershaft then remove the selector forks and shafts

5.11 Withdraw the mainshaft and countershaft together from the bellhousing

5.12 Remove the differential from the bellhousing

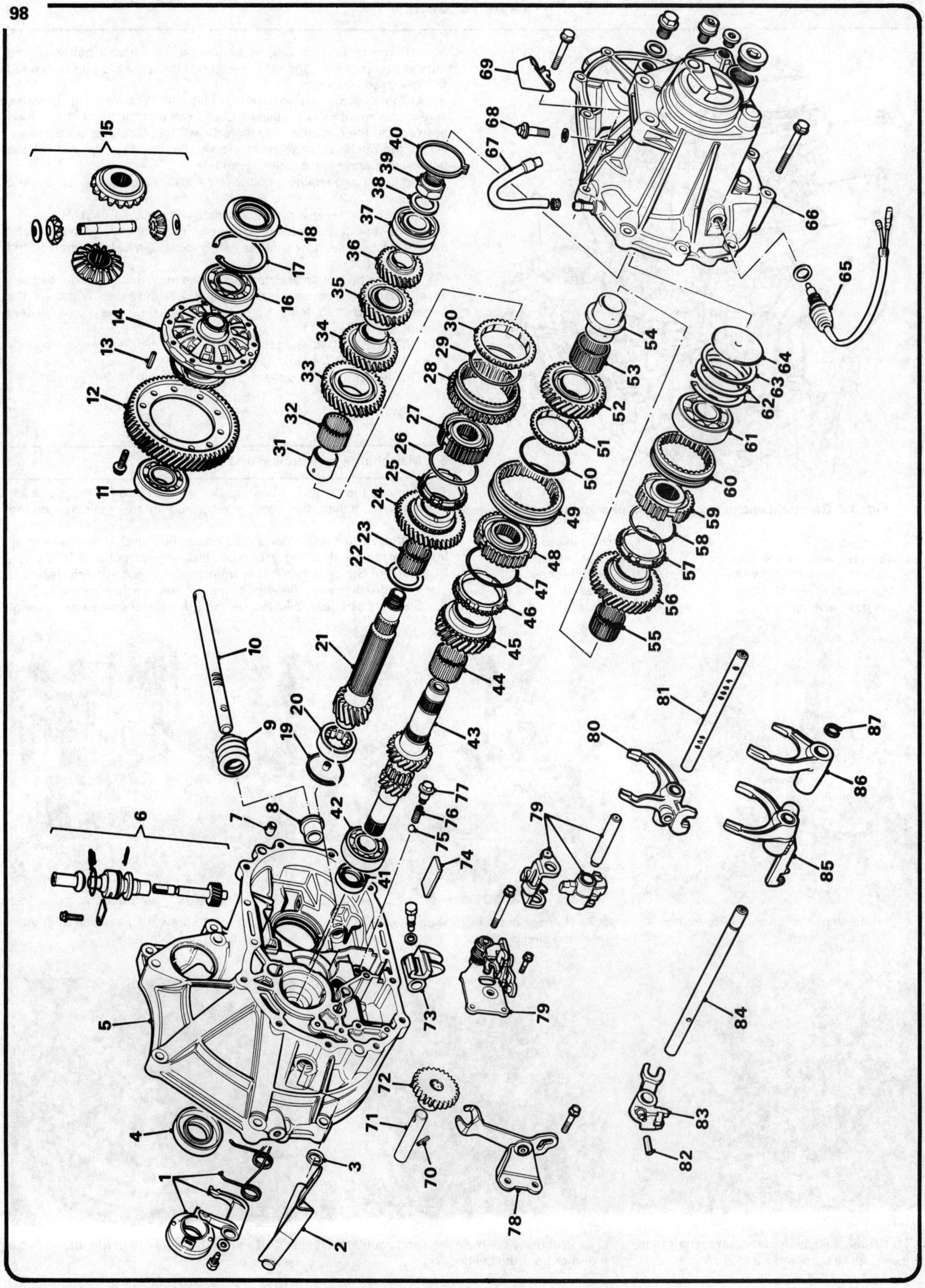

1
2
3
4
5
6
7
8
9
10
11
12
13
14
15
16
17
18
19
20
21
22
23
24
25
26
27
28
29
30
31
32
33
34
35
36
37
38
39
40
41
42
43
44
45
46
47
48
49
50
51
52
53
54
55
56
57
58
59
60
61
62
63
64
65
66
67
68
69
70
71
72
73
74
75
76
77
78
79
80
81
82
83
84
85
86
87

**Fig. 6.3 Exploded view of the gearbox components (Sec 5)**

1 Clutch release bearing, fork and retaining spring wire
2 Clutch operating lever
3 Clutch operating lever oil seal
4 Differential oil seal
5 Bellhousing
6 Speedometer pinion assembly
7 Locating dowel
8 Gearchange shaft oil seal
9 Rubber boot
10 Gearchange shaft
11 Final drive support bearing
12 Final drivegear
13 Roll pin
14 Final drive casing
15 Differential sun and planet gear components
16 Final drive support bearing
17 Differential endfloat circlip shim
18 Differential oil seal
19 Oil guide plate
20 Countershaft roller bearing

21 Countershaft
22 Thrustwasher
23 Needle roller bearing
24 1st gear
25 Baulk ring
26 Spring ring
27 1st/2nd synchro hub
28 1st/2nd synchro sleeve
29 Spring ring
30 Baulk ring
31 Distance collar
32 Needle roller bearing
33 2nd gear
34 3rd gear
35 4th gear
36 5th gear
37 Countershaft ball bearing
38 Tongued washer
39 Retaining nut
40 Circlip
41 Mainshaft oil seal
42 Mainshaft ball bearing
43 Mainshaft

44 Needle roller bearing
45 3rd gear
46 Baulk ring
47 Spring ring
48 3rd/4th synchro hub
49 3rd/4th synchro sleeve
50 Spring ring
51 Baulk ring
52 4th gear
53 Needle roller bearing
54 Distance collar
55 Needle roller bearing
56 5th gear
57 Baulk ring
58 Spring ring
59 5th gear synchro hub
60 5th gear synchro sleeve
61 Mainshaft ball bearing
62 Selective circlips
63 Belleville washer
64 Oil guide plate
65 Reversing lamp switch
66 Gearcase

67 Gearbox breather
68 Reverse idler shaft retaining bolt
69 Gearbox breather bracket
70 Roll pin
71 Reverse idler shaft
72 Reverse idler gear
73 Gearchange arm
74 Magnet
75 Detent ball
76 Detent spring
77 Detent plug
78 Reverse gear fork and bracket
79 Gearchange holder and interlock assembly
80 1st/2nd gear selector fork
81 1st/2nd gear selector shaft
82 Roll pin
83 5th/reverse gear selector
84 5th/reverse gear selector shaft
85 3rd/4th gear selector fork
86 5th gear selector fork
87 Circlip

4   Withdraw the distance collar followed by 4th gear and the needle roller bearing.
5   Remove the 3rd/4th synchro hub and sleeve assembly complete with baulk rings and spring rings.
6   Remove 3rd gear and its needle roller bearing.
7   Carry out a careful inspection of the mainshaft components as described in Section 10, and obtain any new parts as necessary.
8   During reassembly, lightly lubricate all the parts with the specified grade of gear oil as the work proceeds.
9   Slide the 3rd gear needle roller bearing onto the mainshaft, followed by 3rd gear with its flat face towards the other gears on the shaft (photos).
10  Place the 3rd gear baulk ring and spring ring on the cone face of 3rd gear, then fit the 3rd/4th synchro hub and sleeve assembly. Ensure that the lugs on the baulk ring engage with the slots in the synchro hub (photos).
11  Locate the 4th gear spring ring and baulk ring in the synchro unit, then slide on 4th gear with its needle roller bearing (photos).
12  Fit the distance collar with its shoulder towards 4th gear (photo).
13  Place the 5th gear needle roller bearing over the collar then slide 5th gear onto the bearing (photos).
14  Locate the 5th gear baulk ring and spring ring in the 5th gear synchro unit then fit this assembly to the mainshaft (photo).

6.10A Place the baulk ring and spring ring on 3rd gear ...

6.9A Slide the 3rd gear needle roller bearing onto the mainshaft ...

6.10B ... then fit the 3rd/4th synchro hub and sleeve assembly

6.9B ... followed by 3rd gear

6.11A Locate the spring ring and baulk ring on the synchro unit ...

6.11B ... then fit 4th gear with its needle roller bearing

6.13B ... then slide on 5th gear

6.12 Fit the distance collar with its shoulder towards 4th gear

6.14 Fit the 5th gear synchro unit complete with spring ring and baulk ring

6.13A Position the 5th gear needle roller bearing over the collar ...

15 If the mainshaft or any of its components, the gearcase or bellhousing have been renewed, then the mainshaft endfloat must be checked, and if necessary adjusted as follows. To do this it will be necessary to remove the mainshaft ball bearing in the gearcase, remove the selective circlips, Belleville washer and oil guide, then refit the bearing.

16 Position the assembled mainshaft in the bellhousing, fit the gearcase and temporarily secure it with several evenly spaced bolts. Tighten the bolts securely.

17 Support the bellhousing face of the gearbox on blocks, so as to provide access to the protruding mainshaft.

18 Place a straightedge across the bellhousing face, in line with the mainshaft, then accurately measure and record the distance from straightedge to mainshaft.

19 Turn the gearbox over so that the bellhousing is uppermost and gently tap the mainshaft back into the gearcase using a soft-faced mallet. Take a second measurement of the mainshaft to straightedge distance.

20 Subtract the first measurement from the second measurement and identify this as dimension A:

21 Measure the thickness of the Belleville washer and add an

allowance of 0.006 in (0.17 mm) which is the nominal mainshaft endfloat. Identify this as dimension B:

22 Subtract dimension B from dimension A and the value obtained is the thickness of selected circlip(s) required to give the specified mainshaft endfloat.

23 Remove the gearcase and the mainshaft. Remove the bearing from the gearcase, refit the oil guide, Belleville washer and circlips of the required thickness, then refit the bearing.

## 7 Countershaft – dismantling and reassembly

1 Support the pinion gear on the countershaft in a vice between two blocks of wood. Tighten the vice just sufficiently to prevent the countershaft turning as the nut is undone.

2 Using a small punch release the staking on the countershaft nut then undo and remove the nut. Note that the nut has a left-hand thread and must be turned clockwise to unscrew it.

3 Remove the tongued washer then draw off the countershaft bearing using a two or three-legged puller.

4 Slide 5th, 4th, 3rd and 2nd gears off the countershaft noting their fitted directions.

5 Remove the 2nd gear baulk ring and spring ring.

6 Slide off the 2nd gear needle roller bearing followed by the distance collar. Use two screwdrivers to lever off the collar if it is tight.

7 Remove the 1st/2nd synchro hub and sleeve assembly followed by the 1st gear baulk ring and spring ring.

8 Slide off 1st gear followed by the needle roller bearing and thrust washer.

9 Carry out a careful inspection of the countershaft components as described in Section 10, and obtain any new parts as necessary.

10 During reassembly, lightly lubricate all the parts with the specified grade of gear oil as the work proceeds.

11 Fit the thrustwasher to the countershaft followed by the needle roller bearing and 1st gear (photos).

12 Fit the baulk ring and spring ring to the cone face of 1st gear then slide on the 1st/2nd synchro unit. The synchro unit must be fitted with the selector fork groove in the synchro sleeve away from 1st gear. As the unit is fitted ensure that the lugs on the baulk ring engage with the slots in the synchro hub (photos).

13 Warm the distance collar in boiling water then slide it onto the countershaft with the oil hole offset towards 1st gear (photo).

14 Fit the 2nd gear needle roller bearing to the distance collar (photo).

15 Locate the 2nd gear baulk ring and spring ring on the synchro unit then slide 2nd gear into place over the needle roller bearing (photos).

16 Fit 3rd gear to the countershaft with its longer boss away from 2nd gear (photo).

17 Fit 4th gear with its boss towards the 3rd gear boss (photo).

18 Fit 5th gear with its flat face towards 4th gear then tap the countershaft bearing into position using a hammer and suitable tube (photos).

19 Fit the tongued washer followed by a new countershaft nut (photos). Hold the pinion between blocks of wood in the vice as before and tighten the nut to the specified torque.

20 Using feeler gauges, measure the clearance between the rear face of the pinion and 1st gear, and between the 2nd and 3rd gear faces (photos). Compare the measurements with the endfloat dimension given in the Specifications. If the recorded endfloat is outside the tolerance range, dismantle the countershaft again and fit an alternative thrustwasher or distance collar.

21 With the countershaft assembled and the endfloat correctly set, recheck the torque of the countershaft nut then peen its edge into the countershaft groove using a small punch.

7.11A Fit the thrustwasher to the countershaft ...

7.11B ... followed by the needle roller bearing ...

7.11C ... and 1st gear

7.12A Fit the baulk ring and spring ring to 1st gear ...

7.12B ... then slide on the 1st/2nd synchro unit

7.13 Fit the distance collar with its oil hole (arrowed) offset towards 1st gear

7.14 Locate the 2nd gear needle roller bearing over the distance collar

7.15A Fit the baulk ring and spring ring ...

7.15B ... followed by 2nd gear

7.16 Fit 3rd gear with its boss away from 2nd gear

7.17 Fit 4th gear with its boss towards 3rd gear

7.18A Fit 5th gear ...

7.18B ... followed by the countershaft bearing ...

7.18C ... then drive the bearing fully onto the shaft

7.19A Engage the tongued washer with the countershaft groove ...

7.19B ... and screw a new nut onto the shaft

7.20A Check the clearance between the pinion and 1st gear ...

7.20B ... and between 2nd and 3rd gear

## 8  Gearcase – inspection and overhaul

1  Check the gearcase for cracks or any damage to its bellhousing mating face. Renew the case if damaged.

2  Check the condition of the mainshaft bearing in the gearcase and ensure that it spins smoothly with no trace of roughness or harshness. The bearing must be removed if it is worn, if the gearcase is to be renewed, or if it is necessary to gain access to the mainshaft endfloat selective circlips located behind it.

3  Removal of the bearing entails the use of a slide hammer with adaptor consisting of internally expanding flange or legs, to locate behind the inner race. A BL special tool is available for this purpose, but it may be possible to make up a suitable alternative with readily available tools. Whichever option is chosen it is quite likely that the oil guide plate will be damaged or broken in the process. If so a new one must be obtained.

4  If any of the mainshaft components are being renewed during the course of overhaul, do not refit the bearing, circlips, Belleville washer or oil guide plate until after the mainshaft endfloat has been checked and adjusted.

5  When the bearing is fitted this can be done by tapping it squarely into place using a hammer and tube of suitable diameter in contact with the bearing outer race.

6  If there is any sign of leakage, the differential oil seal in the gearcase should be renewed. Drive or hook out the old seal and install the new one with its open side facing inward ie towards the differential (photo). Tap the seal squarely into place using a suitable tube or the old seal. Smear a little grease around the sealing lip to aid refitting of the driveshaft. **Note:** *If the differential or differential bearings have been renewed or disturbed from their original position, do not fit the oil seal until the gearbox has been completely reassembled.* The differential bearing clearances are checked through the oil seal aperture, and cannot be done with the seal in place.

## 9  Bellhousing – inspection and overhaul

1  With the mainshaft and countershaft removed, lift out the magnet from its location in the bellhousing edge (photo).

2  Remove the release bearing as described in Chapter 5.

3  Undo the retaining peg bolt securing the release fork to the clutch operating lever (photo).

4  Withdraw the lever from the fork and lift out the fork noting the fitted position of the release bearing retaining spring wire.

5  Undo the gearchange shaft detent plug and lift out the detent spring and ball (photos).

6  Undo the bolt securing the gearchange arm to the shaft and slide the arm off the shaft (photos).

7  Withdraw the gearchange shaft from the bellhousing and recover the rubber boot.

8  Examine the bellhousing for cracks or any damage to its gearcase mating face. Renew the case if damaged.

9  Check the condition of the ball and roller bearings in the bellhousing ensuring that they spin smoothly with no trace of roughness or harshness.

10  Renewal of the bearings entails the use of a slide hammer with adaptor consisting of an internally expanding flange or legs, to locate through the centre of the bearing. A BL special tool is available for this purpose, but it may be possible to make up a suitable substitute with readily available tools. Another alternative would be to take the bellhousing along to your dealer and have him renew the bearings for you. Whichever option is chosen it is quite likely that the oil guide plate behind the countershaft roller bearing will be damaged or broken in the process (assuming this bearing is to be renewed) and if so a new guide plate must be obtained (photo).

11  Refit the bearings by tapping them squarely into place using a hammer and tube of suitable diameter. Ensure that the oil hole in the countershaft bearing faces the gearbox interior.

8.6 Fit a new differential oil seal with its open side facing inward

9.1 Remove the magnet from the bellhousing

9.3 Undo and remove the clutch release fork peg bolt

9.5A Undo the gearchange shaft detent plug ...

9.5B ... then lift out the detent spring ...

9.5C ... and detent ball

9.6A Undo the gearchange arm retaining bolt ...

9.6B ... slide off the arm and remove the shaft

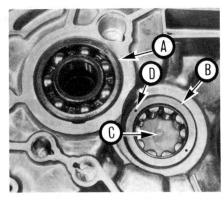

9.10 Mainshaft ball bearing (A) countershaft roller bearing (B) oil guide plate (C) and bearing oil holes (D)

12  Carefully inspect all the oil seals in the bellhousing and renew any that show signs of leakage. The old oil seals can be driven out with a tube or punch and the new seals tapped squarely into place using a block of wood or the old seal. Ensure that in all cases the open side of the seal faces inward. In the case of the mainshaft oil seal it will be necessary to remove the mainshaft bearing to enable a new seal to be fitted.

13  Inspect the gearchange shaft for distortion or wear across the detent grooves and check the gearchange arm for wear of the forks. Renew these components if wear is evident.

14  With the new bearings and seals in position and any other new parts obtained as necessary, refit the gearchange shaft and rubber boot with the detent grooves facing outward, ie towards the gear clusters.

15  Slide on the gearchange arm so that its forked side is facing away from the bellhousing starter motor aperture. Refit the retaining bolt and washer and tighten to the specified torque.

16  Refit the detent ball followed by the spring and plug bolt. Tighten the bolt to the specified torque.

17  Slide the clutch operating lever into the bellhousing and engage the release fork. Ensure that the joined end of the bearing retaining spring wire is positioned behind the release fork arms. Refit and tighten the retaining bolt.

18  Refit the magnet with its forked end down, then refit the clutch release bearing as described in Chapter 5.

---

**10  Mainshaft and countershaft components and synchro units – inspection and overhaul**

1  With the mainshaft and countershaft dismantled, examine the shafts and gears for signs of pitting, scoring, wear ridges or chipped teeth. Check the fit of the gears on the mainshaft and countershaft splines and ensure that there is no lateral free play.

2  Check the smoothness of the bearings and check for any signs of scoring on the needle roller bearing tracks and distance collars.

3  Check the mainshaft and countershaft for straightness, check for damaged threads or splines and ensure that the lubrication holes are clear (photos).

4  Mark one side of each synchro hub and sleeve before separating the two parts, so that they may be refitted in the same position.

5  Withdraw the hub from the sleeve and examine the internal gear teeth for wear or ridges. Ensure that the hub and sleeve are a snug sliding fit with the minimum of lateral movement.

6  Check the fit of the selector forks in their respective synchro sleeve grooves. If the clearance exceeds the figure given in the Specifications, check for wear ridges to give an indication of whether it is the fork or the sleeve groove that has worn. As a general rule the selector fork usually wears first, but not always. If in doubt compare with new parts.

7  Place each baulk ring on the cone face of its respective gear and measure the distance between the baulk ring and the gear face. If the clearance is less than specified renew the baulk ring. Renew them also if there is excessive wear or rounding off of the dog teeth around their periphery, if they are cracked, or if they are in any way damaged. If the gearbox is in reasonable condition and is to be rebuilt it is advisable to renew all the baulk rings as a matter of course. The improvement in the synchromesh action when changing gear will justify the expense.

8  When reassembling the synchro units make sure that the two oversize teeth in the synchro sleeve engage with the two oversize grooves in the hub (photo).

9  If any of the gears on the mainshaft are to be renewed then the corresponding gear on the countershaft must also be renewed and vice versa. This applies to the countershaft and differential final drivegear as well.

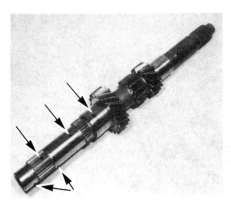

10.3A Gearbox mainshaft showing oil holes (arrowed) ...

10.3B ... and gearbox countershaft

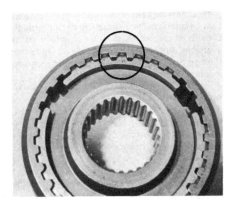

10.8 Oversized teeth in the synchro sleeve engaged with corresponding grooves in the hub

## 11 Selector forks, shafts and gearchange mechanism – inspection and overhaul

1   Visually inspect the selector forks for obvious signs of wear ridges, cracks or deformation.
2   Slide the selector forks off the shafts, noting their fitted position and check the detent action as the fork is removed. Note that the detent balls and springs are located in the selector forks and cannot be removed. If the detent action is weak or if there is evidence of a broken spring or damaged ball, the fork must be renewed.
3   Check the fit of the selector forks in their respective synchro sleeves. If the clearance exceeds the figure given in the Specifications check for wear ridges to give an indication of whether it is the fork or the sleeve groove that is worn. If in doubt compare them with new parts and renew any that are worn.
4   Examine the selector shafts for wear ridges around the detent grooves and for any obvious signs of distortion. Renew any suspect shafts.
5   Examine the gearchange holder and interlock assembly for any visible sign of wear or damage. It is advisable not to dismantle the mechanism unless it is obviously worn and is in need of renewal. If this is the case it can be separated into three main units as shown in Fig. 6.2, and the worn part can be renewed.
6   Having obtained any new parts as required, reassemble the selector forks back onto the shafts and reassemble the gearchange holder and interlock components if these were dismantled for renewal.

## 12 Final drive differential – inspection and overhaul

1   As mentioned earlier the only parts which can be renewed as a practical proposition are the two main support bearings on the final drive casing. The differential unit should be examined for any signs of wear or damage, but if any is found it is recommended that you seek the advice of your dealer. Differential parts are supplied in sets, ie final drive gear and matching countershaft; sun gears and matching planet gears etc, and consequently are extremely expensive. The cost of individual parts may even equal the price of a complete exchange box.
2   Check that the bearings spin freely with no sign of harshness or roughness. If renewal is necessary, remove the bearings by levering them off the differential using two screwdrivers or use a small puller.
3   Fit the new bearings by tapping them into place using a hammer and tube in contact with the bearing inner race.

## 13 Gearbox – reassembly

1   Position the differential in its location in the bellhousing and tap it down gently, using a soft-faced mallet to ensure that the bearing is fully seated.
2   Fit the gearcase to the bellhousing and secure it temporarily with several bolts tightened to the specified torque.
3   Using feeler gauges inserted through the oil seal aperture in the gearcase, measure the clearance between the bearing and the circlip type shim in the bearing recess. If the clearance is not equal to the differential endfloat dimension given in the Specifications, slacken the case retaining bolts, extract the circlip through the oil seal aperture and substitute a thicker or thinner circlip as required from the range available. Repeat this procedure until the correct endfloat is obtained then remove the gearcase. Gearbox reassembly can now proceed as follows.
4   Insert the magnet into its location in the edge of the bellhousing, forked end first.
5   Refit the gearchange shaft and arm as described in Section 9 if this has not already been done.
6   With the bearings in place in the bellhousing, hold the assembled mainshaft and countershaft together and insert them into their locations.
7   With the help of an assistant, lift up the mainshaft and countershaft assemblies together approximately 0.5 in (12.0 mm). Engage the selector forks with their respective synchro sleeves and locate the selector shafts in the bellhousing (photo). Return the mainshaft and countershaft to their original positions ensuring that the selector shafts engage fully with their holes in the bellhousing.

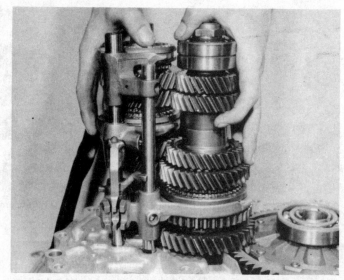

13.7 Lift up the mainshaft and countershaft to allow fitment of the selector forks and shafts

8   Refit the gearchange holder and interlock assembly, noting that the holder locates in a slot in the 1st/2nd selector shaft (photo).
9   Refit the three gearchange holder and interlock retaining bolts and tighten them to the specified torque (photo).
10  Refit the reverse gear idler shaft with reverse gear, noting that the long boss on the gear must face the bellhousing, and the hole in the top of the shaft faces away from the gear clusters (photo).
11  Engage the reverse gear fork over the reverse gear teeth and over the peg on the 5th/reverse selector. Secure the reverse gear fork bracket with the two retaining bolts tightened to the specified torque (photo).
12  Apply a continuous bead of RTV sealant to the gearcase mating face. Lower the gearcase over the gear clusters and engage the shafts and bearings in their locations. Using circlip pliers inserted through the countershaft access plug aperture, spread the circlip and tap the gearcase fully into position using a soft-faced mallet. Release the circlip ensuring that it enters the groove on the countershaft bearing.
13  Refit the gearcase retaining bolts and breather bracket then tighten the bolts progressively to the specified torque.
14  Refit the reverse idler shaft retaining bolt and tighten it to the specified torque.

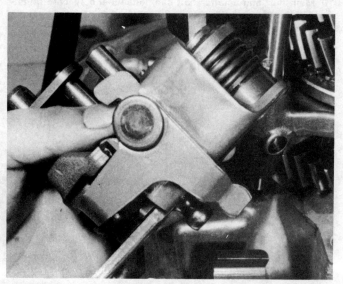

13.8 Fit the gearchange holder and interlock assembly ...

13.9 ... and secure with the three retaining bolts (arrowed)

13.10 Fit the reverse idler shaft and gear

13.11 Engage the reverse gear fork with reverse gear and with the 5th/reverse selector peg (arrowed)

15 Apply RTV sealant to the threads of the countershaft access plug, refit the plug and tighten it to the specified torque.

16 Refit the speedometer drive pinion, if this was not removed when removing the gearbox from the car, and the reversing light switch.

17 Finally refit the final drive differential oil seal to the gearcase if not already done.

18 Check the operation of the gearchange mechanism ensuring that all gears can be engaged, then refit the gearbox to the car as described in Section 3.

## 14 Gear lever and remote control housing – removal and refitting

*Gear lever*

1 Jack up the front of the car and support it on axle stands.

2 From under the car undo the nut and withdraw the bolt securing the gear lever to the gearchange remote control rod (photo).

3 From inside the car unscrew the gear lever knob then remove the gaiter and rubber boot.

4 Remove the retaining circlip and withdraw the gear lever from the remote control housing.

5 Release the sealing washers, extract the bushes and slide out the spacer complete with O-ring seals located at the base of the gear lever. Examine the components for wear.

6 Inspect the O-rings and gear lever seat in the remote control housing for signs of wear or damage.

7 Renew any worn or damaged parts then reassemble and refit the gear lever using the reverse sequence to removal.

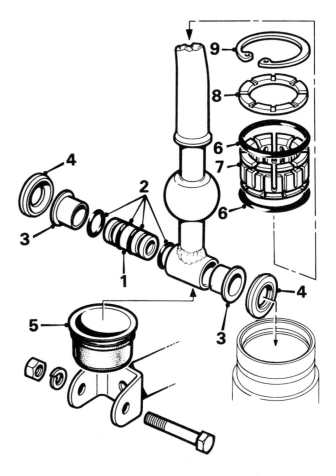

**Fig. 6.4 Exploded view of the gear lever (Sec 14)**

| | | | |
|---|---|---|---|
| 1 | Spacer | 6 | O-ring |
| 2 | O-rings | 7 | Gear lever seat |
| 3 | Bushes | 8 | Retaining ring |
| 4 | Sealing washers | 9 | Circlip |
| 5 | Dust cover | | |

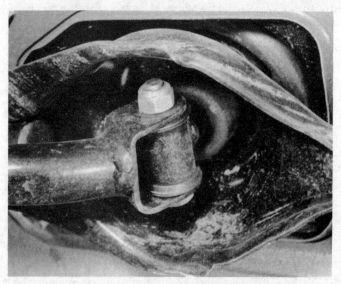

14.2 Gear lever to gearchange remote control rod retaining bolt

11 Undo and remove the small bolt in the centre of the gearbox steady rod, at the gearbox end of the remote control housing.
12 Remove the dished washer, slide off the steady rod and remove the flat washer. Note that the lip on the inner washer faces the steady rod rubber bush.
13 At the remote control housing end, undo the two bolts securing the mounting bracket to the vehicle floor and withdraw the remote control assembly from under the car (photo).
14 Refitting is the reverse sequence of removal.

14.13 Remote control housing bracket retaining bolts (arrowed)

### Remote control housing

8 Remove the gear lever as described above.
9 From under the front of the car remove the spring clip to expose the gearchange rod to gearbox, gearchange shaft retaining roll pin.
10 Using a suitable punch, drive out the roll pin and slide the gearchange rod rearwards off the shaft. Remove the rod from under the car.

---

### 15 Fault diagnosis — gearbox

| Symptom | Reason(s) |
|---|---|
| Gearbox noisy in neutral | Mainshaft bearings worn |
| Gearbox noisy only when moving (in all gears) | Countershaft bearings worn<br>Differential bearings worn<br>Differential final drive gear or countershaft pinion chipped or worn |
| Gearbox noisy in only one gear | Worn, damaged or chipped gear teeth<br>Worn needle roller bearings |
| Gearbox jumps out of gear | Worn synchro hubs or synchro sleeves<br>Weak or broken selector shaft detent spring<br>Weak or broken gearchange shaft detent spring<br>Worn shaft detent grooves<br>Worn selector forks |
| Ineffective synchromesh | Worn baulk rings or synchro hubs |
| Difficulty in engaging gears | Clutch fault<br>Ineffective synchromesh<br>Worn gear lever bushes and linkage<br>Gear linkage retaining clips broken |

# Chapter 7  Driveshafts

*For modifications, and information applicable to later models, see Supplement at end of manual*

## Contents

## Specifications

**Type** ............................................................................................  Unequal length solid steel, splined to inner and outer constant velocity joints

### Lubrication

Overhaul only – see text
Lubricant type:
    Outer constant velocity joint ................................................................  Molycote grease VN2461/C, or equivalent
    Inner constant velocity joint .................................................................  Mobil 525 grease, or equivalent
    Hub bearing water shield ......................................................................  Duckhams Admax, or equivalent
Quantity:
    Outer constant velocity joint ................................................................  90 cc
    Inner constant velocity joint .................................................................  190 to 205 cc

### Torque wrench settings

| | lbf ft | Nm |
|---|---|---|
| Driveshaft retaining nut* ............................................................. | 150 | 203 |
| Roadwheel nuts ........................................................................... | 53 | 72 |
| Swivel hub to strut nuts .............................................................. | 66 | 90 |
| Tie-rod outer balljoint to steering arm ........................................ | 22 | 30 |

*Refer to Chapter 12, Section 11*

## 1  General description

Drive is transmitted from the differential to the front wheels by means of two unequal length, solid steel driveshafts.

Both driveshafts are fitted with constant velocity joints at each end. The outer joints are of the Rzeppa ball and cage type and are splined to accept the driveshaft and wheel hub drive flange. The inner joints are of the sliding tripode type allowing lateral movement of the driveshaft to cater for suspension travel. The inner joints are splined to accept the driveshafts and differential sun gears.

To eliminate driveshaft-induced harmonic vibrations and resonance, a rubber mounted steel damper is attached to the longer right-hand driveshaft.

Driveshaft repair procedures are limited, as only the inner and outer rubber boots and outer constant velocity joints are available separately. The driveshafts and inner joints are supplied as complete assemblies.

## 2  Maintenance and inspection

1  At the intervals given in Routine Maintenance at the beginning of this manual carry out a thorough inspection of the driveshafts and joints as follows.

2  Jack up the front of the car and support it securely on axle stands.
3  Slowly rotate the roadwheel and inspect the condition of the outer joint rubber boots. Check for signs of cracking, splits or deterioration of the rubber which may allow the grease to escape and lead to water and grit entry into the joint. Also check the security and condition of the retaining clips. Repeat these checks on the inner constant velocity joints. If any damage or deterioration is found the boots should be renewed as described in Section 5.
4  Continue rotating the roadwheel and check for any distortion or damage to the driveshaft. Check for any free play in the joints by first holding the driveshaft and attempting to rotate the wheel. Repeat this check by holding the inner joint and attempting to rotate the driveshaft. Any appreciable movement indicates wear in the joints, wear in the driveshaft splines or loose driveshaft retaining nut.
5  Lower the car to the ground, remove the wheel trim, extract the split pin and check the tightness of the driveshaft retaining nut. Fit a new split pin after aligning the holes and slots in the joint and nut. Refit the wheel trim.
6  Road test the car and listen for a metallic clicking from the front as the car is driven slowly in a circle with the steering on full lock. If a clicking noise is heard this indicates wear in the outer constant velocity joint caused by excessive clearance between the balls in the joint and the recesses in which they operate. Remove and inspect the joint, as described in Section 4.
7  If vibration, consistent with road speed is felt through the car

when acceleraing, there is a possibility of wear in the inner constant velocity joint. If so, renewal of the driveshaft complete with inner joint will be necessary.

### 3   Driveshaft – removal and refitting

**Note:** *First refer to Chapter 12, Section 11 'Driveshaft retaining nut – modifications'*

1   While the vehicle is standing on its wheels, firmly apply the handbrake and put the transmission in gear.

2   Remove the wheel trim and extract the driveshaft retaining nut split pin. Using a suitable socket and bar, undo and remove the driveshaft nut. Recover the thrust washer (photo).

3   Slacken the wheel nuts, jack up the front of the car and support it on axle stands. Remove the roadwheel and return the gearlever to neutral.

4   Place a suitable container beneath the gearbox drain plug (see Chapter 6), undo the plug and allow the oil to drain. Refit the plug after draining.

5   Undo and remove the nut securing the tie-rod outer balljoint to the swivel hub steering arm. Release the balljoint from the steering arm using a balljoint separator tool.

6   Undo and remove the nuts and washers then withdraw the two bolts securing the suspension strut to the swivel hub (photo).

7   Pull the upper part of the swivel hub outwards as far as possible without placing undue strain on the flexible brake hose. Push the driveshaft inwards and manoeuvre the outer constant velocity joint from the hub (photo).

8   Using a suitable flat bar or large screwdriver, lever between the inner constant velocity joint and differential housing to release the joint from the differential sun gear.

9   Withdraw the inner joint fully from the differential then remove the driveshaft assembly from under the wheel arch. Recover the plastic bearing water shield from the end of the outer constant velocity joint (photo).

10  To refit the driveshaft place it in position under the car and enter the inner joint splines into the differential sun gear. Push the driveshaft firmly inward to engage the internal spring ring on the sun gear with the groove in the inner joint splines.

11  Position the bearing water shield on the flange of the outer joint and fill the water shield groove with the specified grease.

12  Pull the swivel hub out at the top and enter the outer constant velocity joint into the hub. Refit the thrust washer and driveshaft retaining nut and use the nut to draw the joint fully into place.

13  Refit the swivel hub to the suspension strut and secure with the two bolts, washers and nuts tightened to the specified torque.

14  Refit the steering tie-rod outer balljoint to the swivel hub arm.

3.2 Remove the driveshaft retaining nut and recover the thrustwasher

3.7 Remove the outer constant velocity joint from the hub

3.6 Suspension strut to swivel hub retaining nuts

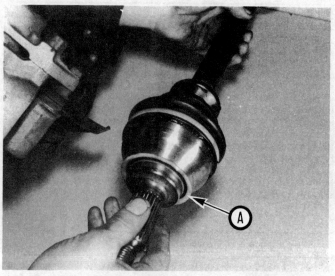

3.9 Withdraw the driveshaft assembly and recover the bearing water shield (A)

Screw on the nut and tighten it to the specified torque.

15 Refit the roadwheel then lower the car to the ground.

16 With the car standing on its wheels tighten the wheel nuts then the driveshaft retaining nut, to the specified torque. Continue tightening the driveshaft nut until the slots in the nut align with the hole in the constant velocity joint. Secure the nut using a new split pin. Refit the wheel trim.

17 Refill the gearbox using the specified grade of oil with reference to Chapter 6.

## 4 Outer constant velocity joint – removal, inspection and refitting

1 Remove the driveshaft from the car as described in the previous Section.

2 With the driveshaft on the bench, release the two rubber boot retaining clips and fold back the boot to expose the outer joint.

3 Firmly grasp the driveshaft, or support it in a vice. Using a hide, or plastic mallet, sharply strike the outer edge of the joint and drive it off the shaft. The outer joint is retained on the driveshaft by an internal circular section circlip and striking the joint in the manner described forces the circlip to contract into a groove, so allowing the joint to slide off.

4 With the constant velocity joint removed from the driveshaft, thoroughly clean the joint using paraffin, or a suitable solvent, and dry it, preferably using compressed air. Carry out a careful visual inspection of the joint, paying particular attention to the following areas.

5 Move the inner splined driving member from side to side to expose each ball in turn at the top of its track. Examine the balls for cracks, flat spots or signs of surface pitting.

6 Inspect the ball tracks on the inner and outer members. If the tracks have widened, the balls will no longer be a tight fit. At the same time check the ball cage windows for wear or for cracking between the balls. Wear in the balls, ball tracks and ball cage windows will lead to the characteristic clicking noise on full lock described previously.

7 If any of the above checks indicate wear in the joint it will be necessary to renew it complete, as the internal parts are not available separately. If the joint is in a satisfactory condition, obtain a repair kit consisting of a new rubber boot, retaining clips, and the correct quantity of grease.

8 The help of an assistant will be necessary whilst refitting the joint to the driveshaft. Ensure that the circlip is undamaged and correctly located in its groove in the driveshaft (photo). Position the new rubber boot over the shaft and locate its end in the shaft groove.

9 Place the retaining clip over the rubber boot and wrap it round until the slot in the clip end can be engaged with the tag. Make sure the clip is as tight as possible using pliers, or a screwdriver, if

4.8 Constant velocity joint retaining circlip (arrowed) fitted to driveshaft groove

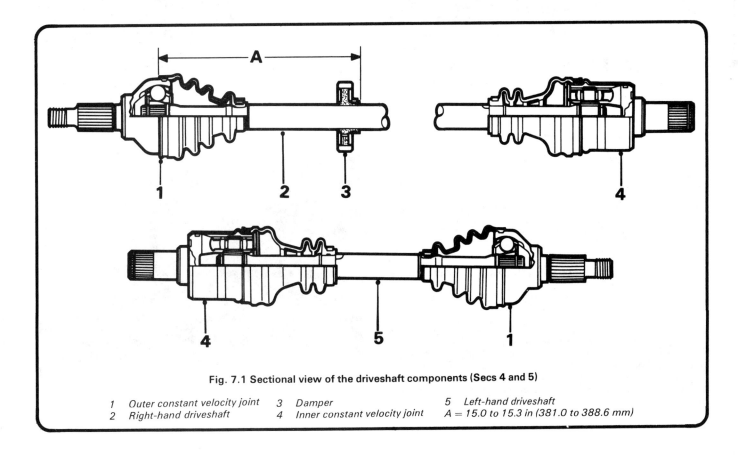

**Fig. 7.1 Sectional view of the driveshaft components (Secs 4 and 5)**

| | | |
|---|---|---|
| 1 Outer constant velocity joint | 3 Damper | 5 Left-hand driveshaft |
| 2 Right-hand driveshaft | 4 Inner constant velocity joint | A = 15.0 to 15.3 in (381.0 to 388.6 mm) |

necessary. Fully tighten the clip by squeezing the raised portion with pliers.

10 Fold back the rubber boot and position the constant velocity joint over the splines on the driveshaft until it abuts the circlip (photo).

11 Using two small screwdrivers placed either side of the circlip, compress the clip and at the same time have your assistant firmly strike the end of the joint with a hide, or plastic mallet.

12 The joint should slide over the compressed circlip and into position on the shaft. It will probably take several attempts until you achieve success. If the joint does not spring into place the moment it is struck, remove it, reposition the circlip and try again. Do not force the joint, otherwise the circlip will be damaged.

13 With the joint in position against the retaining collar, pack it thoroughly with the specified quantity of the grease supplied in the repair kit. Work the grease well into the ball tracks while twisting the joint, and fill the rubber boot with any excess.

14 Ease the rubber boot over the joint and secure it with the retaining clip, as described in paragraph 9.

15 The driveshaft can now be refitted to the car, as described in Section 3.

the three bearing caps in the same way and check for evidence of excessive play between the roller bearing caps and their tracks in the outer member.

9 If any of the above checks indicate wear in the joint it will be necessary to renew the driveshaft and inner joint as an assembly; they are not available separately. If the joint is in a satisfactory condition obtain a repair kit consisting of the new rubber boot, retaining clips and the specified grease.

10 Slide the new rubber boot onto the driveshaft followed by the rubber retaining ring. Position the small end of the boot in the driveshaft groove then slip the rubber ring over it to secure it in place (photos).

11 Fold back the boot and pack the constant velocity joint with the specified quantity of the grease supplied in the kit (photo). Work the grease well into the joint while moving it from side to side. Fill the boot with any excess.

12 Position the large end of the boot over the joint outer member so that it locates squarely in the groove.

13 Position the retaining clip over the boot and engage one of the slots in the clip end over the small tag. Make sure the clip is as tight as possible and, if necessary use a screwdriver to ease the slot over the tag (photo).

4.10 Fitting the outer constant velocity joint to the driveshaft

5.10A Fit the inner joint rubber boot ...

## 5 Constant velocity joint rubber boots – removal and refitting

1 Remove the driveshaft from the car as described in Section 3.

### Outer joint rubber boot

2 Remove the outer constant velocity joint from the driveshaft as described in Section 4. Renewal of the rubber boot is also covered in Section 4 as it is an integral part of the outer joint removal and refitting procedures.

### Inner joint rubber boot

3 Remove the outer constant velocity joint and rubber boot as described in Section 4.

4 If working on the right-hand driveshaft, mark the position of the damper on the shaft then release the retaining clip and slide off the damper.

5 Release the rubber retaining ring on the small end of the inner joint boot and slide the ring off the driveshaft.

6 Release the retaining clip, slip the boot off the inner joint and withdraw it from the driveshaft.

7 Clean out as much of the grease in the inner constant velocity joint as possible using a wooden spatula and old rags. As the joint cannot be dismantled it is advisable not to clean it using paraffin or solvents.

8 Examine the bearing tracks in the joint outer member for signs of scoring, wear ridges or evidence of lack of lubrication. Also examine

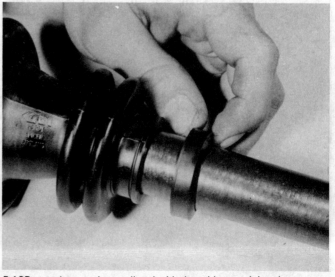

5.10B ... and secure its small end with the rubber retaining ring

16 Secure the damper in position with the retaining clip as described in paragraphs 13 and 14 (photo).
17 Refit the outer constant velocity joint and rubber boot as described in Section 4, then refit the driveshaft as described in Section 3.

5.11 Pack the joint with the special grease

5.13 Use a screwdriver to stretch the retaining clip slot over the tag

14 Fully tighten the clip by squeezing the raised portion with pincers (photo).
15 Refit the damper to the driveshaft (where applicable) and position it against the marks made on the shaft during removal. If a new driveshaft has been fitted or if the marks have been lost, use the setting dimension given in Fig. 7.1.

5.14 Tighten the clip fully by squeezing the raised portion

5.16 Secure the damper in place with the retaining clip

## 6 Fault diagnosis – driveshafts

| Symptom | Reason(s) |
| --- | --- |
| Vibration and/or noise on turns | Worn constant velocity outer joint(s) |
| Vibration when accelerating | Worn constant velocity inner joint(s)<br>Bent or distorted driveshaft |
| Noise on taking up drive | Worn driveshaft or constant velocity joint splines<br>Loose driveshaft retaining nut<br>Worn constant velocity joints |

*See also Fault diagnosis – suspension and steering*

# Chapter 8 Braking system

*For modifications, and information applicable to later models, see Supplement at end of manual*

## Contents

## Specifications

### System type ...................................................
Diagonally split dual circuit hydraulic with pressure regulating valve in rear hydraulic circuit. Cable operated handbrake on rear wheels. Servo assistance on all models

### Front brakes
| | |
|---|---|
| Type ................................................................................... | Ventilated disc with single piston sliding calipers |
| Disc diameter ...................................................................... | 9.5 in (241.3 mm) |
| Disc thickness ..................................................................... | 0.866 to 0.871 in (22.0 to 22.13 mm) |
| Maximum disc run-out .......................................................... | 0.006 in (0.15 mm) |
| Minimum pad thickness ........................................................ | 0.125 in (3.1 mm) |

### Rear brakes
| | |
|---|---|
| Type ................................................................................... | Single leading shoe drum, self-adjusting |
| Drum diameter ..................................................................... | 8.0 in (203.0 mm) |
| Lining width ........................................................................ | 1.5 in (38.1 mm) |
| Minimum lining thickness ..................................................... | 0.0625 in (1.6 mm) |
| Wheel cylinder diameter ....................................................... | 0.687 in (17.45 mm) |
| Handbrake lever stop endfloat .............................................. | 0 to 0.080 in (0 to 2.03 mm) |

### General
| | |
|---|---|
| Master cylinder bore diameter .............................................. | 0.81 in (20.57 mm) |
| Servo unit boost ratio .......................................................... | 3:1 |
| Brake fluid type/specification ............................................... | Hydraulic fluid to FMVSS 116 DOT 4 or SAE J1703C (Duckhams Universal Brake and Clutch Fluid) |

### Torque wrench settings
| | lbf ft | Nm |
|---|---|---|
| Bleed screws ....................................................................... | 7 | 10 |
| Brake caliper to swivel hub bolts .......................................... | 53 | 72 |
| Twin GP valve mounting bolts .............................................. | 9 | 12 |
| Guide pin bolts .................................................................... | 24 | 33 |
| Master cylinder mounting nuts ............................................. | 9 | 12 |
| Servo unit mounting nuts ..................................................... | 9 | 12 |
| Wheel cylinder to backplate bolts ......................................... | 5 | 7 |
| Rear hub retaining nut ......................................................... | 50 | 68 |
| Wheel nuts .......................................................................... | 53 | 72 |
| Brake disc to drive flange .................................................... | 8 | 11 |

## 1 General description

The braking system is of the servo-assisted, dual circuit hydraulic type with disc brakes at the front and drum brakes at the rear. A diagonally split dual circuit hydraulic system is employed in which each circuit operates one front and one diagonally opposite rear brake from a tandem master cylinder. Under normal conditions both circuits operate in unison; however, in the event of hydraulic failure in one circuit, full braking force will still be available at two wheels. A pressure regulating device or 'twin GP' (Gravity Pressure) valve is incorporated in the rear brake hydraulic circuit. This valve regulates the pressure applied to each rear brake and reduces the possibility of the rear wheels locking under heavy braking.

The ventilated front disc brakes are operated by single piston sliding type calipers. At the rear, leading and trailing brake shoes are operated by twin piston wheel cylinders and are self-adjusting by footbrake application.

The handbrake provides an independent mechanical means of rear brake application.

Driver warning lights are provided for brake pad wear, low brake hydraulic fluid level and handbrake applied.

## 2 Maintenance and inspection

1 At the intervals given in Routine Maintenance at the beginning of this Manual the following service operations should be carried out on the braking system components.
2 Check the brake hydraulic fluid level and if necessary top up with the specified fluid to the MAX mark on the reservoir. Any need for frequent topping up indicates a fluid leak somewhere in the system which must be investigated and rectified immediately.
3 Check the front disc pads and the rear brake shoe linings for wear and inspect the condition of the discs and drums. Details will be found in Sections 3 and 6 respectively.
4 Check the condition of the hydraulic pipes and hoses as described in Section 14. At the same time check the condition of the handbrake cables, lubricate the exposed cables and linkages and if necessary adjust the handbrake as described in Section 16.
5 The three braking system warning lights should be tested as follows. Check the operation of the handbrake warning light by applying the handbrake with the ignition switched on. The light should illuminate when the handbrake is applied. To test the low brake hydraulic fluid level warning light, place the car in gear, release the handbrake and switch on the ignition. The light should illuminate when the flexible contact cover in the centre of the brake fluid reservoir filler cap is depressed. To check the disc pad wear warning indicator, locate the twin terminal black plastic socket which is in the wiring harness adjacent to the right-hand brake caliper. Switch on the ignition and connect a bridging wire between the terminals on one socket; the pad wear warning light should be illuminated on the instrument panel when the bridging wire is earthed. If any of the lights fail to illuminate in the test condition, then either the bulb is blown, a fuse is at fault, or there is a fault in the circuit.
6 Renew the brake hydraulic fluid at the specified intervals by draining the system and refilling with fresh fluid as described in Section 15.
7 The flexible brake hoses and rubber seals in the brake calipers, wheel cylinders and master cylinder should also be renewed at the less frequent intervals given, as should the air filter in the vacuum servo unit. Details of these operations will be found in the relevant Sections of this Chapter.

## 3 Front disc pads – inspection and renewal

1 Apply the handbrake, prise off the front wheel trim and slacken the wheel nuts. Jack up the front of the car and support it securely on axle stands. Remove the roadwheels.
2 The thickness of the disc pads can now be checked by viewing through the slot in the front of the caliper body. If the lining on any of the pads is at, or below, the minimum specified thickness all four pads must be renewed as a complete set.
3 To renew the pads, first remove the protective cap over the caliper bleed screw (photo). Obtain a plastic or rubber tube of suitable diameter to fit snugly over the bleed screw and submerge the free end in a jar containing a small quantity of brake fluid.
4 Open the bleed screw half a turn and pull the caliper body toward you. This will push the piston back into its bore to facilitate removal and refitting of the pads. When the piston has moved in as far as it will go, close the bleed screw, remove the tube and refit the protective cap.
5 Disconnect the pad wear indicator wiring connector (right-hand caliper only) and, using a suitable spanner, unscrew the lower guide pin bolt while holding the guide pin with a second spanner (photo).
6 Pivot the caliper body upwards (photo), withdraw the two disc pads and, where fitted, the anti-squeal shim(s).
7 Brush the dust and dirt from the caliper, piston, disc and pads, but **do not inhale, as it is injurious to health.**
8 Rotate the brake disc by hand and scrape away any rust and scale. Carefully inspect the entire surface of the disc and if there are any signs of cracks, deep scoring or severe abrasions, the disc must be renewed. Also inspect the caliper for signs of fluid leaks around the piston, corrosion, or other damage. Renew the piston seals or the caliper body as necessary.
9 To refit the pads, first attach the anti-squeal shim(s) to the pad(s). If a single shim is fitted, it must be fitted to the outer pad. Later models are fitted with an anti-squeal shim on both pads; in this instance the inner pad can be identified by the cut-out section to facilitate the location of the pad wear warning leads. Very lightly smear the pad and caliper contact areas with a silicone grease, ensuring that no grease is allowed to come into contact with the friction material.
10 Place the pads in position against the disc, noting that the pad with the wear indicator lead is fitted to the inner position on the right-hand caliper.
11 Place the caliper body over the pads and refit the guide pin bolt. Tighten the bolt to the specified torque.
12 Reconnect the wear indicator wiring connector (where applicable), and refit the roadwheels – do not tighten the wheel nuts fully until the weight of the car is on the wheels.
13 Depress the brake pedal several times to bring the piston into contact with the pads and then lower the car to the ground. Check and, if necessary, top up the fluid in the master cylinder reservoir.

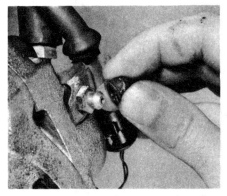

3.3 Remove the protective cap over the caliper bleed screw

3.5 Disconnect the pad wear indicator wiring connector

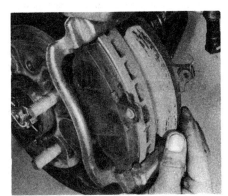

3.6 Pivot the caliper body upwards and withdraw the disc pads

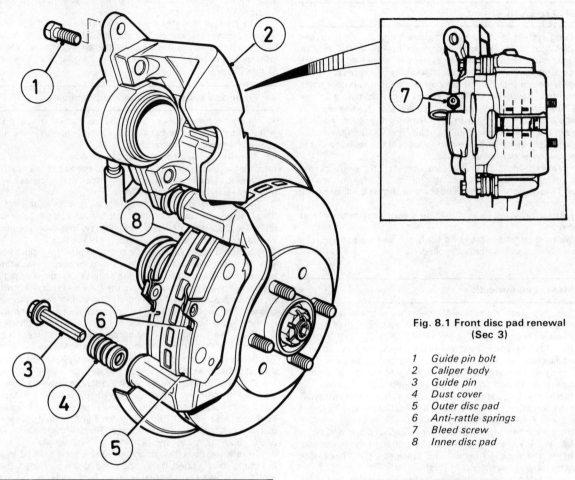

Fig. 8.1 Front disc pad renewal
(Sec 3)

1　Guide pin bolt
2　Caliper body
3　Guide pin
4　Dust cover
5　Outer disc pad
6　Anti-rattle springs
7　Bleed screw
8　Inner disc pad

## 4　Front brake caliper – removal, overhaul and refitting

1　Apply the handbrake, prise off the front wheel trim and slacken the wheel nuts. Jack up the front of the car and support it securely on axle stands. Remove the roadwheel.

2　Using a suitable spanner, unscrew the guide pin bolts while holding the guide pins with a second spanner.

3　Disconnect the pad wear indicator wiring connector (right-hand side caliper only) and lift away the caliper body, leaving the disc pads and anchor bracket still in position. It is not necessary to remove the anchor bracket unless it requires renewal because of accident damage or severe corrosion.

4　With the flexible brake hose still attached to the caliper body, very slowly depress the brake pedal until the piston has been ejected just over halfway out of its bore.

5　Using a brake hose clamp, or self-locking wrench with protected jaws, clamp the flexible brake hose. This will minimise brake fluid loss during subsequent operations.

6　Slacken the brake hose-to-caliper body union, and then, while holding the hose, rotate the caliper to unscrew it from the hose. Lift away the caliper and plug or tape over the end of the hose to prevent dirt entry.

7　With the caliper on the bench wipe away all traces of dust and dirt, but **avoid inhaling the dust as it is injurious to health.**

8　Withdraw the partially ejected piston from the caliper body and remove the dust cover.

9　Using a suitable blunt instrument, such as a knitting needle, carefully extract the piston seal from the caliper bore.

10　Clean all the parts in methylated spirit, or clean brake fluid, and wipe dry using a lint free cloth. Inspect the piston and caliper bore for signs of damage, scuffing or corrosion and if these conditions are evident renew the caliper body assembly. Also renew the guide pins if bent or damaged.

11　If the components are in a satisfactory condition, a repair kit consisting of new seals and dust cover should be obtained.

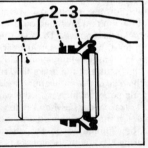

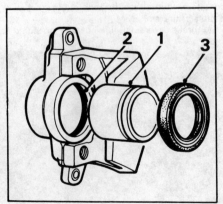

Fig. 8.2 Front brake caliper components (Sec 4)

1　Piston　　　　　　　　　3　Dust cover
2　Piston seal

12 Thoroughly lubricate the components and new seals with clean brake fluid and carefully fit the seal to the caliper bore.

13 Position the dust cover over the innermost end of the piston so that the caliper bore sealing lip protrudes beyond the base of the piston. Using a blunt instrument, if necessary, engage the sealing lip of the dust cover with the groove in the caliper. Now push the piston into the bore until the other sealing lip of the dust cover can be engaged with the groove in the piston. Having done this, push the piston fully into its bore. Ease the piston out again slightly, and make sure that the cover lip is correctly seating in the piston groove.

14 Remove the guide pins from the anchor bracket and smear them with a high melting-point brake grease. Fit new dust covers to the guide pins and refit them to the anchor bracket.

15 Hold the flexible brake hose and screw the caliper body back onto the hose.

16 With the piston pushed fully into its bore, refit the caliper and secure it with the guide pin bolts. Tighten the bolts to the specified torque.

17 Fully tighten the brake hose union and remove the clamp. If working on the right-hand caliper, reconnect the pad warning light wiring connector.

18 Refer to Section 15 and bleed the brake hydraulic system; noting that, if precautions were taken to minimise fluid loss, it should only be necessary to bleed the relevant front brake.

19 Refit the roadwheel and lower the car to the ground before fully tightening the wheel nuts and refitting the wheel trim.

## 5 Front brake disc – removal and refitting

1 Apply the handbrake, remove the front wheel trim and slacken the wheel nuts. Jack the front of the car up and support it securely on axle stands. Remove the roadwheel.

2 Rotate the disc by hand and examine it for deep scoring, grooving or cracks. Light scoring is normal, but if excessive, the disc must be renewed. Any loose rust and scale around the outer edge of the disc can be removed by lightly tapping it with a small hammer while rotating the disc.

3 To remove the disc undo and remove the two bolts securing the brake caliper to the swivel hub. Withdraw the caliper, complete with pads, off the disc and support it to one side. Avoid straining the flexible brake hose.

4 Undo and remove the two screws securing the disc to the drive flange and withdraw the disc.

5 Refitting is the reverse sequence to removal. Ensure that the mating face of the disc and drive flange are thoroughly clean and tighten all retaining bolts to the specified torque.

## 6 Rear brake shoes – inspection and renewal

1 Chock the front wheels, remove the rear wheel trim and slacken the rear wheel nuts. Jack up the rear of the car and support it securely on axle stands. Remove the roadwheel and release the handbrake.

2 By judicious tapping and levering remove the hub cap and extract the split pin from the hub retaining nut.

3 Using a large socket and bar, undo and remove the hub retaining nut and flat washer (photo). *Note that the left-hand nut has a left-hand thread and the right-hand nut has a conventional right-hand thread.* **Take care not to tip the car from the axle stands.** If the hub nuts are particularly tight, temporarily refit the roadwheel and lower the car to the ground. Slacken the nut in this more stable position and then raise and support the car before removing the nut.

4 Withdraw the hub and brake drum assembly from the stub axle. If it is not possible to withdraw the hub due to the brake drum binding on the brake shoes, the following procedure should be adopted. Refer to Section 16, if necessary, and slacken off the handbrake cable at the adjuster. From the rear of the brake backplate, prise out the handbrake lever stop, which will allow the brake shoes to retract sufficiently from the hub assembly to be removed. It will, however, be necessary to remove the brake shoes and fit a new handbrake lever stop to the backplate.

5 With the brake drum assembly removed, brush or wipe the dust from the brake drum, brake shoes and backplate. **Take great care not to inhale the dust, as it is injurious to health.**

6 Measure the brake shoe lining thickness. If it is worn down to the

6.3 Rear hub retaining nut and flat washer

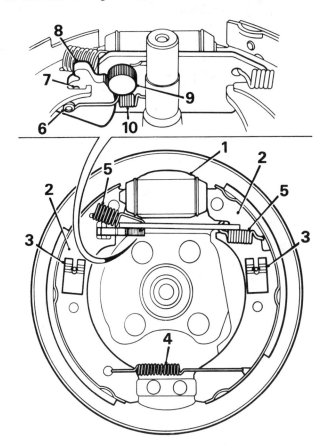

**Fig. 8.3 Rear brake assembly – left-hand side (Sec 6)**

| | | | |
|---|---|---|---|
| 1 | Wheel cylinder | 6 | Self-adjusted quadrant |
| 2 | Brake shoes | | pivot pin |
| 3 | Hold-down springs | 7 | Hollow pin |
| 4 | Lower return springs | 8 | Quadrant |
| 5 | Upper return springs | 9 | Ratchet wheel |
| | | 10 | Operating spring |

specified minimum amount, or if it is nearly worn down to the rivets, renew all four rear brake shoes. The shoes must also be renewed if any are contaminated with brake fluid or grease, or show signs of cracking or glazing. If contamination is evident, the cause must be traced and cured before fitting new brake shoes.

7   If the brake shoes are in a satisfactory condition proceed to paragraph 19; if removal is necessary, proceed as follows.

8   First make a careful note of the location and position of the various springs and linkages as an aid to refitting (photo).

9   Depress the brake shoe hold-down springs (photo) while supporting the hold-down pin from the rear of the backplate with your finger. Turn the springs through 90° and lift off. Withdraw the hold-down pins.

10  Using a screwdriver, if necessary, release the shoes from their lower pivots and disengage the lower return spring (photo).

11  Using long-nosed pliers, release the upper return spring from the leading shoe and the arm of the self-adjust mechanism (photo). Detach the leading shoe from the wheel cylinder and self-adjust mechanism, and remove it from the backplate.

12  Withdraw the trailing shoe from the backplate, move the

handbrake operating lever away from the shoe and unhook the handbrake cable (photo). Release the upper return spring and detach the self-adjust mechanism from the trailing brake shoe.

13  Before fitting the new brake shoes clean off the brake backplate with a rag and apply a trace of silicone grease to the brake shoe contact areas (photo). Clean the self-adjust mechanism and make sure that it is free to move in its elongated slot (photo).

14  Refit the self-adjust mechanism to the trailing brake shoe and secure with the return spring.

15  Position the self-adjust mechanism in the fully retracted position by moving the quadrant away from the ratchet wheel and turning it until the hollow pin locates the inner cutaway (photo).

16  Connect the handbrake cable to the operating lever and position the trailing shoe on the backplate.

17  Refit the leading shoe to the backplate, ensuring that the shoe is

6.8 Layout and position of the rear brake components

6.9 Brake shoe hold-down springs and pins

6.10 Brake shoe lower return spring

6.11 Brake shoe upper return spring removal

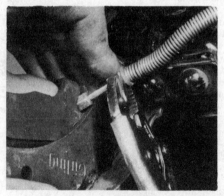
6.12 Detach the handbrake cable from the trailing shoe operating lever

6.13A Apply a trace of silicone grease to the brake shoe contact areas (arrowed)

6.13B Brake shoe self-adjust mechanism

6.15 Self-adjust mechanism pin (arrowed) in fully retracted position

6.17 Brake shoe correctly located around self-adjust mechanism hollow pin (arrowed)

properly located around the pin on the self-adjust mechanism (photo).

18  Refit the upper and lower return springs and the hold-down springs and pins.

19  Refit the brake drum and hub assembly, flat washer and hub retaining nut. Tighten the hub retaining nut to the specified torque and then tighten further until a split pin hole is aligned. Fit a new split pin and tap on the hub cap.

20  Refit the roadwheel, but do not fully tighten the wheel nuts.

21  If it was necessary to slacken the handbrake cable to enable the brake drum to be removed, adjust the cable as described in Section 16.

22  Lower the car to the ground, tighten the wheel nuts to the specified torque and refit the wheel trim. Depress the footbrake two or three times to operate the brake adjusters.

### 7  Rear wheel cylinder – removal and refitting

1  Begin by removing the rear hub and brake drum assembly, as described in Section 6, paragraphs 1 to 5 inclusive.

2  Using a screwdriver, or other suitable tool, carefully ease the upper end of the leading brake shoe (the one nearest the front of the car) away from the wheel cylinder piston. Take care not to damage the wheel cylinder rubber boot as you do this. As the leading shoe is moved away from the wheel cylinder, the self-adjust mechanism will expand and hold both brake shoes in the expanded position. This will provide sufficient clearance to allow removal of the wheel cylinder.

3  Using a brake hose clamp, or self-locking wrench with protected jaws, clamp the flexible brake hose just in front of the rear axle pivot mounting. This will minimise brake fluid loss during subsequent operations.

4  At the rear of the brake backplate, unscrew the union nut securing the brake pipe to the wheel cylinder. Carefully ease the pipe out of the cylinder and plug or tape over its end to prevent dirt entry.

5  Undo and remove the two bolts securing the wheel cylinder to the backplate and withdraw the cylinder from between the brake shoes.

6  To refit the wheel cylinder, place it in position on the backplate and engage the brake pipe and union. Screw in the union nut two or three turns to ensure the thread has started.

7  Refit the wheel cylinder retaining bolts and tighten them to the specified torque. Now fully tighten the brake pipe union nut.

8  Using a screwdriver or other suitable tool, as before, ease the leading brake shoe away from the wheel cylinder and retract the self-adjust mechanism. To do this move the quadrant, in its elongated slot, away from the ratchet wheel and turn it until the hollow pin contacts the cutaway in the operating lever. As you do this, ease the brake shoes back into their upper locations in the wheel cylinder pistons.

9  Refit the brake drum and hub assembly, flat washer and hub retaining nut. Tighten the hub retaining nut to the specified torque and then tighten further until a split pin hole is aligned. Fit a new split pin and tap on the hub cap.

10  Remove the clamp from the brake hose and bleed the brake hydraulic system, as described in Section 15. Providing suitable precautions were taken to minimise loss of fluid, it should only be necessary to bleed the relevant rear brake.

11  If it was necessary to slacken the handbrake cable to remove the brake drum, adjust the cable, as described in Section 16.

12  Refit the roadwheel, but do not tighten the wheel nuts fully.

13  Lower the car to the ground and tighten the wheel nuts. Depress the footbrake two or three times to operate the brake adjusters.

### 8  Rear wheel cylinder – overhaul

1  Remove the wheel cylinder from the car, as described in the previous Section.

2  With the wheel cylinder on the bench, remove the dust cover retainers, where fitted, and withdraw the dust covers from the ends of the piston, and cylinder body.

3  Withdraw the pistons and piston spring. Remove the seals from the pistons and unscrew the bleed screw from the cylinder body.

4  Thoroughly clean all the components in methylated spirit or clean brake fluid and dry with a lint-free rag.

5  Carefully examine the surfaces of the pistons and cylinder bore for wear, score marks or corrosion and, if evident, renew the complete

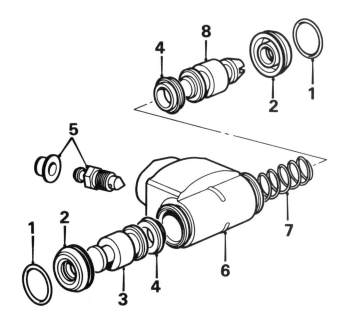

**Fig. 8.4 Exploded view of the rear wheel cylinder (Sec 8)**

| | |
|---|---|
| 1  Dust cover retainers (where fitted) | 5  Bleed screw and dust cap |
| 2  Dust cover | 6  Wheel cylinder body |
| 3  Trailing shoe piston | 7  Piston spring |
| 4  Piston seal | 8  Leading shoe piston |

wheel cylinder. If the components are in a satisfactory condition, obtain a repair kit consisting of new seals and dust covers.

6  Dip the new seals and pistons in clean brake fluid and assemble the components wet, as follows.

7  Using your fingers, fit the new seals to the pistons with their sealing lips facing inwards.

8  Lubricate the cylinder bore with clean brake fluid and insert the spring, followed by the two pistons. Note that the piston with the slot in its end must face toward the front of the car when fitted.

9  Place the dust covers over the pistons and cylinder and, where applicable, refit the dust cover retainers.

10  Screw the bleed screw into the cylinder body and then refit the wheel cylinder to the car, as described in the previous Section.

### 9  Rear brake backplate – removal and refitting

The rear brake backplate is removed in conjunction with the stub axle and details of this procedure will be found in Chapter 10.

### 10  Master cylinder – removal and refitting

1  Working under the front of the car, remove the dust cover from the bleed screw on each front brake caliper. Obtain two plastic or rubber tubes of suitable diameter to fit snugly over the bleed screws, and place the other ends in a suitable receptacle.

2  Open the bleed screws half a turn and operate the brake pedal until the master cylinder reservoir is empty. Tighten the bleed screws and remove the tubes. Discard the expelled brake fluid.

3  Disconnect the wiring connectors from the reservoir filler cap terminals.

4  Unscrew the two brake pipe union nuts and carefully withdraw the pipes from the master cylinder. Plug or tape over the ends of the pipes to prevent dirt entry, and place rags beneath the master cylinder to protect the surrounding paintwork.

5  Undo and remove the two retaining nuts and washers and withdraw the master cylinder from the servo unit.

6  Refitting the master cylinder is the reverse sequence to removal. Tighten the retaining nuts to the specified torque and, on completion, bleed the brake hydraulic system, as described in Section 15.

## 11 Master cylinder – overhaul

1 Remove the master cylinder from the car, as described in the previous Section. Drain any fluid remaining in the reservoir and prepare a clean uncluttered working surface, ready for dismantling.

2 Remove the primary piston assembly and then mount the cylinder horizontally in a soft-jawed vice. Using a parallel pin punch, drive out the two roll pins securing the reservoir to the cylinder body.

3 Lift off the reservoir and withdraw the two reservoir sealing washers from the master cylinder inlet ports.

4 Using a blunt instrument, push the secondary piston in as far as it will go and withdraw the secondary piston stop pin from its location in the secondary inlet port.

5 Slowly release the secondary piston, remove the cylinder from the vice and tap it on a block of wood to release the secondary piston from the cylinder bore.

6 Note the location and position of the components on the secondary piston and then remove the piston spring. Withdraw the seal retainer, followed by the seal and washer. Now remove the remaining seal from the other end of the secondary piston.

7 The primary piston should not be dismantled, as parts are not

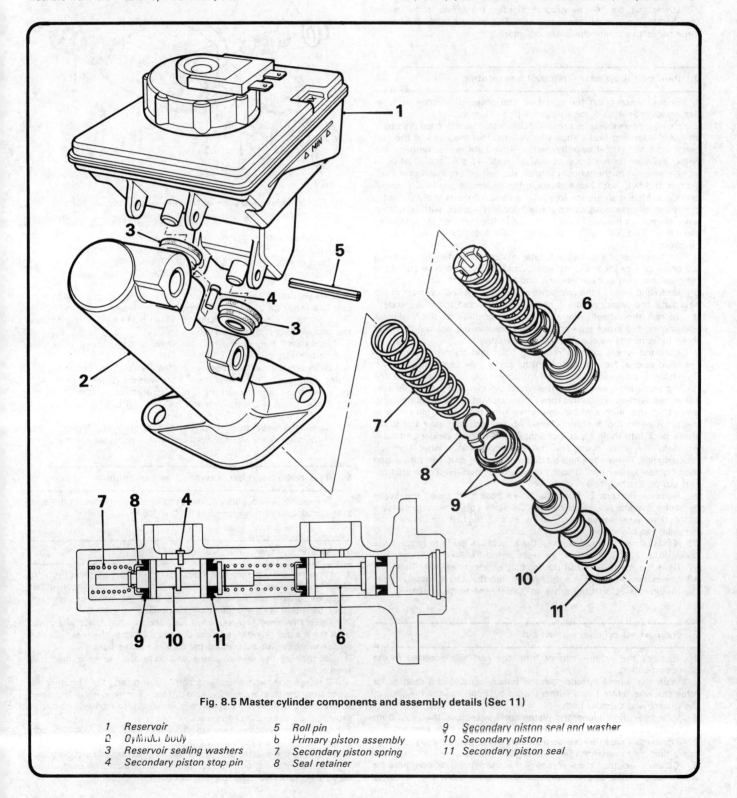

**Fig. 8.5 Master cylinder components and assembly details (Sec 11)**

| | | |
|---|---|---|
| 1 Reservoir | 5 Roll pin | 9 Secondary piston seal and washer |
| 2 Cylinder body | 6 Primary piston assembly | 10 Secondary piston |
| 3 Reservoir sealing washers | 7 Secondary piston spring | 11 Secondary piston seal |
| 4 Secondary piston stop pin | 8 Seal retainer | |

available separately. If the master cylinder is in a serviceable condition, and is to be reused, a complete new primary piston assembly is included in the repair kit.

8 With the master cylinder completely dismantled, clean all the components in methylated spirit, or clean brake fluid, and dry with a lint-free rag.

9 Carefully examine the cylinder bore and secondary piston for signs of wear, scoring or corrosion, and if evident renew the complete cylinder assembly.

10 If the components are in a satisfactory condition, obtain a repair kit consisting of new seals, springs and primary piston assembly.

11 Lubricate the master cylinder bore, pistons and seals thoroughly in clean brake fluid, and assemble them wet.

12 Fit the washer and seal onto the inner end of the secondary piston using your fingers only. Use the notes made during dismantling and the accompanying illustrations as a guide to the direction of fitment of the seal. Place the seal retainer over the seal and refit the spring. Fit the remaining seal to the other end of the secondary piston.

13 Insert the secondary piston into the cylinder bore, taking care not to turn over the lips of the seals as the piston is inserted.

14 Push the secondary piston down the cylinder bore as far as it will go and refit the stop pin into its hole in the inlet port.

15 Insert the new primary piston assembly into the cylinder bore, again taking care not to turn over the seal lips as they enter the bore.

16 Place the two reservoir seals into the inlet ports and refit the reservoir. Secure the reservoir with the two roll pins.

17 The master cylinder can now be refitted to the car, as described in the previous Section.

## 12 Twin GP valve – description and testing

1 The twin GP valve is mounted in the engine compartment and contains two gravity/pressure valves, one for each hydraulic circuit (photo).

12.1 Location of the twin GP valve in the engine compartment

2 The purpose of the valve is to distribute brake fluid to the front and rear brakes, and to limit the fluid pressure supplied to the rear brakes under heavy braking.

3 The operation of the valve may be suspect if one or both rear wheels continually lock during normal braking. It is essential, however, before condemning the valve to ensure the fault does not lie with the brake shoe assemblies or wheel cylinders, and that adverse road conditions are also not responsible.

4 In the event of failure of the valve it must be renewed as an assembly, as parts are not available separately.

## 13 Twin GP valve – removal and refitting

1 Remove the master cylinder reservoir filler cap and place a piece of polythene over the filler neck. Secure the polythene in place with an elastic band ensuring that an airtight seal is obtained. This will minimise brake fluid loss during subsequent operations.

2 Place some rags around the valve to protect the paintwork from brake fluid spillage. Wash off any brake fluid that comes into contact with the body immediately, using copious amounts of cold water. Brake fluid is a very effective paint stripper.

3 Clean the area around the brake pipe unions thoroughly and unscrew the union nuts. Note that the nuts securing the primary brake pipes are of a smaller diameter than the secondary nuts. Very carefully withdraw the brake pipes from the valve.

4 Undo and remove the two bolts securing the valve to the inner member and withdraw the unit from the engine compartment.

5 Refitting is the reverse sequence to removal. Refer to Fig. 8.6 if in any doubt about the pipe locations, and bleed the brake hydraulic system, as described in Section 15, after refitting.

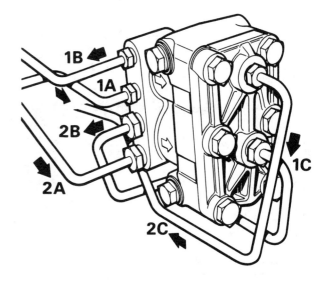

Fig. 8.6 Hydraulic pipe arrangement at twin GP valve (Sec 13)

Primary circuit hydraulic pipes (11 mm unions)
*1A From master cylinder*
*1B To right-hand front brake*
*1C To left-hand rear brake*
Secondary circuit hydraulic pipes (13 mm unions)
*2A From master cylinder*
*2B To left-hand front brake*
*2C To right-hand rear brake*

## 14 Hydraulic pipes and hoses – inspection, removal and refitting

1 At intervals given in Routine Maintenance carefully examine all brake pipes, hoses, hose connections and pipe unions.

2 First check for signs of leakage at the pipe unions. Then examine the flexible hoses for signs of cracking, chafing and fraying.

3 The brake pipes must be examined carefully and methodically. They must be cleaned off and checked for signs of dents, corrosion or other damage. Corrosion should be scraped off and, if the depth of pitting is significant, the pipes renewed. This is particularly likely in those areas underneath the vehicle body where the pipes are exposed and unprotected.

4 If any section of pipe or hose is to be removed, first unscrew the master cylinder reservoir filler cap and place a piece of polythene over the filler neck. Secure the polythene with an elastic band ensuring that an airtight seal is obtained. This will minimise brake fluid loss when the pipe or hose is removed.

5   Brake pipe removal is usually quite straightforward. The union nuts at each end are undone, the pipe and union pulled out and the centre section of the pipe removed from the body clips. Where the unions nuts are exposed to the full force of the weather they can sometimes be quite tight. As only an open-ended spanner can be used, burring of the flats on the nuts is not uncommon when attempting to undo them. For this reason a self-locking wrench is often the only way to separate a stubborn union.

6   To remove a flexible hose, wipe the unions and brackets free of dirt and undo the union nut from the brake pipe end(s).

7   Next extract the hose retaining clip, or unscrew the nut, and lift the end of the hose out of its bracket (photos). If a front hose is being removed, it can now be unscrewed from the brake caliper.

8   Brake pipes can be obtained individually, or in sets, from most accessory shops or garages with the end flares and union nuts in place. The pipe is then bent to shape, using the old pipe as a guide, and is ready for fitting to the car.

9   Refitting the pipes and hoses is a reverse of the removal sequence. Make sure that the hoses are not kinked when in position and also make sure that the brake pipes are securely supported in their clips. After refitting, remove the polythene from the reservoir and bleed the brake hydraulic system, as described in Section 15.

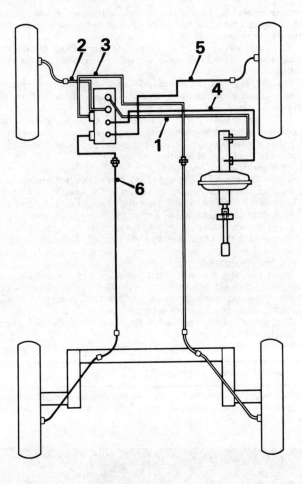

**Fig. 8.7 Layout of brake hydraulic pipes and hoses (Sec 14)**

Secondary circuit
1   *Master cylinder to GP valve*
2   *GP valve to left-hand front hose*
3   *GP valve to right-hand rear hose*
Primary circuit
4   *Master cylinder to GP valve*
5   *GP valve to right-hand front hose*
6   *GP valve to left-hand rear hose*

14.7A Rear brake hose and retaining clip (arrowed)

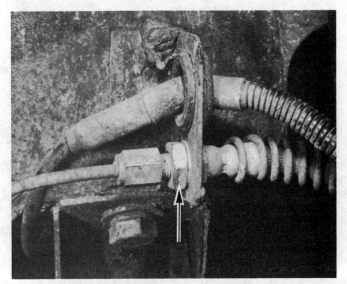

14.7B Front brake hose and retaining nut (arrowed)

## 15  Hydraulic system – bleeding

1   The correct functioning of the brake hydraulic system is only possible after removal of all air from the components and circuit; this is achieved by bleeding the system. Note that only clean unused brake fluid, which has remained unshaken for at least 24 hours, must be used.

2   If there is any possibility of incorrect fluid being used in the system, the brake lines and components must be completely flushed with uncontaminated fluid and new seals fitted to the components.

3   *Never reuse brake fluid which has been bled from the system.*

4   During the procedure, do not allow the level of brake fluid to drop more than halfway down the reservoir.

5   Before starting work, check that all pipes and hoses are secure, unions tight and bleed screws closed. Take great care not to allow brake fluid to come into contact with the car paintwork, otherwise the finish will be seriously damaged. Wash off any spilled fluid immediately with cold water.

6   There are a number of one-man, do-it-yourself, brake bleeding kits currently available from motor accessory shops. Always follow the instructions supplied with the kit. It is recommended that one of these kits is used wherever possible, as they greatly simplify the bleeding operation and also reduce the risk of expelled air and fluid being drawn back into the system. If one of these kits is not available, it will be

necessary to gather together a clean jar and a suitable length of clear plastic tubing which is a tight fit over the bleed screw, and also to engage the help of an asisstant.

7 If brake fluid has been lost from the master cylinder due to a leak in the system, ensure that the cause is traced and rectified before proceeding further.

8 If the hydraulic system has only been partially disconnected and suitable precautions were taken to prevent further loss of fluid it should only be necessary to bleed that part of the system (ie primary or secondary circuit).

9 If the complete system is to be bled then it should be done in the following sequence.

*Secondary circuit: Left-hand front then right-hand rear*
*Primary circuit: Right-hand front then left-hand rear*

10 To bleed the system, first clean the area around the bleed screw (photo) and fit the bleed tube. If necessary, top up the master cylinder reservoir with brake fluid.

11 *If a one-man brake bleeding kit is being used,* open the bleed screw half a turn and position the unit so that it can be viewed from the car (photo). Depress the brake pedal to the floor and slowly release it; the one-way valve in the kit will prevent expelled air from returning to the system. Repeat the procedure then top up the brake fluid level. Continue bleeding until clean brake fluid, free from air bubbles, can be seen coming through the tube. Now tighten the bleed screw and remove the tube.

15.11 Front brake caliper bleed screw (arrowed)

12 *If a one-man brake bleeding kit is not available,* immerse the free end of the bleed tube in the jar and pour in sufficient brake fluid to keep the end of the tube submerged. Open the bleed screw half a turn and have your assistant depress the brake pedal to the floor and then slowly release it. Tighten the bleed screw at the end of each downstroke to prevent the expelled air and fluid from being drawn back into the system. Repeat the procedure, then top up the brake fluid level. Continue bleeding until clean brake fluid, free from bubbles, can be seen coming through the tube. Now tighten the bleed screw and remove the tube.

13 Repeat the procedure described in paragraphs 10 to 12 on the remaining wheels, in the correct sequence.

14 When completed, recheck the fluid level in the reservoir, top up if required and refit the cap. Depress the brake pedal several times; it should feel firm and free from 'sponginess' which would indicate air is still present in the system.

## 16 Handbrake cable – adjustment

1 Chock the front wheels, jack up the rear of the car and support it securely on axle stands. Release the handbrake.

2 Apply the footbrake firmly two or three times to ensure full movement of the self-adjust mechanism on the rear brake shoes. This is particularly important if the brake drums have recently been removed.

3 Apply the handbrake sharply and then release it to equalise the cable loads.

4 From under the rear of the car measure the endfloat of the handbrake lever stops located on the rear of each brake backplate. If the movement of the stops is not as given in the Specifications, adjustment is necessary and is carried out as follows.

5 Turn the cable adjuster, located just in front of the rear axle (photo), clockwise to increase the cable tension and anti-clockwise to decrease it. The direction of rotation assumes the adjuster is being viewed from the rear of the car, ie facing forwards.

6 Turn the cable adjuster until the endfloat of the handbrake lever stops is as specified and the travel of the handbrake lever necessary to lock the wheels is between 5 and 8 clicks of the ratchet. Do not overtighten the cable.

7 After adjustment, ensure that the wheels are free to turn, without binding when the handbrake is released, and then lower the car to the ground.

16.5 Location of the handbrake cable adjuster (arrowed) in front of the rear axle

## 17 Handbrake cable (front) – removal and refitting

1 Chock the front wheels, jack up the rear of the car and support it securely on axle stands.

2 From under the rear of the car slacken the handbrake cable adjuster to remove all tension from the cable. Refer to Section 16, if necessary.

3 From inside the car remove the centre console, as described in Chapter 11.

4 Slide both the front seats fully forward then undo and remove the rear bolts securing the inner seat rails to the floor. Now remove the centre console rear mounting bracket.

5 Undo and remove the seat belt centre stalk mounting bolts and remove the two stalks.

6 Undo and remove the bolt securing the left-hand seat belt to the inner sill. Remove the carpet retaining screws and lift up the rear carpet sufficiently to gain access to the handbrake cable cover plate. Remove the cover plate and seal.

7 Extract the split pin and withdraw the clevis pin securing the handbrake cable to the handbrake lever. Remove the cable from the lever and feed it through the cover plate aperture.

8 Disconnect the front cable from the intermediate cable at the connector and withdraw it from under the car.

9 Refitting the cable is the reverse sequence of removal. Ensure that the seat belt mounting bolts are tightened to the specified torque (Chapter 11) and adjust the cable, as described in Section 16, after refitting.

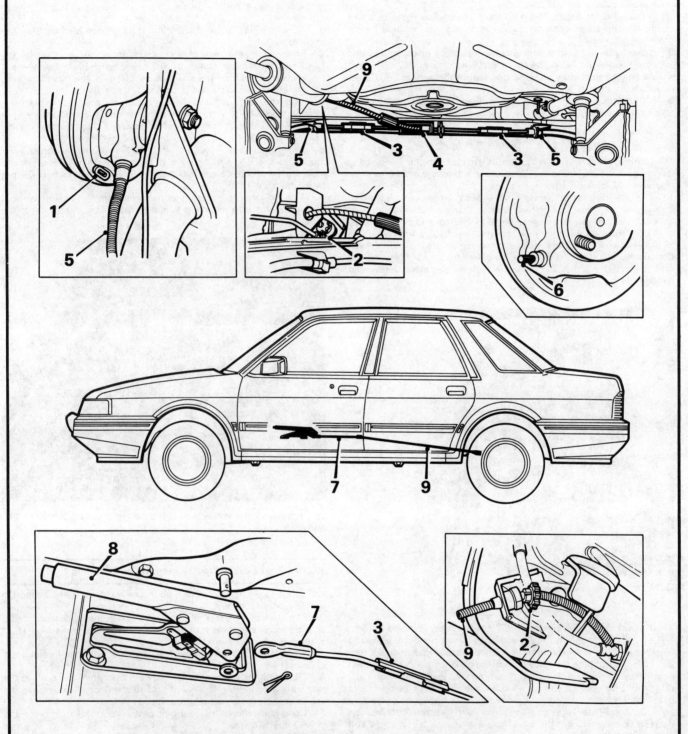

**Fig. 8.8 Layout of the handbrake mechanism (Secs 16 to 20)**

| | | | | | |
|---|---|---|---|---|---|
| 1 | Brake shoe inspection aperture | 4 | Compensator assembly | 8 | Handbrake lever |
| 2 | Handbrake cable adjuster | 5 | Handbrake cable – rear | 9 | Handbrake cable – intermediate |
| 3 | Cable connector | 6 | Handbrake lever stop | | |
| | | 7 | Handbrake cable – front | | |

## 18 Handbrake cable (intermediate) – removal and refitting

1   Chock the front wheels, jack up the rear of the car and support it securely on axle stands.

2   From under the rear of the car, slacken the handbrake cable adjuster to remove all tension from the cable. Refer to Section 16 if necessary.

3   Disconnect the intermediate cable from the front cable at the connector (photo), release the cable adjuster from its bracket and pull the cable through the bracket.

4   Disconnect both rear cables from the connectors located to the rear of the rear axle transverse member. Release the intermediate cable from its mounting bushes and withdraw it from under the car.

5   Refitting is the reverse sequence to removal but adjust the cable, as described in Section 16, after fitting.

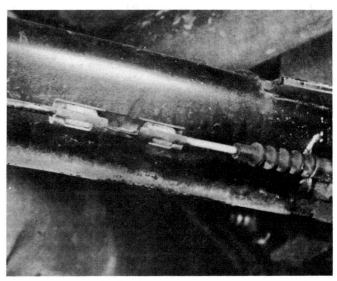

19.2 Handbrake rear cable connector

18.3 Handbrake intermediate cable to front cable connector located just forward of the fuel tank

## 19 Handbrake cable (rear) – removal and refitting

1   Chock the front wheels, prise off the rear wheel trim and slacken the wheel nuts. Jack the car up and support it securely on axle stands. Remove the roadwheel.

2   Disconnect the rear cable at the connector just behind the rear axle transverse member (photo).

3   Using circlip pliers, extract the retaining circlip and withdraw the rear cable from its mounting bracket on the rear axle member.

4   Refer to Section 6, paragraphs 2 to 4 inclusive, and remove the rear hub and brake drum assembly.

5   With the drum removed, ease the handbrake operating lever on the trailing brake shoe forward and disengage the inner cable nipple from the elongated slot on the lever.

6   Withdraw the cable from the brake backplate and remove the spring collar and felt seal.

7   Refitting is the reverse sequence to removal. Adjust the cable, as described in Section 16, after refitting.

## 20 Handbrake lever and switch – removal and refitting

1   Refer to Chapter 11, and remove the centre console.

2   If only the warning light switch is to be removed, disconnect the wires from the switch terminals, remove the securing screws and withdraw the switch.

3   To remove the lever assembly, extract the split pin, withdraw the clevis pin and detach the handbrake cable from the lever.

4   Undo and remove the two mounting bolts and lift away the handbrake lever assembly.

5   Refitting is a straightforward reversal of the removal sequence.

## 21 Footbrake pedal – removal and refitting

1   From inside the car extract the split pin and withdraw the clevis pin and washer securing the servo unit pushrod to the brake pedal.

2   Undo and remove the nut and washer at the clutch pedal end of the pedal pivot bolt.

3   Draw the pivot bolt toward the side of the car to release the clutch pedal. Recover the washer from the pivot shaft (where fitted), unhook the top of the clutch pedal from the cable and withdraw the pedal.

4   Fully remove the pivot bolt and withdraw the brake pedal after detaching the return spring. Where fitted, recover the pivot bolt washers.

5   With the pedal removed, examine the pivot bushes and, if worn, renew them.

6   Refitting the pedals is the reverse sequence to removal. Smear the pivot bolt and bushes with a little lithium-based grease prior to fitting.

## 22 Stop-light switch – removal, refitting and adjustment

1   To remove the switch, first disconnect the battery negative terminal.

2   Undo and remove the bolts on the right-hand side of the pedal securing the switch bracket in position.

3   Disconnect the wiring connector and remove the switch. Unscrew the switch from the plastic insert.

4   To refit the switch, screw it fully into the plastic insert and position the switch with the terminal in the vertical position.

5   Refit the wiring connector and secure the switch bracket to the pedal bracket with the bolts, finger tight only at this stage.

6   To adjust the switch, reconnect the battery negative terminal and switch on the ignition.

7   Depress the brake pedal by 0.25 in (6 mm) and adjust the switch position so that the stop-lights just come on.

8   Tighten the switch bracket retaining bolts and recheck the stop-light operation.

## 23 Vacuum servo unit – description

A vacuum servo unit is fitted into the brake hydraulic circuit in series with the master cylinder, to provide assistance to the driver when the brake pedal is depressed. This reduces the effort required by the driver to operate the brakes under all braking conditions.

The unit operates by vacuum obtained from the inlet manifold and basically consists of a booster diaphragm, control valve, and a non-return valve.

The servo unit and hydraulic master cylinder are connected together so that the servo unit piston rod acts as the master cylinder pushrod. The driver's braking effort is transmitted through another pushrod to the servo unit piston and its built-in control system. The servo unit piston does not fit tightly into the cylinder, but has a strong diaphragm to keep its edges in constant contact with the cylinder wall, so ensuring an airtight seal between the two parts. The forward chamber is held under vacuum conditions created in the inlet manifold of the engine and, during periods when the brake pedal is not in use, the controls open a passage to the rear chamber so placing it under vacuum conditions as well. When the brake pedal is depressed, the vacuum passage to the rear chamber is cut off and the chamber opened to atmospheric pressure. The consequent rush of air pushes the servo piston forward in the vacuum chamber and operates the main pushrod to the master cylinder.

## 24 Vacuum servo unit – removal and refitting

1   Undo and remove the two nuts and washers securing the brake master cylinder to the servo unit. Carefully withdraw the master cylinder, taking great care not to strain the brake pipes, and tie it to one side, just clear of the servo.

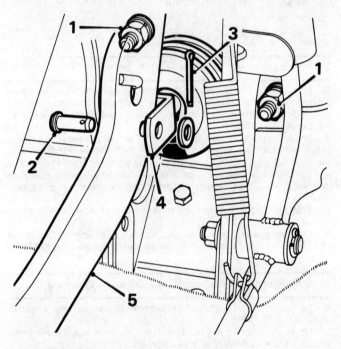

Fig. 8.9 Vacuum servo unit attachments (Sec 24)

1   Servo unit retaining nuts     4   Servo pushrod
2   Clevis pin                    5   Brake pedal
3   Split pin

2   Using a wide-bladed screwdriver, carefully prise the vacuum hose elbow out of the grommet on the front face of the servo.
3   From inside the car, extract the split pin and withdraw the clevis pin securing the servo pushrod to the brake pedal.
4   Undo and remove the two nuts and washers securing the servo to the bulkhead and withdraw the unit from the engine compartment.
5   Refitting the servo unit is the reverse sequence to removal. Tighten the servo and master cylinder retaining nuts to the specified torque and use a new split pin in the servo pushrod clevis pin.

## 25 Vacuum servo unit – servicing

1   At the intervals given in Routine Maintenance the servo unit air filter should be renewed as follows. Note that this is the only work that can be carried out on the servo and no attempt should be made to dismantle the unit or alter the setting of the domed nut in the output rod (where fitted).
2   From inside the car, extract the split pin and withdraw the clevis pin securing the servo pushrod to the brake pedal.
3   Pull back the rubber dust cover, release the end cap and extract the old filter. Cut the new filter, as shown in Fig. 8.10, place it over the pushrod and push the filter into the neck of the servo. Refit the end cap and dust cover.
4   To test the operation of the servo, depress the footbrake and then start the engine. As the engine starts there should be a noticeable 'give' in the brake pedal. Allow the engine to run for at least two minutes and then switch it off. If the brake pedal is now depressed again, it should be possible to detect a hiss from the servo when the pedal is depressed. After about four or five applications no further hissing will be heard and the pedal will feel considerably firmer.

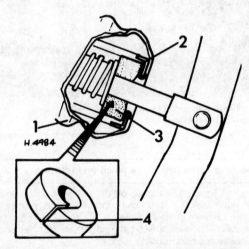

Fig. 8.10 Servo unit air filter renewal (Sec 25)

1   Dust cover      3   Filter
2   End cap         4   Position of cut in filter

## 26 Fault diagnosis – braking system

| Symptom | Reason(s) |
| --- | --- |
| Excessive pedal travel | Rear brake self-adjust mechanism inoperative<br>Air in hydraulic system<br>Faulty master cylinder |
| Brake pedal feels spongy | Air in hydraulic system<br>Faulty master cylinder |
| Judder felt through brake pedal or steering wheel when braking | Excessive run-out or distortion of front discs or rear drums<br>Brake pedals or linings worn<br>Brake backplate or disc caliper loose<br>Wear in suspension or steering components or mountings – see Chapter 10 |
| Excessive pedal pressure required to stop car | Faulty servo unit, disconnected, damaged or insecure vacuum hose<br>Wheel cylinder(s) or caliper piston seized<br>Brake pads or brake shoe linings worn or contaminated<br>Brake shoes incorrectly fitted<br>Incorrect grade of pads or linings fitted<br>Primary or secondary hydraulic circuit failure |
| Brakes pull to one side | Brake pads or linings worn or contaminated<br>Wheel cylinder or caliper piston seized<br>Seized rear brake self-adjust mechanism<br>Brake pads or linings renewed on one side only<br>Faulty twin GP valve<br>Tyre, steering or suspension defect – see Chapter 10 |
| Brakes binding | Wheel cylinder or caliper piston seized<br>Handbrake incorrectly adjusted<br>Faulty master cylinder |
| Rear wheels locking under normal braking | Rear brake shoe linings contaminated<br>Faulty twin GP valve |

# Chapter 9 Electrical system

*For modifications, and information applicable to later models, see Supplement at end of manual*

## Contents

## Specifications

### System type ........................................................ 12 volt, negative earth

### Battery
Type ............................................................ Unipart 'Sealed for life'
Performance – cold start current (Amps)/reserve capacity
   (minutes) .................................................. 360/70

### Alternator
Type ............................................................ Lucas 127/55
Maximum output .......................................... 55 Amps
Brush length:
   New ..................................................... 0.8 in (20.0 mm)
   Minimum ............................................... 0.4 in (10.0 mm)
Brush spring tension (brush face flush with brushbox) .... 4.7 to 9.8 oz f (1.3 to 2.7N)

## Starter motor

| | |
|---|---|
| Type | Lucas 9M90 pre-engaged |
| Minimum brush length | 0.4 in (10.0 mm) |
| Brush spring tension | 32.0 to 36.0 oz f (9.0 to 10.0N) |
| Commutator minimum skimming thickness | 0.08 in (2.0 mm) |

## Relays and control units

| Component | Location |
|---|---|
| Direction indicator/hazard flasher unit | In fusebox on right-hand side of facia |
| Starter solenoid relay | In fusebox on right-hand side of facia |
| Ignition auxiliary relay | In fusebox on right-hand side of facia |
| Heated rear window relay | In fusebox on right-hand side of facia |
| Headlamp relay | In fusebox on right-hand side of facia |
| Inlet manifold heater (if fitted) | In fusebox on right-hand side of facia |
| Fuel system electronic control unit – ECU (models with carburettor induction) | Attached to panel on glovebox roof |
| Fuel system electronic control unit – ECU (models with electronic fuel injection) | Under carpet in passenger's foot well |
| Wipe/washer program control unit | Attached to panel on glovebox roof |
| Courtesy light delay control unit (if fitted) | Attached to panel on glovebox roof |
| Programmed ignition system control unit | Attached to engine compartment left-hand valance |
| Electric window lift relay and control unit (if fitted) | Behind the driver's door inner trim panel |
| Central locking control unit (if fitted) | Behind the driver's door inner trim panel |
| Electronic fuel injection main and fuel pump relays (if fitted) | Attached to bracket on engine compartment left-hand valance |

## Fuses

| Fuse colour coding | Current rating |
|---|---|
| Violet | 3 Amp |
| Tan | 5 Amp |
| Red | 10 Amp |
| Blue | 15 Amp |
| Natural | 25 Amp |
| Green | 30 Amp |

## Fusible links

| | |
|---|---|
| Link A | 150 mm long, 28/0.3 wire |
| Link B | 200 mm long, 28/0.3 wire |
| Link C | 450 mm long, 14/0.3 wire |

| | |
|---|---|
| **Wiper blades** | Champion X-5103 (front) and X-3603 (rear) |
| **Wiper arms** | Champion CCA6 |

## Bulbs

| | Wattage |
|---|---|
| Headlamps | 60/55 |
| Sidelamps | 4 |
| Direction indicators | 21 |
| Reverse lamps | 21 |
| Stop/tail lamps | 21/5 |
| Tail lamps | 5 |
| Rear fog guard lamps | 21 |
| Number plate lamps | 5 |
| Interior and luggage compartment lamps | 10 |
| Glovebox lamp | 5 |
| Switch illumination | 0.36 |
| Instrument panel illumination and warning lamps | 1.2 |
| Ignition warming lamp | 2 |
| Heater control illumination | 0.36 and 1.2 |

## Torque wrench settings

| | lbf ft | Nm |
|---|---|---|
| Alternator adjustment arm to alternator | 18 | 25 |
| Alternator adjustment arm to engine | 38 | 51 |
| Alternator mounting nuts | 13 | 18 |
| Alternator pulley nut | 28 | 38 |
| Battery tray to body | 18 | 25 |
| Starter motor retaining bolts | 66 | 89 |
| Wiper arm retaining nuts | 7 | 9 |
| Wiper motor spindle nuts: | | |
|    M18 nut | 3 | 4 |
|    Plastic nut | 2 | 3 |

## 1  General description

The electrical system is of the 12 volt negative earth type, and consists of a 12 volt battery, alternator, starter motor and related electrical accessories, components and wiring. The battery is of the maintenance-free, 'sealed for life' type and is charged by an alternator which is belt-driven from the crankshaft pulley. The starter motor is of the pre-engaged type incorporating an integral solenoid. On starting, the solenoid moves the drive pinion into engagement with the flywheel ring gear before the starter motor is energised. Once the engine has started, a one-way clutch prevents the motor armature being driven by the engine until the pinion disengages from the flywheel.

Further details of the major electrical systems are given in the relevant Sections of this Chapter.

**Caution**: *Before carrying out any work on the vehicle electrical* *system, read through the precautions given in Safety First! at the beginning of this manual and in Section 2 of this Chapter.*

## 2 Electrical system – precautions

It is necessary to take extra care when working on the electrical system to avoid damage to semi-conductor devices (diodes and transistors), and to avoid the risk of personal injury. In addition to the precautions given in Safety First! at the beginning of this manual, observe the following items when working on the system.
1   *Always remove rings, watches, etc before working on the electrical system.* Even with the battery disconnected, capacitive discharge could occur if a component live terminal is earthed through a metal object. This could cause a shock or nasty burn.
2   *Do not reverse the battery connections.* Components such as the alternator or any other having semi-conductor circuitry could be irreparably damaged.
3   If the engine is being started using jump leads and a slave battery, connect the batteries *positive to positive* and *negative to negative*. This also applies when connecting a battery charger.
4   Never disconnect the battery terminals, or alternator multi-plug connector, when the engine is running.
5   The battery leads and alternator multi-plug must be disconnected before carrying out any electric welding on the car.
6   Never use an ohmmeter of the type incorporating a hand cranked generator for circuit or continuity testing.
7   Models equipped with carburettor induction incorporate an inlet manifold heater which will operate whenever the ignition is switched on and the coolant temperature is below 90°F (30°C). Owing to the high current consumption of this unit, always disconnect the wiring connector if the ignition is to be left on for any length of time with the engine stopped. Further details will be found in Chapter 3.

## 3 Maintenance and inspection

1   At regular intervals (see Routine Maintenance) carry out the following maintenance and inspection operations on the electrical system components.
2   Check the operation of all the electrical equipment, ie wipers, washers, lights, direction indicators, horn etc. Refer to the appropriate Sections of this Chapter if any components are found to be inoperative.
3   Visually check all accessible wiring connectors, harnesses and retaining clips for security, or any signs of chafing or damage. Rectify any problems encountered.
4   Check the alternator drivebelt for cracks, fraying or damage. Renew the belt if worn or, if satisfactory, check and adjust the belt tension. These procedures are covered in Chapter 2.
5   Check the condition of the wiper blades and if they are cracked or show signs of deterioration, renew them, as described in Section 39. Check the operation of the windscreen and headlamp washers (if fitted). Adjust the nozzles using a pin, if necessary.
6   Check the battery terminals, and if there is any sign of corrosion disconnect and clean them thoroughly. Smear the terminals and battery posts with petroleum jelly before refitting the plastic covers. If there is any corrosion on the battery tray, remove the battery, clean the deposits away and treat the affected metal with an anti-rust preparation. Repaint the tray in the original colour after treatment.
7   Top up the windscreen washer reservoir and check the security of the pump wires and water pipes.
8   It is advisable to have the headlight aim adjusted using optical beam setting equipment.
9   While carrying out a road test check the operation of all the instruments and warning lights, and the operation of the direction indicator self-cancelling mechanism.
10  At less frequent intervals (see Routine Maintenance) renew the alternator drivebelt, as described in Chapter 2.

## 4 Battery – removal and refitting

1   The sealed for life battery is located on the left-hand side of the engine compartment.

4.2 Battery negative terminal disconnected

2   To remove the battery, slacken the negative (–) terminal clamp bolt and lift the terminal off the battery post (photo).
3   Lift the plastic cover from the positive (+) terminal, loosen the clamp bolt and lift the terminal off the battery post.
4   Undo and remove the retaining bolt and lift off the battery clamp plate.
5   Withdraw the battery from the carrier tray.
6   Refitting is the reverse sequence of removal, but make sure that the polarity is correct before connecting the leads, and do not overtighten the clamp bolts.

## 5 Battery – charging

1   In winter when a heavy demand is placed on the battery, such as when starting from cold and using more electrical equipment, it may be necessary to have the battery fully charged from an external source.
2   Charging is best done overnight at a 'trickle' rate of 1 to 1.5 amps. Owing to the design of certain maintenance-free batteries, rapid or boost charging is not recommended. If in any doubt about the suitability of certain types of charging equipment for use with maintenance-free batteries, consult a BL dealer or automotive electrical specialist.
3   Battery charging must be carried out with the battery removed from the vehicle and in a well ventilated area.
4   *The terminals of the battery and the leads of the charger must be connected positive to positive and negative to negative.*

## 6 Alternator – removal and refitting

1   Disconnect the battery negative terminal.
2   Jack up the front of the car and support it on axle stands. Remove the right-hand front roadwheel and the access panel from under the wheel arch.
3   From under the wheel arch, slacken the alternator adjustment arm bolt (photo) and the two mounting nuts. Push the alternator in towards the engine and slip the drivebelt off the alternator pulley.
4   From under the car undo the two bolts securing the heat shield to the engine and remove the shield (photo).
5   Release the retaining clip and disconnect the wiring multi-plug from the rear of the alternator (photo).
6   Undo and remove the alternator adjustment arm bolt and the two nuts and washers on the mounting stud. Withdraw the mounting stud and remove the alternator from under the car.
7   Refitting is the reverse sequence of removal, but before tightening the adjustment arm bolt and the mounting nuts, adjust the drivebelt tension as described in Chapter 2.

6.3 Alternator adjustment arm bolt (arrowed)

6.4 Heat shield retaining bolts (arrowed)

6.5 Alternator wiring multi-plug (A) and retaining clip (B)

## 7 Alternator – fault tracing and retification

Due to the specialist knowledge and equipment required to test or repair an alternator, it is recommended that, if the performance is suspect, the car be taken to an automobile electrician who will have the facilities for such work. Because of this recommendation, information is limited to the inspection and renewal of the brushes. Should the alternator not charge, or the system be suspect, the following points should be checked before seeking further assistance:

(a) *Check the drivebelt condition and tension*
(b) *Ensure that the battery is fully charged*
(c) *Check the ignition warning light bulb, and renew it if blown*

## 8 Alternator brushes – removal, inspection and refitting

1   Remove the alternator, as described in Section 6.
2   Disconnect the electrical lead and remove the suppression capacitor from the rear of the alternator.
3   Undo the retaining screws, lift off the regulator and brushbox assembly and disconnect the electrical lead.
4   Check the brush length and the brush spring tension against the figures given in the Specifications. New brushes are supplied as an assembly complete with brushbox and regulator.
5   Refitting is the reverse sequence to removal.

## 9 Starter motor – testing in the car

1   If the starter motor fails to operate, first check the condition of the battery by switching on the headlamps. If they glow brightly then gradually dim after a few seconds, the battery is in an uncharged condition.
2   If the battery is satisfactory, check the starter motor main terminal and the engine earth cable for security. Check the terminal connections on the starter solenoid – located on top of the starter motor.
3   If the starter still fails to turn, use a voltmeter, or 12 volt test lamp and leads, to ensure that there is battery voltage at the solenoid main terminal (containing the cable from the battery positive terminal).
4   With the ignition switched on and the ignition key in position III, check that voltage is reaching the solenoid terminal with the Lucar connector, and also the starter main terminal.
5   If there is no voltage reaching the Lucar connector there is a wiring, relay, or ignition switch fault. If voltage is available, but the starter does not operate, then the starter or solenoid is likely to be at fault.

## 10 Starter motor – removal and refitting

1   Disconnect the battery negative terminal then refer to Chapter 3 and remove the air cleaner assembly.
2   Disconnect the electrical leads at the solenoid terminals noting their respective locations (photo). **Note:** *Take care when disconnecting the leads from the Lucar type terminal connections as they are easily broken. Pull back the cover along the lead to expose the connector, then use a small screwdriver to depress the lock tag in the centre of the connector and release it.*
3   Undo the starter motor retaining bolts, washers and nut, then withdraw the unit from the gearbox.
4   Refitting is the reverse sequence of removal, but tighten the retaining bolts to the specified torque.

## 11 Starter motor – overhaul

1   Remove the starter motor from the car, as described in the previous Section.
2   At the rear of the solenoid, unscrew the nut and lift away the lead from the solenoid 'STA' terminal.
3   Undo and remove the two screws securing the solenoid to the drive end housing. Disengage the solenoid plunger from the engaging lever and withdraw the solenoid.
4   Withdraw the end cap from the commutator end housing and then prise off the 'spire' retaining washer from the armature shaft.

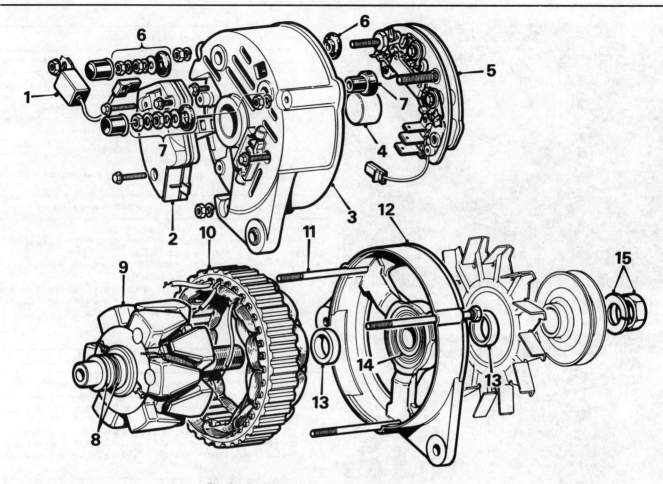

Fig. 9.1 Exploded view of the alternator (Sec 8)

1  Suppression capacitor
2  Regulator and brushbox
   assembly
3  Slip ring end bracket
4  Bearing

5  Rectifier
6  Phase terminal and washer
   assembly
7  Main terminal and washer
   assembly

8  Slip rings
9  Rotor assembly
10 Stator
11 Retaining bolts
12 Drive end bracket

13 Spacers
14 Bearing
15 Pulley retaining nut and
   washer

10.2 Disconnect the leads at the solenoid terminals (arrowed)

5   Undo and remove the two through-bolts and withdraw the end housing sufficiently to gain access to the brushes.
6   Release the field brushes from their brushbox locations and then remove the end housing.
7   Mark the relationship of the field coil assembly to the drive end housing, unscrew the two housing retaining screws and withdraw the drive end housing.
8   Slide the armature and drive assembly out of the field coil assembly.
9   Mount the armature in a vice and, using a suitable tubular drift, tap the thrust collar on the end of the shaft toward the pinion, to expose the jump ring. Prise the jump ring out of its groove and slide it off the shaft. Withdraw the thrust collar and the drive assembly.
10  With the starter motor now completely dismantled, clean all the components in paraffin or a suitable solvent, and wipe dry.
11  Check the length of the brushes and the tension of the brush springs. If the length and tension are not as given in the Specifications renew the brushes and springs. Note that the field brushes must be soldered in place.
12  Check that the brushes move freely in their holders, and clean the holders and brushes with a petrol-moistened rag if they show any tendency to stick. If necessary use a fine file on the brushes if they still stick after cleaning.
13  Check the armature shaft for distortion, and examine the commutator for excessive scoring, wear or burns. If necessary, the commutator may be skimmed in a lathe and then polished with fine glass paper. Make sure that it is not reduced below the specified minimum thickness and do not undercut the mica insulation.

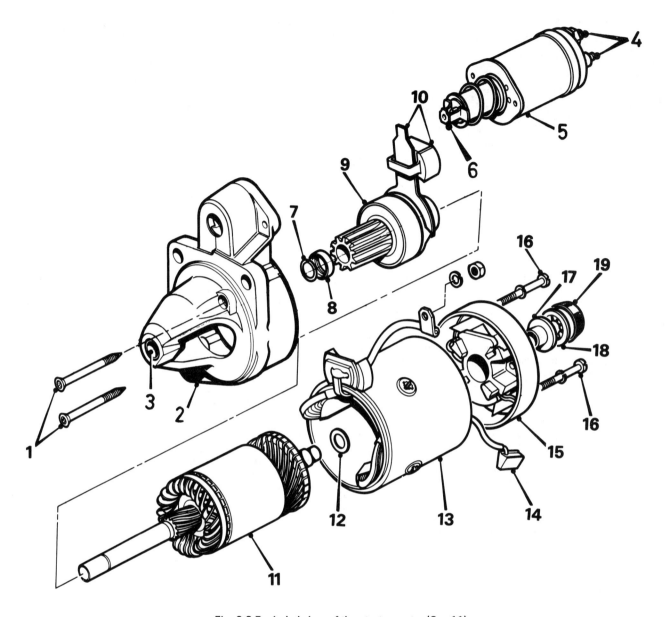

**Fig. 9.2 Exploded view of the starter motor (Sec 11)**

| | |
|---|---|
| 1 Housing retaining screws | 6 Solenoid plunger |
| 2 Drive end housing | 7 Jump ring |
| 3 Armature support bush | 8 Thrust collar |
| 4 Solenoid terminals | 9 Drive assembly |
| 5 Solenoid | 10 Engaging lever and bush |

| | |
|---|---|
| 11 Armature | 16 Through-bolt |
| 12 Thrust washer | 17 Bush |
| 13 Field coil assembly | 18 'Spire' washer |
| 14 Field brush | 19 End cap |
| 15 Commutator end housing | |

14 Check the taping of the field coils, check all joints for security and check the coils and commutator for signs of burning.

15 Check the drive pinion assembly, drive end housing, engaging lever and solenoid for wear or damage. Make sure that the drive pinion one-way clutch permits movement of the pinion in one direction only and renew the complete assembly, if necessary.

16 Check the condition of the brush in the commutator end housing and, if necessary, renew it.

17 Accurate checking of the armature, commutator and field coil windings and insulation requires the use of special test equipment. If the starter motor was inoperative when removed from the car and the previous checks have not highlighted the problem, then it can be assumed that there is a continuity or insulation fault and the unit should be renewed.

18 If the starter is in a satisfactory condition, or if a fault has been traced and rectified, the unit can be reassembled using the reverse of the dismantling procedure.

## 12 Fuses, relays and control units – general

### Fuses

1 The fusebox is situated below the facia on the right-hand side, adjacent to the steering column. To gain access to the fusebox release the two turnbuckles using a coin and lift off the fusebox cover (photo). The fuse locations, current rating and circuits protected are shown on the fusebox cover (photo). Each fuse is colour coded and has its rating stamped on it.

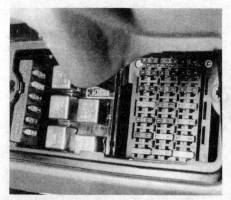

12.1A Fuse and relay locations in the fusebox ...

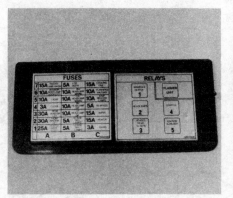

12.1B ... and circuit details on the fusebox cover

12.7 Control unit locations on glovebox roof panel

2    To remove a fuse from its location, hook it out using the small plastic removal tool provided. This is located, together with the spare fuses, in a holder on the left-hand side of the fusebox. Hook the tool over the fuse and pull up to remove. Refit the fuse by pressing it frmly into its location.

3    Always renew the fuse with one of an identical rating. Never renew a fuse more than once without finding the source of the trouble.

4    Circuits protected by fuses C1, C2 and C3 operate when the ignition switch is at positions 1 or 11. Circuits A1, A2, C5 and C6 only operate when the ignition switch is at position 11. All other circuits operate irrespective of ignition switch position.

*Relays and control units*

5    The relays are located in the fusebox under the right-hand side of the facia.

6    The relays can be removed by simply pulling them from their respective locations. If a system controlled by a relay becomes inoperative, and the relay is suspect, operate the system and if the relay is functioning it should be possible to hear it click as it is energised. If this is the case, the fault lies with the components of the system. If the relay is not being energised then the relay is not receiving a main supply voltage, a switching voltage or the relay itself is faulty.

7    Control units for the wiper delay, courtesy light delay and electronic mixture control are located on a panel attached to the roof of the glovebox. Release the two turnbuckles using a coin and lower the panel to gain access (photo). On MG EFi models the control unit for the electronic fuel injection system is located under the carpet above the front passenger's foot well.

8    The programmed ignition system control unit will be found on the left-hand valance in the engine compartment. Where applicable the electric window lift relay and control unit together with the central locating control unit are located inside the driver's door behind the trim panel.

9    On models equipped with a driver information system, the location and operation of the various sensors and relays associated with this system are described in later Sections of this Chapter.

## 13  Fusible links – general

1    Certain electrical circuits on Montego models are protected by three fusible links which form the first part of the battery positive (+) cable. The links consist of fusible cable of different thicknesses and lengths according to the circuits they protect. The wire size and fusible link lengths are given in the Specifications.

2    If power to a number of related electrical circuits is lost, the relevant fusible link may be suspect. The circuits protected by the fuslble links are as follows.

| Link A | *Ignition circuit* |
| Link B | *Window lift circuit* |
| Link C | *Interior lights, lighter, clock circuit* |

3    To renew a fusible link, disconnect the battery leads then remove the leads from the battery '+' terminal.

4    Carefully cut back the binding and pull the sleeve from the cables. Identify the link to be renewed by referring to the Specifications.

5    Remove the joint cover and unsolder the fusible link at both ends.

6    Solder a new fusible link into place and seal the joints with insulating tap. Re-tape the sleeve to the cables and reconnect the battery terminals.

## 14  Direction indicator and hazard flasher system – general

1    The flasher unit is located in the fusebox situated under the right-hand side of the facia.

2    Should the flashers become faulty in operation, check the bulbs for security and make sure that the contact surfaces are not corroded. If one bulb blows or is making a poor connection due to corrosion, the system will not flash on that side of the car.

3    If the flasher unit operates in one direction and not the other, the fault is likely to be in the bulbs, or wiring to the bulbs. If the system will not flash in either direction, operate the hazard flashers. If these function, check for a blown fuse in position C6. If the fuse is satisfactory renew the flasher unit.

4    On models equipped with a driver information system, the location of a failed bulb will be shown on the vehicle map. This information is provided by a bulb fail monitor located adjacent to the lamp clusters at each corner of the car. The failure of all the bulbs in a particular lamp cluster may indicate a possible fault in the monitor itself and this should be checked before renewing the bulbs. Further information will be found in Section 27.

## 15  Ignition switch/steering column lock – removal and refitting

The ignition switch is an integral part of the steering column lock, and removal and refitting procedures are given in Chapter 10.

## 16  Steering column switches – removal and refitting

1    Disconnect the battery negative terminal.

2    Remove the steering wheel as described in Chapter 10.

3    Undo the retaining screws and lift off the left-hand and right-hand steering column cowls.

4    Carefully lift up the retaining clip and slip the fibre optic guide out of the bulb housing (photo).

5    Depress the retainers at the top and bottom of the switch then pull the switch out of the steering column boss (photo).

6    Disconnect the wiring multi-plug(s) as applicable and remove the switch (photo).

7    Refitting is the reverse sequence of removal. Position the switch striker bush with the arrow pointing towards the direction indicator switch before refitting the steering wheel.

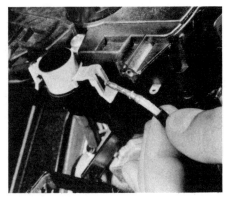

16.4 Removing the fibre optic guide from the bulb housing

16.5 Depress the retainers to release the switch ...

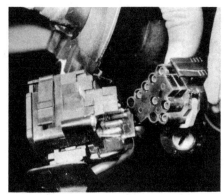

16.6 ... then disconnect the wiring multi-plugs

## 17 Facia switches – removal and refitting

1   Disconnect the battery negative terminal.
2   Undo the retaining screws and withdraw the right-hand steering column cowl from the column (photo). Position the steering wheel so that the spokes are facing away from the cowl to provide sufficient clearance for removal.
3   Undo the two retaining screws and withdraw the switch panel from the facia (photos).
4   Disconnect the wiring multi-plug and push the switch out of the panel.
5   Refitting is the reverse sequence to removal.

## 18 Instrumentation – description

Two completely different instrument packs are available, according to model, to present information to the driver. The first pack available on all 2.0 litre Montego models has conventional electro-mechanical instruments and consists of a speedometer, tachometer, fuel gauge, temperature gauge and driver warning lamps. The instruments are arranged in a panel in front of the driver and all electrical interconnections are by a printed circuit at the rear of the unit. A multi-function unit attached to the rear of the instrument panel provides a stabilized 10 volt supply for the instruments and warning lamps and also controls the operation of the tachometer.

MG versions of the Montego range are available with sophisticated solid-state instrumentation arranged in two modules to form a comprehensive driver information centre. Module 1 is the instrument panel mounted centrally in the facia consisting of a logic board and display board. The display board contains three liquid crystal plates which display road speed, engine speed, coolant temperature and fuel level together with a service interval counter, warning lights, vehicle map and attention display. Module 2 is located to the left of module 1 and consists of a voice synthesis unit, message centre display, computer and keyboard. Information on the vehicle operating conditions are obtained from sensors located around the car whose inputs are processed by the vehicle condition monitor. This unit utilizes the message centre, vehicle map, voice synthesis unit, attention display and certain warning lights to relay the information from the various sensors to the driver.

When in operation the vehicle condition monitor together with the driver information centre form the complete driver information system. **Note:** *The instrumentation fitted to Montego models is extremely delicate and must be handled with great care.* The solid-state instrumentation consists of fragile electronic circuit boards which can be easily damaged. Removal and refitting procedures are given for the solid-state instrumentation, but it is recommended that the two modules are not dismantled in any way. Fault diagnosis on these instruments entails the use of a sophisticated electronic tester, and, if a fault develops, repair should be left to a suitably equipped BL dealer.

## 19 Instrument panel and instruments (electro-mechanical instruments) – removal and refitting

1   Disconnect the battery negative terminal.
2   Undo the two retaining screws and ease the instrument panel

17.2 Remove the steering column cowl ...

17.3A ... undo the two screws (arrowed) ...

17.3B ... then withdraw the switch panel

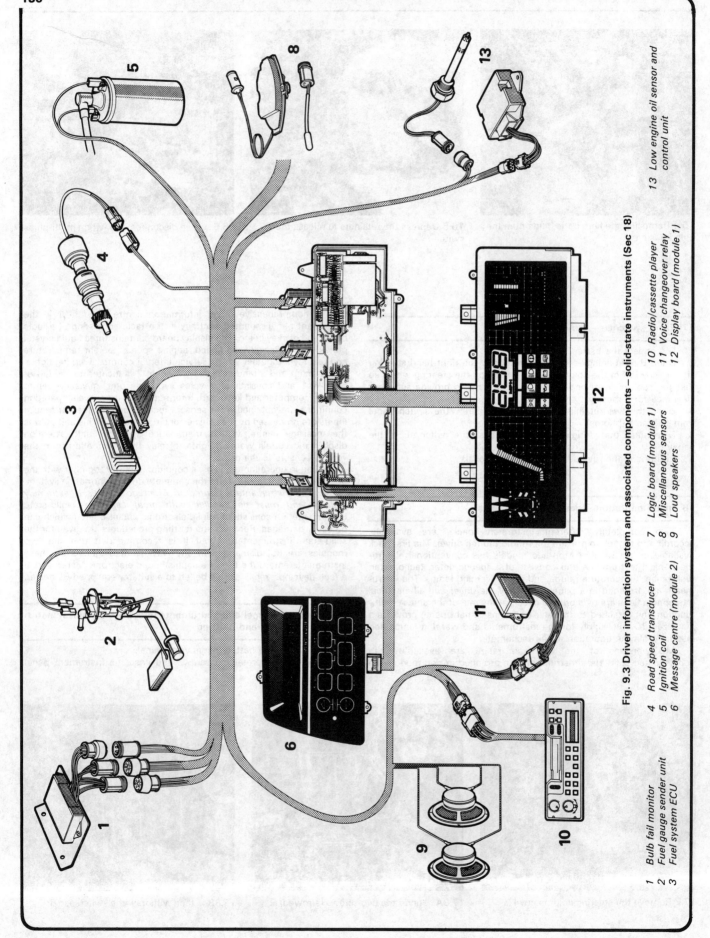

Fig. 9.3 Driver information system and associated components – solid-state instruments (Sec 18)

| | |
|---|---|
| 1 Bulb fail monitor | 7 Logic board (module 1) |
| 2 Fuel gauge sender unit | 8 Miscellaneous sensors |
| 3 Fuel system ECU | 9 Loud speakers |
| 4 Road speed transducer | 10 Radio/cassette player |
| 5 Ignition coil | 11 Voice changeover relay |
| 6 Message centre (module 2) | 12 Display board (module 1) |
| | 13 Low engine oil sensor and control unit |

19.2A Undo the retaining screws ...

19.2B ... then withdraw the instrument panel surround and disconnect the clock wiring

19.4 Disconnect the instrument panel wiring multi-plugs

surround from its location (photo). Disconnect the clock wiring harness plug and remove the surround (photo).
3   Undo the five screws securing the instrument panel mounting brackets to the facia. Ease the panel down slightly and tip it towards the steering wheel at the top.
4   Disconnect the wiring harness multi-plugs (photo) then reach round behind the panel and disconnect the speedometer cable by pulling the retainer straight from the speedometer. It may be necessary to push the cable through the bulkhead grommet from the engine compartment side to facilitate removal.
5   Withdraw the instrument panel from the facia.
6   Refitting is the reverse sequence to removal.

---

**20  Instrument panel and instruments (electro-mechanical instruments) – dismantling and reassembly**

---

1   Remove the instrument panel from the car as described in the previous Section.

### Panel illumination and warning lamp bulbs
2   The bulbholders are secured to the rear of the instrument panel by a bayonet fitting and are removed by turning the holders anticlockwise (photo). Note that only the no-charge warning lamp bulb can be separated from its bulbholder (coloured red). All the other illumination and warning lamp bulbs are renewed complete with their bulbholders.
3   To gain access to the panel front illumination lamp bulb, remove the upper instrument panel mounting bracket (photo).
4   Refit the bulbholders by turning clockwise to lock.

### Multi-function unit
5   Undo the three nuts and lift off the multi-function unit cover (photo).
6   Undo the four nuts securing the fuel and temperature gauges and the three nuts securing the tachometer.
7   Undo the multi-function unit retaining screws and withdraw the unit from the instrument panel.
8   Refit the multi-function unit using the reverse sequence of removal.

### Printed circuit
9   Remove all the instrument panel warning and illumination bulbholders as previously described.
10  Undo the three nuts and remove the multi-function unit cover.
11  Carefully extract the pegs securing the printed circuit to the instrument panel. Remove the tape, ease the printed circuit off the pegs and studs and lift off.
12  Refitting the printed circuit is the reverse sequence of removal.

### Speedometer
13  Release the retaining wires and lift off the instrument panel shroud and window glass assembly on the front illuminated panel. Note that the two parts will be connected by the printed circuit.
14  Ease the face plate from the case noting that it is secured between the fuel and temperature gauges with double sided tape.
15  Undo the retaining screws and withdraw the speedometer from the instrument panel (photo).
16  Refitting is the reverse sequence of removal, but use new double sided tape to secure the face plate.

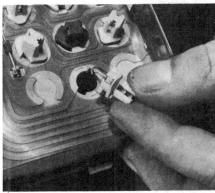
20.2 Remove the instrument panel bulbholders by turning them anti-clockwise

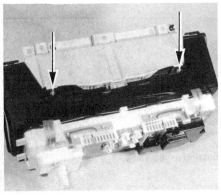
20.3 Panel mounting bracket retaining nuts (arrowed)

20.5 Multi-function unit cover retaining nuts (arrowed)

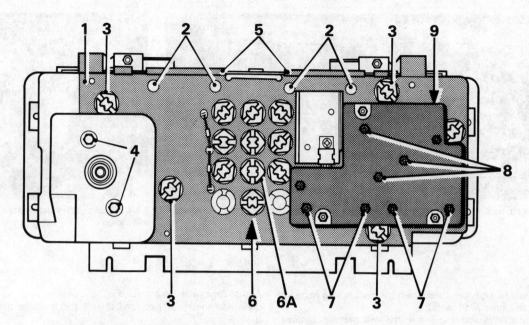

**Fig. 9.4 Instrument panel connections and component locations – electro-mechanical instruments (Sec 20)**

| | | | |
|---|---|---|---|
| 1 | Printed circuit | 3 | Panel illumination bulbs |
| 2 | Printed circuit retaining pegs | 4 | Speedometer retaining screws |
| | | 5 | Wiring multi-plug connections |

| | | | |
|---|---|---|---|
| 6 | Warning lamp bulbholders | 7 | Fuel and temperature gauge retaining nuts |
| 6A | Ignition no-charge warning lamp bulbholder | 8 | Tachometer retaining nuts |
| | | 9 | Multi-function unit |

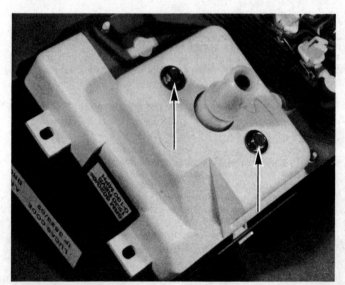

20.15 Speedometer retaining screws (arrowed)

### Fuel gauge, temperature gauge and tachometer

17 Carry out the operations described in paragraphs 13 and 14.

18 Undo the retaining nuts and remove the cover from the multi-function unit.

19 Undo the nuts securing the fuel and temperature gauges and the tachometer, to the printed circuit and instrument panel.

20 With the panel held face down, carefully remove the instruments. Recover the wavy washers from the gauge studs.

21 Refitting is the reverse sequence of removal bearing in mind the following points:

    (a) Ensure that an equal number of wavy washers are positioned on each of the fuel and temperature gauge studs

    (b) Position all the instruments in the panel before tightening the retaining nuts then tighten the tachometer retaining nuts first

### 21 Clock (electro-mechanical instruments) – removal and refitting

1 Disconnect the battery negative terminal.

2 Undo the two retaining screws and ease the instrument panel surround from its location.

3 Disconnect the clock wiring harness plug and remove the surround.

4 Undo the two screws securing the clock to the surround and carefully lift the clock from its location (photo).

5 The clock time set push rods should not be disturbed. If they become dislodged, the longest legs must be positioned horizontally.

6 Refitting the clock is the reverse sequence of removal. Reset the hour and minute display by depressing first the hour then the minute control buttons using a pencil or ball point pen. To zero the display press both control buttons simultaneously.

21.4 Clock retaining screws (arrowed)

## 22 Message centre (solid-state instruments – module 2) – removal and refitting

1   Disconnect the battery negative terminal.
2   Undo the two retaining screws and ease the instrument panel surround complete with message centre from the facia.
3   Carefully prise apart the lugs at each end of the message centre multi-plug ribbon connector and release the connector. Lift away the panel surround and message centre.
4   Undo the three retaining screws and remove the message centre from the surround.
5   Refitting is the reverse sequence to removal.

## 23 Message centre bulbs (solid-state instruments – module 2) – renewal

1   Remove the message centre as described in the previous Section.
2   Undo and remove the message centre rear cover right-hand retaining screw then slacken the left-hand screw. Move the cover aside taking care not to strain the wires.
3   Turn the bulbholder anti-clockwise and withdraw the holder complete with bulb. Do not touch the bulb glass with your fingers; if touched, clean the bulb with methylated spirit.
4   Refit the bulb and holder and turn clockwise to lock.
5   Refit the rear cover and message centre using the reverse sequence to removal.

## 24 Instrument panel (solid-state instruments – module 1) – removal and refitting

1   Remove the instrument panel surround and message centre as described in Section 22.

2   Undo and remove the five screws securing the instrument panel to the facia.
3   Withdraw the instrument panel and disconnect the three multi-plugs and earth terminal from the rear of the panel. Remove the instrument panel from the car.
4   Refitting is the reverse sequence to removal.

## 25 Instrument panel bulbs (solid-state instruments – module 1) – renewal

1   Remove the instrument panel as described in the previous Section.
2   Turn the bulbholder at the rear of the panel anti-clockwise and remove the bulbholder. Withdraw the bulb from the holder using a clean cloth. Do not touch the bulb glass with your fingers; if touched, clean the bulb with methylated spirit.
3   Refit the bulb to the holder then insert the holder into the panel and turn clockwise to lock.
4   Refit the instrument panel using the reverse sequence to removal.

## 26 Vehicle condition monitor sensors (solid-state instruments) – general

As described earlier, sensors located at various positions around the car monitor 34 different vehicle operating conditions. Inputs from the sensors are relayed to the vehicle condition monitor which if necessary acts on the information received in the form of an audible and visual warning to the driver. The sensors and their locations are shown in Fig. 9.6.
Some of the sensors are an integral part of many of the car's mechanical components (ie brake pad wear sensor) and others are separate individual units (ie bulb fail monitor). Of those sensors that can be removed and refitted separately, details are given in the

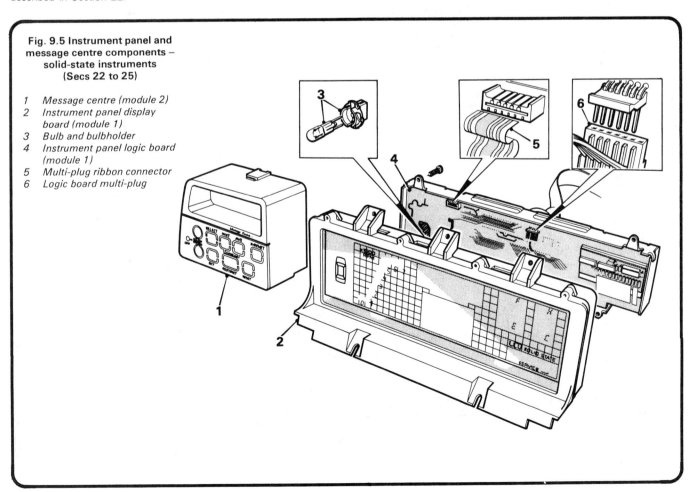

**Fig. 9.5 Instrument panel and message centre components – solid-state instruments (Secs 22 to 25)**

1   Message centre (module 2)
2   Instrument panel display board (module 1)
3   Bulb and bulbholder
4   Instrument panel logic board (module 1)
5   Multi-plug ribbon connector
6   Logic board multi-plug

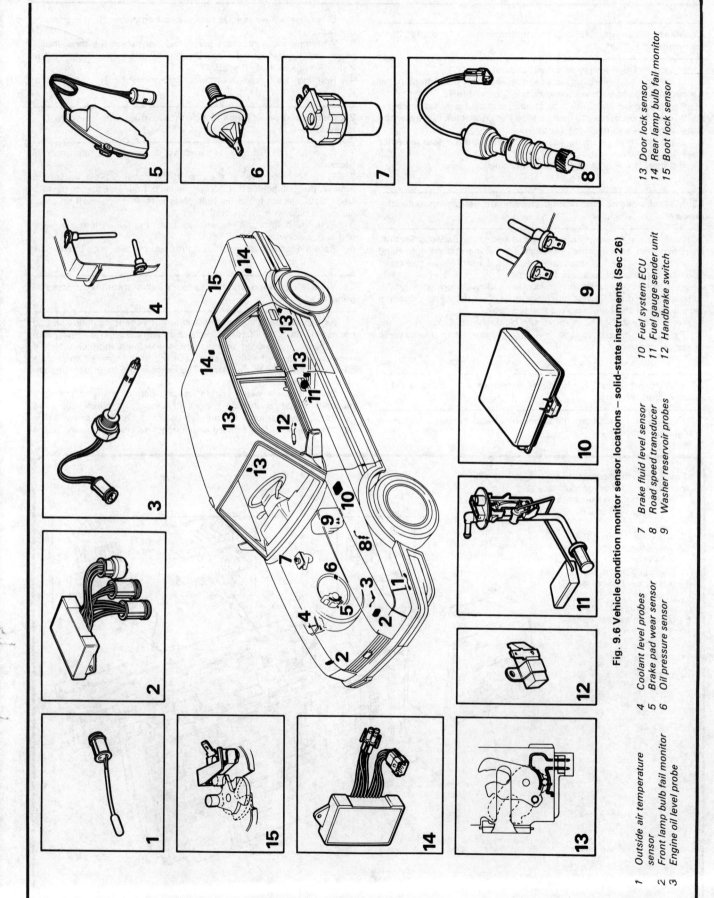

**Fig. 9.6 Vehicle condition monitor sensor locations — solid-state instruments (Sec 26)**

1  Outside air temperature sensor
2  Front lamp bulb fail monitor
3  Engine oil level probe
4  Coolant level probes
5  Brake pad wear sensor
6  Oil pressure sensor
7  Brake fluid level sensor
8  Road speed transducer
9  Washer reservoir probes
10  Fuel system ECU
11  Fuel gauge sender unit
12  Handbrake switch
13  Door lock sensor
14  Rear lamp bulb fail monitor
15  Boot lock sensor

following Sections of this Chapter and in other applicable Chapters of this Manual. In all cases each sensor is a sealed unit and cannot be dismantled.

### 27 Bulb fail monitor (solid-state instruments) – checking, removal and refitting

1    Four bulb fail monitors are used, one at each corner of the car to monitor the bulbs in its adjacent lamp cluster.

2    In the event of a fault developing in one of the monitors, it is likely that all the bulbs in the relevant lamp cluster will fail to operate. To enable the car to be driven until such time as a new monitor can be fitted, the unit may be temporarily bypassed as follows.

3    Observe the group of wiring multi-plugs adjacent to the lamp cluster and, with reference to Fig. 9.7, identify the two 5-pin multi-plugs.

4    Disconnect the two multi-plugs and connect the male member of one of the plugs to the female member of the other. Now also connect the two remaining members together.

5    If the bulbs in the lamp cluster now operate correctly the monitor is proved faulty and must be renewed. Note that the relevant segments on the vehicle map will not function whilst the bulb fail monitor is bypassed.

6    To remove the monitor, disconnect the wiring multi-plugs and

27.6 Right-hand front bulb fail monitor location and retaining screws (arrowed)

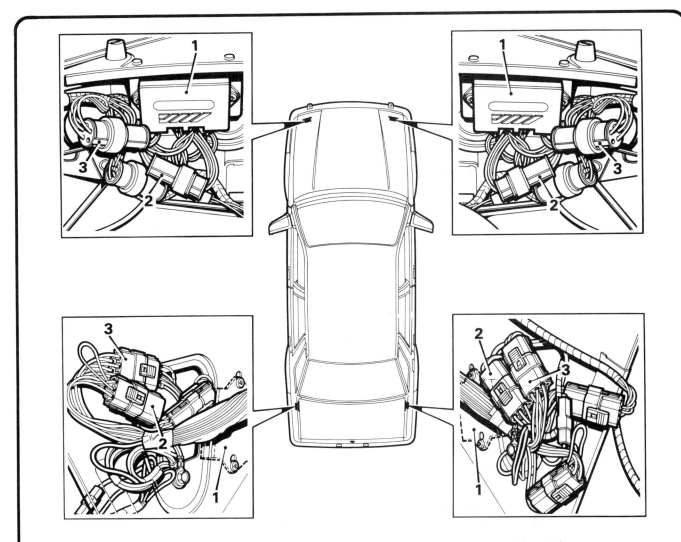

Fig. 9.7 Bulb fail monitor locations and connections – solid-state instruments (Sec 27)

| | | |
|---|---|---|
| *1    Bulb fail monitor* | *2    5-pin multi-plug male member* | *3    5-pin multi-plug female member* |

unscrew the two retaining screws (photo). Remove the unit from its location.

7    Refitting is the reverse sequence of removal, but remember to reconnect the multi-plugs in their original positions if the unit was bypassed prior to renewal.

### 28  Road speed transducer (solid-state instruments) – removal and refitting

1    Disconnect the wiring harness multi-plug then undo the bolt securing the transducer retaining plate to the gearbox.
2    Lift the speed transducer and pinion assembly out of its location at the rear of the gearbox.
3    Refitting is the reverse sequence of removal, but renew the O-ring seal if necessary.

### 29  Headlamp and sidelamp bulbs – renewal

1    From within the engine compartment pull off the wiring connector at the rear of the headlamp bulb (photo).
2    Slip off the rubber cover to gain access to the bulbs (photo).

29.1 Disconnect the headlamp bulb wiring connector ...

29.2 ... then slip off the rubber cover

### Headlamp bulb

3    Release the bulb retaining clip and withdraw the bulb from its location in the headlamp (photos). Take care not to touch the bulb glass with your fingers; if touched, clean the bulb with methylated spirit.

29.3A Release the headlamp bulb retaining clip ...

29.3B ... and withdraw the bulb

4    Fit the bulb ensuring that the lugs on the bulb engage with the slots in the headlamp.
5    Refit the retaining clip, rubber cover and wiring connector.

### Sidelamp bulb

6    Withdraw the bulbholder from the headlamp and remove the bulb from the holder by turning anti-clockwise (photo).
7    Fit the new bulb to the holder, push the holder into the headlamp and refit the rubber cover.

29.6 The sidelamp bulbholder is a push fit in the lens assembly

## 30 Headlamp lens assembly – removal and refitting

1    From within the engine compartment pull off the wiring connector at the rear of the headlamp bulb.
2    Slip off the rubber cover then withdraw the sidelamp bulbholder.
3    Undo the two nuts securing the top of the lens unit to the body, release the two lower plastic ball pegs from their sockets and remove the assembly from the car.

4    Remove the headlamp bulb from the lens assembly if required as described in the previous Section.
5    Refitting is the reverse sequence of removal, but have the headlamp aim adjusted using optical beam setting equipment (see Section 38).

## 31 Front direction indicator bulb – renewal

1    Working in the engine compartment, release the spring retainer (photo) and ease the lamp body and seal from the body panel (photo). Release the bulbholder by turning anti-clockwise.
2    Remove the bulb from the holder by turning anti-clockwise.
3    Fit the bulb and holder to the lamp body, locate the lamp flange behind the panel and push the unit into position. Make sure the seal seats correctly.
4    Secure the lamp body with the spring retainer.

## 32 Rear lamp cluster bulbs – renewal

1    Working in the luggage compartment, press and disengage the retainers then withdraw the bulb panel from the lens unit (photos).
2    The bulbs can now be removed from the panel as required. All the bulbs except the tail lamp bulb are removed by turning them anti-clockwise. The tail lamp bulb is a push fit.
3    Fit the bulbs and bulb panel using the reverse sequence to removal.

## 33 Number plate lamp bulb – renewal

1    Push the lens housing forward and lift up to release it from the bumper.
2    Turn the bulbholder anti-clockwise and withdraw it from the lens housing (photo). The bulb is a push fit in the holder.
3    Refit using the reverse sequence to removal.

31.1A Release the direction indicator lamp body spring retainer ...

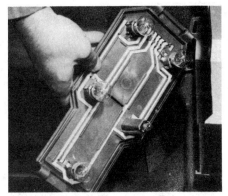

31.1B ... to gain access to the bulbholder and bulb

32.1A Depress the plastic retainers ...

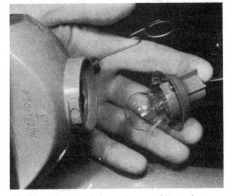

32.1B ... then withdraw the rear lamp cluster bulb panel

33.2 Turn the number plate bulbholder anti-clockwise to remove it from the lens housing

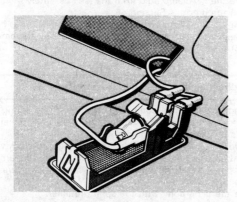

Fig. 9.8 Interior courtesy lamp bulb and holder arrangement
(Sec 34)

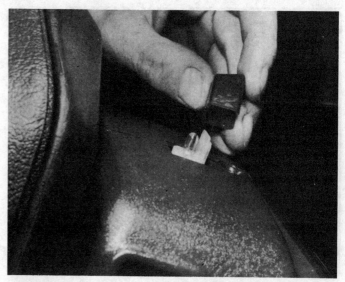

36.3 Lift off the switch lens to renew the hazard switch bulb

## 34 Interior courtesy lamp bulbs – renewal

1   Carefully ease the lamp lens from its location and then withdraw
the bulb by turning anti-clockwise.
2   Fit the bulb and push the unit into place.

## 35   Instrument panel illumination and warning lamp bulbs (electro-mechanical instruments) – renewal

1   Refer to Section 20.

## 36 Switch illumination bulbs – renewal

1   Remove the facia switch panel as described in Section 17 to gain
access to the illumination bulbs.
2   With the wiring multi-plug disconnected, remove the relevant bulb
which is a push fit in the holder (photo).
3   To renew the hazard warning switch bulb, lift off the switch lens
and remove the push fit bulb (photo).
4   Refitting is the reverse sequence to removal.
5   To renew the steering column switch bulb first remove the
steering wheel as described in Chapter 10.

6   Undo the retaining screws and lift off the steering column left-
hand and right-hand cowls.
7   Carefully lift up the retaining clips and slip the fibre optic guides
out of the bulb housing (photo).
8   Undo the bulb housing cover retaining screw, lift off the cover and
pull the bulb from its holder.
9   Refitting is the reverse sequence to removal. Position the switch
striker bush with the arrow pointing towards the direction indicator
switch before refitting the steering wheel.

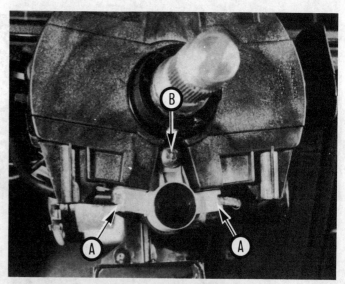

36.7 Fibre optic guide retaining clips (A) and bulb housing cover
retaining screw (B)

## 37 Heater control illumination bulbs – renewal

1   Carefully prise off the switch blanking plate on the right-hand side
of the heater control knobs.
2   Undo the retaining screw located behind the switch blanking
plate.
3   Centralize the heater control knobs and ease the control panel,
complete with knobs, off the facia.
4   Disconnect the switch multi-plug to gain access to the push fit
bulbs.

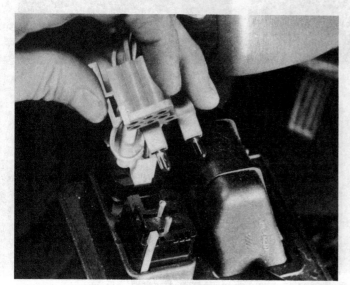

36.2 Switch panel bulbs located in wiring multi-plug

3   Release the spring retaining catch and separate the blade from the wiper arm (photo).
4   Insert the new blade into the arm, making sure that the spring retainer catch is engaged correctly.

### Wiper arms

5   To remove a wiper arm, open the bonnet, lift the hinged cover and unscrew the retaining nut (photo).
6   Carefully prise the arm off the spindle.
7   Before refitting the arm, switch the wipers on and off, allowing the mechanism to return to the 'park' position.
8   Noting that the longer arm is fitted to the driver's side, refit the arm to the spindle with the blade positioned on the finisher at the bottom edge of the windscreen.
9   Refit and tighten the retaining nut then close the hinged cover.

37.5 Heater panel illumination bulbs located in wiring multi-plug (A) and at rear of panel (B)

5   The panel illumination bulbholder and bulb are removed by turning anti-clockwise (photo).
6   Refitting is the reverse sequence of removal.

## 38  Headlamp aim – adjustment

1   At regular intervals (see Routine Maintenance) headlamp aim should be checked and, if necessary, adjusted.
2   Due to the light pattern of the homofocal headlamp lenses fitted to Montegò models, optical beam setting equipment must be used to achieve satisfactory aim of the headlamps. It is recommended therefore that this work is entrusted to a BL dealer.
3   Holts Amber Lamp is useful for temporarily changing the headlight colour to conform with the normal usage on Continental Europe.

## 39  Wiper blades and arms – removal and refitting

### Wiper blades

1   The wiper blades should be renewed when they no longer clean the windscreen effectively.
2   Lift the wiper arm away from the window.

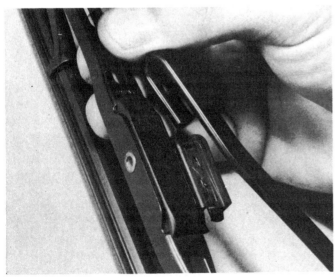

39.3 Release the catch and separate the wiper blade from the arm

39.5 Lift the hinged cover to gain access to the wiper arm retaining nut

## 40  Windscreen wiper motor – removal and refitting

1   Disconnect the battery negative terminal.
2   Undo the screws securing the washer reservoir in place and move the reservoir to one side.
3   Pull the rubber edge seal off the top of the closure panel, undo the retaining screws and lift away the panel (photos).
4   Remove the wiper arms as described in Section 39.
5   Undo the retaining nut and remove the wiper linkage rotary link from the motor spindle (photo).
6   Undo the three bolts securing the wiper motor mounting bracket to the bulkhead.
7   Ease the motor assembly out of its fitted position and prise the wiring harness grommet out of the bulkhead.
8   From inside the car, open the glovebox lid and lower the relay panel from the roof of the glovebox.
9   Identify the wiper motor wiring harness (coloured grey) and disconnect the multi-plugs.
10  Pull the wiring harness through the bulkhead and remove the wiper motor assembly from its location.
11  Refitting is the reverse sequence of removal bearing in mind the following points:

(a)  *Before reconnecting the linkage to the motor spindle, switch the wipers on and then off again to set the motor in the 'park' position. Apply silicone grease to the rotary link and gimbal, hold the primary link and turn the rotary link anti-clockwise to set the cam in its extended position. Hold the primary link and rotary link in a straight line as shown in Fig. 9.9 and refit the link to the wiper motor spindle.*

(b)  *Switch the wipers on and off once more so that the linkage is in the 'park' position then refit the wiper arms as described in Section 39.*

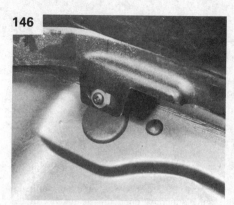

40.3A Undo the retaining screws ...

40.3B ... and lift out the closure panel

40.5 Wiper linkage rotary link retaining nut (A) and motor bracket retaining bolts (B)

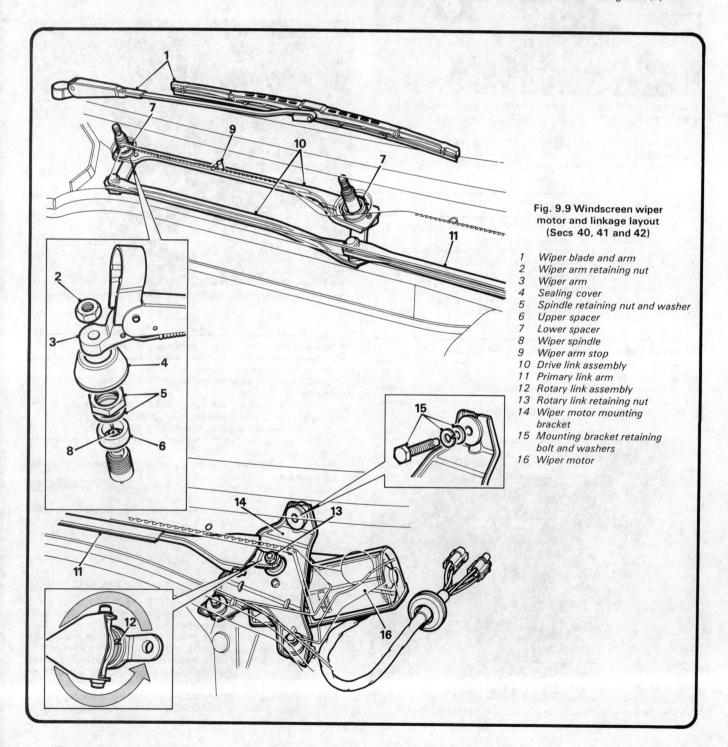

Fig. 9.9 Windscreen wiper motor and linkage layout (Secs 40, 41 and 42)

1   Wiper blade and arm
2   Wiper arm retaining nut
3   Wiper arm
4   Sealing cover
5   Spindle retaining nut and washer
6   Upper spacer
7   Lower spacer
8   Wiper spindle
9   Wiper arm stop
10  Drive link assembly
11  Primary link arm
12  Rotary link assembly
13  Rotary link retaining nut
14  Wiper motor mounting bracket
15  Mounting bracket retaining bolt and washers
16  Wiper motor

## 41 Windscreen wiper linkage – removal and fitting

1   Disconnect the battery negative terminal.
2   Undo the screws securing the washer reservoir in place and move the reservoir to one side.
3   Pull the rubber edge seal off the top of the closure panel, undo the retaining screws and lift away the panel.
4   Remove the wiper arms as described in Section 39.
5   Carefully prise the primary link arm socket off the rotary link assembly.
6   Remove the wiper spindle sealing covers followed by the retaining nuts, washers and upper spacers.
7   Push the wiper spindles through the body then withdraw the linkage assembly and central water shield from the engine compartment.
8   Remove the central water shield and lower spindle spacers.
9   Refitting is the reverse sequence of removal bearing in mind the following points:

  (a)  *Apply silicone grease to the primary link bolt and socket then press the two parts together using a lightly adjusted self-locking wrench*
  (b)  *Switch the wipers on and off to set the linkage in the 'park' position then refit the wiper arms as described in Section 39.*

## 42 Windscreen wiper primary link – removal and refitting

1   Disconnect the battery negative terminal.
2   Undo the screws securing the washer reservoir in place and move the reservoir to one side.
3   Pull the rubber edge seal off the top of the closure panel, undo the retaining screws and lift away the panel.
4   Carefully prise the primary link arm socket off the rotary link assembly.
5   Undo the retaining nut and remove the rotary link from the wiper motor spindle.
6   To refit the primary link, reconnect the battery then switch the wipers on and off again to set the motor in the 'park' position.
7   Hold the primary link and turn the rotary link anti-clockwise to position the cam in the fully extended position.
8   Hold the primary and rotary link together in a straight line as shown in Fig. 9.9, then refit the rotary link to the wiper motor spindle. Refit and tighten the retaining nut.
9   Apply silicone grease to the primary link ball and socket then press the two parts together using a lightly adjusted self-locking wrench.
10  With the windscreen wet, check the operation of the wipers ensuring that they 'park' on the finisher at the bottom edge of the windscreen. If necessary adjust the wiper arm position as described in Section 39.
11  Refit the closure panel, edge seal and washer reservoir, using the reverse sequence of removal.

## 43 Windscreen washer reservoir and pump – removal and refitting

1   Undo the retaining screws and withdraw the reservoir from its location.
2   Disconnect the wiring connectors and water hoses then remove the reservoir from the car.
3   To remove the pump(s) from the reservoir, lift the pump and pull the inlet nozzle from the seal. Withdraw the pump from the reservoir.
4   Refitting is the reverse sequence to removal.

## 44 Headlamp washer jet – removal and refitting

1   From within the front bumper panel disconnect the hose from the washer extension tube.
2   Undo the retaining nut, remove the seating collar and withdraw the washer jet assembly.
3   Refitting is the reverse sequence to removal.

## 45 Horn – removal and refitting

1   Disconnect the battery negative terminal.
2   Working under the left-hand front wing, disconnect the electrical leads, undo the retaining nut and remove the horn from its bracket.
3   The horn is a sealed unit which cannot be dismantled or adjusted. In the event of horn failure, renewal will be necessary.

## 46 Speedometer cable – removal and refitting

1   Refer to Section 19 paragraphs 1 to 4 inclusive, and detach the speedometer cable from the instrument panel.
2   Release the grommet from the bulkhead and pull the cable through into the engine compartment.
3   Undo the bolt securing the cable retaining plate to the gearbox and withdraw the cable and pinion.
4   Release the cable clips and remove the cable from the car.
5   Refitting the cable is the reverse sequence of removal.

## 47 Radio, radio/cassette player – removal and refitting

1   Disconnect the battery negative terminal.
2   Pull off the knobs and bezels from the radio, radio/cassette player console, unscrew the retaining nuts and lift off the finisher and masking plate.
3   Push the unit back into its aperture and remove the mounting plate from inside the panel.
4   Withdraw the unit, disconnect the wiring, speaker and aerial leads then remove it from the car.
5   Refitting is the reverse sequence of removal.

## 48 Electronic radio/cassette player – removal and refitting

1   Disconnect the battery negative terminal.
2   Reach behind the facia and disconnect the radio multi-plug and the aerial lead.
3   Using 0.10 in (2.5 mm) diameter welding rod or similar material, make up four probes each approximately 3.0 in (76.0 mm) long. As a depth gauge wrap some tape around each probe, 1.0 in (25.4 mm) from one end.
4   Insert the probes into the holes at each end of the radio panel. Move the ends of the probes outwards to release the radio retaining clips and withdraw the unit from its location. Remove the probes.
5   Refitting is the reverse sequence of removal, but simply push the radio into place to engage the retaining clips.

## 49 Electrically operated windows – description

  Certain Montego models covered by this manual are available with electrically operated front windows as standard or optional equipment. The system enables both front windows to be raised or lowered independently by two switches on the driver's door armrest. A single switch on the passenger's door armrest allows independent operation of the passenger's window. A 'one-touch' facility incorporated in the driver's door switch circuitry allows the window to be raised or lowered *fully* when the switch is pressed fully then released.
  Each window is operated by an electric motor acting on the window regulator. A relay mounted within the driver's door controls the electrical supply to the circuit and a thermal cut-out in each motor isolates the circuit should any objects jam the window during operation.

## 50 Window lift motor – removal and refitting

1   Remove front door inner trim panel as described in Chapter 11.
2   If the door window is not closed, temporarily reconnect the switch multi-plugs and the battery, and close the window.

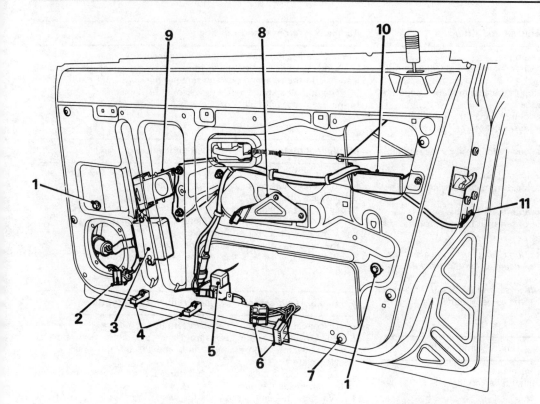

**Fig. 9.10 Electric window and central locking door component locations (Secs 50 to 53)**

1  *Window auxiliary channel retaining screws*
2  *Speaker wiring connector*
3  *Window lift 'one-touch' control unit*
4  *Inner trim panel retainers*
5  *Window lift relay*
6  *Window lift multi-plugs*
7  *Inner trim panel retaining studs*
8  *Inner trim panel mounting bracket*
9  *Window lift motor*
10 *Central locking switch unit*
11 *Interior courtesy lamp door switch*

3  Undo the retaining screws and remove the inner trim panel mounting bracket.
4  Carefully remove the polythene condensation barrier from the door.
5  Support the window in the raised position using wooden wedges then undo the three nuts securing the motor unit to the door (photo).
6  Release the three nylon wheels of the lift arms from the lifting channel and auxiliary channel.
7  Operate the motor so that the lift arms are set in the fully lowered position. Disconnect the switch wiring multi-plug and manoeuvre the motor unit out of the door aperture.
8  Refitting is the reverse sequence of removal, bearing in mind the following points.

(a)  *Lubricate the lifting and auxiliary channels with graphite grease*
(b)  *Install the motor unit with the lifting arms in the lowered position, then engage the nylon wheels of the arms into the channels, with the arms in the raised position*
(c)  *With the motor unit installed check the operation of the window and adjust the lower ends of the window channels if necessary*
(d)  *Ensure that all wiring is neatly secured in its clips*

### 51 Window lift 'one-touch' control unit – removal and refitting

1  Remove front door inner trim panel as described in Chapter 11. Carefully peel back the polythene condensation barrier as necessary, starting at the lower front corner of the door.
2  Disconnect the wiring multi-plug, undo the retaining screws and withdraw the control unit from inside the door.
3  Refitting is the reverse sequence of removal.

### 52 Central door locking – description

A central door locking system is available on certain Montego models covered by this manual. The system enables the passenger's front door lock, both rear door locks and the boot lid lock to be operated simultaneously by the action of the driver's door interior lock button or exterior private lock.

50.5 Window lift motor retaining nuts

The passenger's door, both rear doors and the boot lid each incorporate a solenoid which is connected to the lock mechanism.
The driver's door is equipped with the normal private lock and interior lock button arrangement and is also fitted with a control unit. Operation of the driver's door lock causes the control unit to supply an electric current to each of the door lock solenoids thus locking or unlocking the doors in unison with the driver's door.

### 53 Central locking solenoid – removal and refitting

1  To remove a passenger door solenoid or driver's door control unit, first remove the inner trim panel from the relevant door as described in Chapter 11.
2  If the door window is not closed, temporarily reconnect the switch

multi-plugs and battery if electrically operated windows are fitted, and close the window.

3  Undo the retaining screws and remove the inner trim panel mounting bracket.

4  Carefully remove the polythene condensation barrier from the door.

5  Undo the retaining screws, disconnect the wiring multi-plug and cable clips then lift the solenoid or control unit off the door panel (photo). Detach the lock lever link and remove the unit.

6  Refitting is the reverse sequence of removal.

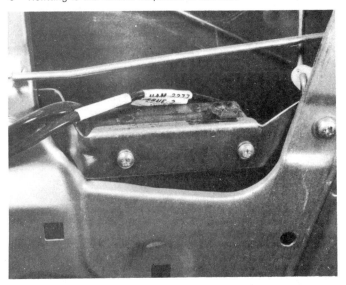

53.5 Central locking control unit location in driver's door

## 54 Boot lock solenoid – removal and refitting

1  Disconnect the battery negative terminal.

2  Open the boot lid and undo the nut securing the solenoid mounting plate to the lock barrel. Remove the lock barrel.

3  Disconnect the wiring multi-plug then position the solenoid mounting plate in the left-hand boot lid aperture.

4  Undo the retaining screws and withdraw the solenoid from behind the mounting plate.

5  Refitting is the reverse sequence of removal.

## 55 Accessory wiring – general

1  If an electrical accessory is to be fitted, electrical connections should be made at the fusebox, on the feed side of the following fuses.

   (a)  *If the accessory is to operate through the ignition switch, connect to fuse C1, C2 or C3 (light green/white wire)*

   (b)  *If the accessory is to operate independent of the ignition switch, connect to fuse A6 (brown wire)*

2  Always use a separate line fuse of the appropriate rating to protect the accessory being fitted.

3  To avoid the risk of damage resulting from overloading the vehicle electrical circuits, do not plug any accessory into the cigarette lighter socket unless it is approved by BL for such fitment.

4  For towing bracket installations, connections for the towing socket can be made at the cables located behind the rear light clusters. To gain access, fold back the floor covering and release the rear of the side panel covers. The cable colours, locations and their respective circuits are shown in Fig. 9.11; note that a special relay is required. Always disconnect the battery negative terminal before making any wiring connections. **Note:** *On vehicles with solid-state instruments consult your BL dealer – incorrect connections could damage the bulb fail monitors.*

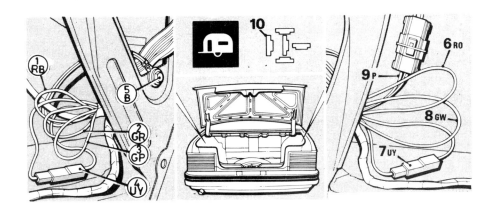

Fig. 9.11 Wiring harness connections for towing bracket installation – vehicles with electro-mechanical instrumentation (Sec 55)

| | | |
|---|---|---|
| *1  Left-hand tail lamps* | *4  Left-hand rear fog guard* | *7  Right-hand rear fog guard* |
| *2  Left-hand indicator* | *5  Earth point* | *8  Right-hand indicator* |
| *3  Stop lamps* | *6  Right-hand tail lamps* | *9  Interior lamp (live feed)* |
| | | *10  Flasher unit relay connector* |

### Cable colour code

| | | | | | |
|---|---|---|---|---|---|
| B | Black | O | Orange | U | Blue |
| G | Green | P | Purple | W | White |
| N | Brown | R | Red | Y | Yellow |

*The second code letter indicates the tracer colour*

## 56 Fault diagnosis – electrical system

| Symptom | Reason(s) |
| --- | --- |
| Starter fails to turn engine | Battery discharged or defective |
| | Battery terminal and/or earth leads loose |
| | Starter motor connections loose |
| | Starter solenoid faulty |
| | Starter brushes worn or sticking |
| | Starter commutator dirty or worn |
| | Starter field coils earthed |
| | Starter solenoid relay faulty |
| Starter turns engine very slowly | Battery discharged |
| | Starter motor connections loose |
| | Starter brushes worn or sticking |
| Starter spins but does not turn engine | Pinion or flywheel ring gear teeth broken or badly worn |
| | Starter pinion sticking |
| Starter noisy | Pinion or flywheel ring gear teeth badly worn |
| | Mounting bolts loose |
| Battery will not hold charge for more than a few days | Battery defective internally |
| | Battery terminals loose |
| | Alternator drivebelt slipping |
| | Alternator or regulator faulty |
| | Short circuit |
| Ignition light stays on | Alternator faulty |
| | Alternator drivebelt broken |
| Ignition light fails to come on | Warning bulb blown |
| | Indicator light open circuit |
| | Alternator faulty |
| Instrument readings increase with engine speed | Faulty instrument multi-function unit |
| Fuel or temperature gauge gives no reading | Wiring open circuit |
| | Sender or thermister faulty |
| | Faulty instrument multi-function unit |
| Fuel or temperature gauge gives continuous maximum reading | Wiring short circuit |
| | Sender or thermister faulty |
| | Faulty instrument multi-function unit |
| | Faulty gauge |
| Lights inoperative | Bulb blown |
| | Fuse blown |
| | Fusible link blown |
| | Battery discharged |
| | Switch faulty |
| | Relay faulty |
| | Wiring open circuit |
| | Bad connection due to corrosion |
| Failure of component motor | Commutator dirty or burnt |
| | Armature faulty |
| | Brushes sticking or worn |
| | Armature bearings dry or misaligned |
| | Field coils faulty |
| | Fuse blown |
| | Relay faulty |
| | Fusible link blown |
| | Poor or broken wiring connections |
| Failure of an individual component | Fuse blown |
| | Relay faulty |
| | Fusible link blown |
| | Poor or broken wiring connections |
| | Switch faulty |
| | Component faulty |

# Chapter 10 Suspension and steering

*For modifications, and information applicable to later models, see Supplement at end of manual*

## Contents

## Specifications

### Front suspension

| | |
|---|---|
| Type | Independent by MacPherson struts with coil springs and integral telescopic shock absorber. Anti-roll bar on all models |
| Coil spring free length | 15.16 in (385.1 mm) |
| Trim height (measured from the centre of the front hub to the edge of the wheel arch) | 14.3 to 15.3 in (363.0 to 389.0 mm) |
| Trim height permissible side-to-side difference | 1.02 in (26.0 mm) |

### Rear suspension

| | |
|---|---|
| Type | Trailing twist axle with coil springs and telescopic shock absorbers |
| Coil spring free length | 13.5 in (343.0 mm) |
| Trim height (measured from the centre of the rear hub to the edge of the wheel arch) | 14.5 to 15.7 in (368.0 to 398.0 mm) |
| Trim height permissible side-to-side difference | 1.02 in (26.0 mm) |
| Rear wheel toe setting | 0° 20′ to 0° 40′ toe-in |
| Rear wheel camber angle | 0° to 1° negative |

## Steering

| | |
|---|---|
| Type ....................................................................................... | Rack and pinion, manual or power-assisted |
| Turns lock-to-lock: | |
|     Manual steering ............................................................... | 4.3 |
|     Power-assisted steering ................................................... | 2.94 |
| Steering wheel diameter ...................................................... | 15.0 in (381.0 mm) |
| Camber angle ...................................................................... | 0° 5′ negative to 0° 26′ negative |
| Castor angle ........................................................................ | 0° 30′ positive to 1° 30′ positive |
| Steering axis inclination .................................................... | 12° 0′ to 13° 0′ |
| Toe setting .......................................................................... | Parallel ± 0° 8′ |
| Steering gear lubricant (overhaul only) ............................. | BP Energrease FGL Fluid Grease |
| Power-assisted steering fluid ............................................. | Dexron TQII type ATF (Duckhams Uni-Matic or D-Matic) |

## Roadwheels

| | |
|---|---|
| Wheel size: | |
|     All models except MG EFi ............................................. | 120 x 365 mm TD type |
|     MG EFi models ................................................................ | 135 x 365 mm TD type |

## Tyres

| | |
|---|---|
| Tyre size: | |
|     All models except MG EFi ............................................. | 180/65 R365 TD type steel braced radial ply |
|     MG EFi models ................................................................ | 180/65 HR365 TD type steel braced radial ply |

## Tyre pressures – cold (all models)

| | lbf/in² | bar |
|---|---|---|
| Front ................................................................................... | 26 | 1.8 |
| Rear .................................................................................... | 28 | 2.0 |

## Torque wrench settings

### Front suspension

| | lbf ft | Nm |
|---|---|---|
| Anti-roll bar bush-to-lower arm nut (10 mm) .......................... | 13 | 18 |
| Anti-roll bar clamp bolts ......................................................... | 33 | 45 |
| Balljoint-to-lower arm (service replacement bolts) ................. | 22 | 30 |
| Balljoint-to-swivel hub clamp nut .......................................... | 33 | 45 |
| Driveshaft nut* ....................................................................... | 150 | 203 |
| Suspension strut-to-swivel hub nuts ..................................... | 66 | 90 |
| Suspension strut upper retaining nut ..................................... | 41 | 55 |
| Suspension strut bearing retaining nut .................................. | 22 | 30 |
| Lower arm rear mounting bolts ............................................... | 33 | 45 |
| Lower arm front mounting bolt ............................................... | 53 | 72 |
| Crossmember-to-body ............................................................ | 67 | 90 |
| Crossmember struts to body ................................................... | 55 | 75 |
| Crossmember struts to crossmember ..................................... | 67 | 90 |
| *Refer to Chapter 12, Section 11 | | |

### Rear suspension

| | lbf ft | Nm |
|---|---|---|
| Rear axle pivot bolts (marked '8.8') ....................................... | 41 | 55 |
| Rear axle pivot bolts (marked '10.9') ..................................... | 63 | 85 |
| Rear hub retaining nut ............................................................ | 50 | 68 |
| Stub axle-to-trailing arm nuts ................................................ | 33 | 45 |
| Suspension strut lower mounting ........................................... | 53 | 72 |
| Suspension strut upper mounting nut ..................................... | 33 | 45 |
| Suspension strut spring retainer nut ...................................... | 30 | 40 |

### Steering

| | lbf ft | Nm |
|---|---|---|
| Steering column mounting bolts .............................................. | 18 | 25 |
| Intermediate shaft clamp nuts ................................................ | 18 | 25 |
| Rack and pinion steering gear mounting bolts ........................ | 33 | 45 |
| Steering gear pinion locknut .................................................. | 30 | 40 |
| Steering wheel nut ................................................................. | 33 | 48 |
| Tie-rod inner balljoint to rack ................................................ | 60 | 80 |
| Tie-rod outer balljoint to steering arm .................................. | 22 | 30 |
| Power steering pump mounting and adjustment bolts ............ | 18 | 25 |

### Roadwheels

| | lbf ft | Nm |
|---|---|---|
| Wheel nuts (all models) .......................................................... | 53 | 72 |

## 1  General description

The independent front suspension is of the MacPherson strut type, incorporating coil springs and integral telescopic shock absorbers. Lateral and longitudinal location of each strut assembly is by pressed steel lower suspension arms utilizing rubber inner mounting bushes and incorporating a balljoint at their outer ends. Both lower suspension arms are interconnected by an anti-roll bar. The front swivel hubs, which carry the wheel bearings, brake calipers and the hub/disc assemblies, are bolted to the MacPherson struts and connected to the lower arms via the balljoints.

The rear suspension is of the trailing twist axle type, whereby the two trailing arms are welded to an 'L' section transverse member. This arrangement allows considerable independent up-and-down movement of each trailing arm whilst providing lateral rigidity and anti-roll

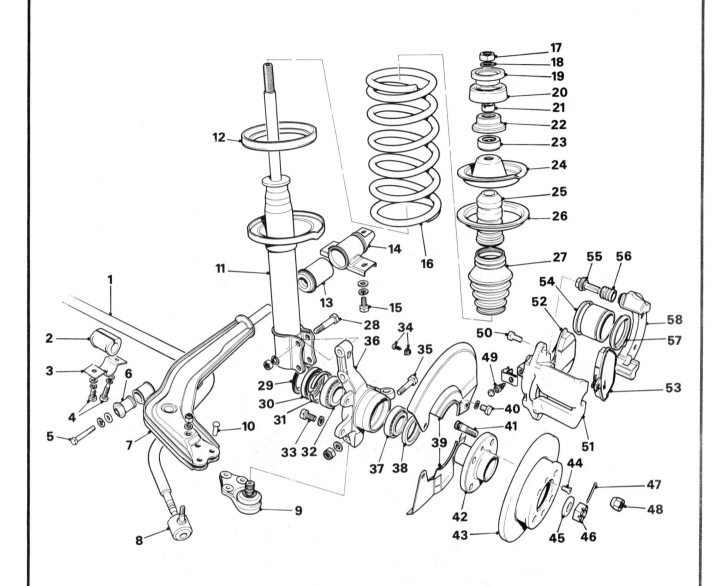

**Fig. 10.1 Exploded view of the front suspension (Sec 1)**

1  Anti-roll bar
2  Mounting block
3  Clamp
4  Clamp retaining bolts
5  Lower arm front mounting
   bolt
6  Rubber bush
7  Lower suspension arm
8  Anti-roll bar mounting bush
9  Lower suspension arm
   balljoint
10 Rivet (replaced by nut and
   bolt after balljoint renewal)
11 Front suspension strut
12 Lower insulator ring
13 Rubber bush
14 Rear mounting bracket

15 Mounting bolt
16 Coil spring
17 Upper mounting nut
18 Washer
19 Cup and washer assembly
20 Upper mounting rubber
21 Bearing retainer nut
22 Bearing housing
23 Upper bearing
24 Spring seat
25 Spring aid
26 Upper insulator ring
27 Rubber boot
28 Strut-to-swivel hub
   retaining bolt
29 Bearing water shield
30 Inner oil seal

31 Oil seal spacer
32 Hub inner bearing
33 Brake caliper mounting
   bolt
34 Disc shield retaining
   screw and clip
35 Lower balljoint clamp
   bolt
36 Swivel hub
37 Hub outer bearing
38 Outer oil seal
39 Disc shield halves
40 Disc shield retaining bolt
41 Wheel stud
42 Drive flange
43 Brake disc
44 Disc retaining screw

45 Flat washer
46 Driveshaft nut (staked
   on later models)
47 Split pin
48 Wheel nut
49 Bleed screw
50 Guide pin bolt
51 Caliper body
52 Inner brake pad
53 Outer brake pad
54 Piston
55 Guide pin
56 Dust cover
57 Piston dust cover
58 Anchor bracket

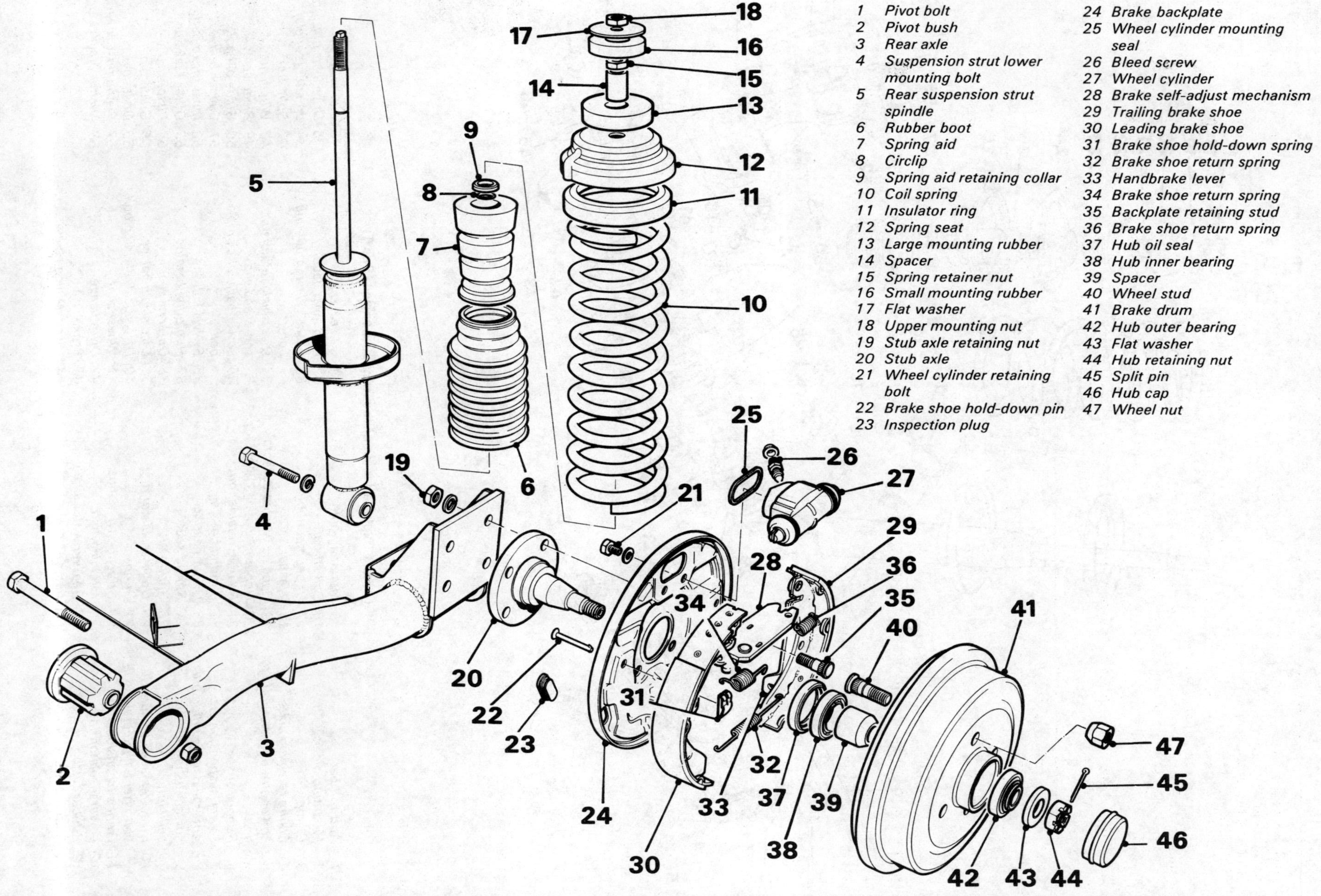

Fig. 10.2 Exploded view of the rear suspension (Sec 1)

1  Pivot bolt
2  Pivot bush
3  Rear axle
4  Suspension strut lower
   mounting bolt
5  Rear suspension strut
   spindle
6  Rubber boot
7  Spring aid
8  Circlip
9  Spring aid retaining collar
10 Coil spring
11 Insulator ring
12 Spring seat
13 Large mounting rubber
14 Spacer
15 Spring retainer nut
16 Small mounting rubber
17 Flat washer
18 Upper mounting nut
19 Stub axle retaining nut
20 Stub axle
21 Wheel cylinder retaining
   bolt
22 Brake shoe hold-down pin
23 Inspection plug
24 Brake backplate
25 Wheel cylinder mounting
   seal
26 Bleed screw
27 Wheel cylinder
28 Brake self-adjust mechanism
29 Trailing brake shoe
30 Leading brake shoe
31 Brake shoe hold-down spring
32 Brake shoe return spring
33 Handbrake lever
34 Brake shoe return spring
35 Backplate retaining stud
36 Brake shoe return spring
37 Hub oil seal
38 Hub inner bearing
39 Spacer
40 Wheel stud
41 Brake drum
42 Hub outer bearing
43 Flat washer
44 Hub retaining nut
45 Split pin
46 Hub cap
47 Wheel nut

capability in a structure of minimum unsprung weight. Suspension and damping is by strut assemblies containing coil springs and integral telescopic shock absorbers.

The steering gear is of the conventional rack and pinion type available in either manual or power-assisted configuration according to model or options fitted. Movement of the steering wheel is transmitted to the steering gear by an intermediate shaft containing two universal joints. The front wheels are connected to the steering gear by tie-rods, each having an inner and outer balljoint. Where power-assisted steering is fitted, fluid pressure is provided by a pump mounted on the engine and driven by a vee-belt from the crankshaft pulley.

## 2 Maintenance and inspection

1   At the intervals given in Routine Maintenance at the beginning of this manual a thorough inspection of all suspension and steering components should be carried out using the following procedure as a guide.

### Front suspension and steering

2   Apply the handbrake, jack up the front of the car and support it securely on axle stands.
3   Visually inspect the lower balljoint dust covers and the steering rack and pinion gaiters for splits, chafing, or deterioration. Renew the rubber gaiters or the balljoint assembly, as described in the appropriate Sections of this Chapter, if any damage is apparent.
4   On vehicles equipped with power-assisted steering, check the fluid hoses for chafing or deterioration, and the pipe and hose unions for fluid leaks. Also check for any signs of fluid leakage from the rubber gaiters which would indicate failed fluid seals within the steering gear. Rectify any faults found using the procedures described later in this Chapter.
5   Check and if necessary top up the fluid in the power-assisted steering pump reservoir, as described in Section 29. Check the condition of the pump drivebelt, adjust the tension or if worn renew the belt using the procedure described in Section 27.
6   Grasp the roadwheel at the 12 o'clock and 6 o'clock positions and try to rock it. Very slight free play may be felt, but if the movement is appreciable further investigation is necessary to determine the source. Continue rocking the wheel while an assistant depresses the footbrake. If the movement is now eliminated or significantly reduced, it is likely that the hub bearings are at fault. If the free play is still evident with the footbrake depressed, then there is wear in the suspension joints or mountings. Pay close attention to the lower balljoint and lower arm mounting bushes. Renew any worn components, as described in the appropriate Sections of this Chapter.
7   Now grasp the wheel at the 9 o'clock and 3 o'clock positions and try to rock it as before. Any movement felt now may again be caused by wear in the hub bearings or the steering tie-rod inner or outer balljoints. If the outer balljoint is worn the visual movement will be obvious. If the inner joint is suspect it can be felt by placing a hand over the rack and pinion rubber gaiter and gripping the tie-rod. If the wheel is now rocked, movement will be felt at the inner joint if wear has taken place. Repair procedures are described in Section 20 and 23 or 26 respectively.
8   Using a large screwdriver or flat bar check for wear in the anti-roll bar mountings and lower arm mountings by carefully levering against these components. Some movement is to be expected, as the mountings are made of rubber, but excessive wear should be obvious. Renew any bushes that are worn.
9   With the car standing on its wheels have an assistant turn the steering wheel back and forth about one eighth of a turn each way. There should be no lost movement whatever between the steering wheel and roadwheels. If this is not the case, closely observe the joints and mountings previously described, but in addition check the intermediate shaft universal joints for wear and also the rack and pinion steering gear itself. Any wear should be visually apparent and must be rectified, as described in the appropriate Sections of this Chapter.

### Rear suspension

10   Chock the front wheels, jack up the rear of the car and support it securely on axle stands.
11   Visually inspect the rear suspension components, attachments and linkages for any visible signs of wear or damage.

12   Grasp the roadwheel at the 12 o'clock and 6 o'clock positions and try to rock it. Any excess movement here indicates wear in the hub bearings which may also be accompanied by a rumbling sound when the wheel is spun. Repair procedures are described in Section 11.

### Wheels and tyres

13   Carefully inspect each tyre, including the spare, for signs of uneven wear, lumps, bulges or damage to the sidewalls or tread face. Refer to Section 32 for further details.
14   Check the condition of the wheel rims for distortion, damage and excessive run-out. Also make sure that the balance weights are secure with no obvious signs that any are missing. Check the torque of the wheel nuts and check the tyre pressures.

### Shock absorbers

15   Check for any signs of fluid leakage around the shock absorber body or from the rubber boot around the piston rod. Should any fluid be noticed the shock absorber is defective internally and renewal is necessary.
16   The efficiency of the shock absorber may be checked by bouncing the car at each corner. Generally speaking the body will return to its normal position and stop after being depressed. If it rises and returns on a rebound, the shock absorber is probably suspect. Examine also the shock absorber upper and lower mountings for any sign of wear. Renewal procedures are contained in Sections 6 and 14.

## 3 Front swivel hub assembly – removal and refitting

1   Securely apply the handbrake, chock the rear wheels and remove the wheel trim from the front roadwheel.
2   Extract the split pin, then, using a socket and long bar, slacken, but do not remove, the driveshaft nut.
3   Slacken the roadwheel retaining nuts then jack up the front of the car and support it securely on axle stands. Remove the roadwheel. Remove the driveshaft retaining nut and washer (photo).
4   Undo and remove the two bolts securing the disc brake caliper to the anchor bracket (photo). Slide the caliper, complete with brake pads, off the anchor bracket and suspend it from a convenient place under the wheel arch using string or wire. Take care not to strain the flexible hydraulic hose.
5   Undo and remove the nut securing the tie-rod balljoint to the steering arm on the swivel hub. Release the balljoint from the steering arm using a suitable extractor (see Section 20).
6   At the base of the swivel hub unscrew and remove the nut and washer, then withdraw the lower balljoint clamp bolt (photo).
7   Undo and remove the nuts and washers, then withdraw the two bolts securing the suspension strut to the upper part of the swivel hub (photo).

3.3 Remove the driveshaft nut and washer

3.4 Brake caliper to anchor bracket upper retaining bolt (arrowed)

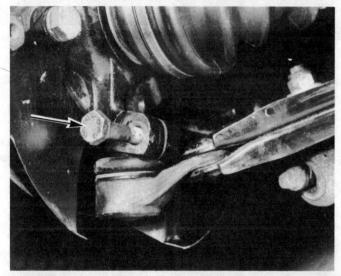

3.6 Undo the nut and withdraw the lower balljoint clamp bolt (arrowed)

3.7 Remove the two nuts and bolts (arrowed) securing the swivel hub to the suspension strut

8    Release the swivel hub from the strut and then lift the hub, while pushing down on the suspension arm, to disengage the lower balljoint. Withdraw the swivel hub assembly from the driveshaft and remove it from the car.

9    Secure the swivel hub in a vice, undo the two bolts and remove the brake caliper anchor bracket. Separate the drive flange and disc from the hub using a hammer and suitable tube or socket, if necessary (photos).

3.9A Remove the brake caliper anchor bracket

3.9B Use a hammer and tube to drift out the drive flange if tight

10  If required, the two halves of the disc shield can now be removed after unscrewing the retaining bolt and screw.

11  Refitting the swivel hub is the reverse of the removal sequence, bearing in mind the following points:

(a)  *Ensure that the bearing water shield is in position on the driveshaft (photo) before fitting the swivel hub. Fill the groove in the water shield with a general purpose grease*

(b)  *Where applicable, tighten all retaining nuts and bolts to the specified torque.* **Do not attempt to fully tighten the driveshaft nut until the weight of the car is on the roadwheels**

(c)  *Tighten the driveshaft nut to the specified torque, but first refer to Chapter 12, Section 11*

3.11 Position the water shield on the driveshaft flange before fitting the hub

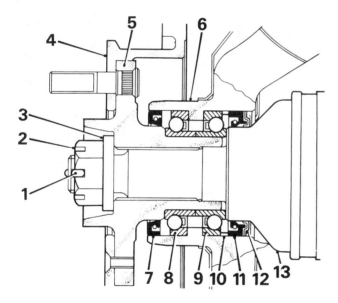

**Fig. 10.3 Cross-sectional view of the front hub (Sec 4)**

| | |
|---|---|
| 1 Split pin | 8 Outer bearing |
| 2 Hub nut | 9 Inner bearing |
| 3 Flat washer | 10 Spacer |
| 4 Brake disc | 11 Inner oil seal |
| 5 Drive flange | 12 Bearing water shield |
| 6 Hub | 13 Drive shaft |
| 7 Outer oil seal | |

### 4  Front hub bearings – removal and refitting

1   Remove the swivel hub assembly from the car, as described in the previous Section.
2   With the hub assembly on the bench, prise out the inner and outer oil seals using a screwdriver or suitable flat bar. Note that there is a spacer fitted between the inner oil seal and the bearing.
3   With the hub supported on blocks use a hammer and drift to drive out one of the bearing inner races from the centre of the hub. Take care not to lose the balls which will be dislodged as the inner race is released. Turn the hub over and repeat the procedure for the other inner race. The outer races can now be driven out in the same way.
4   Wipe away any surplus grease from the bearings and swivel hub

and then clean these components thoroughly using paraffin, or a suitable solvent. Dry with a lint-free rag. Remove any burrs or score marks from the hub bore with a fine file or scraper.
5   Carefully examine the bearing inner and outer races, the balls and ball cage for pitting, scoring or cracks, and if at all suspect renew the bearings as a pair. It will also be necessary to renew the oil seals as they will have been damaged during removal.
6   If the old bearings are in satisfactory condition and are to be reused, reassemble the ball cage, holding the balls in position with grease, and then place this assembly in the outer race. Lay the inner race over the balls and push it firmly into place.
7   Before refitting the bearings to the hub, pack them thoroughly with a high melting-point grease.
8   Place one of the bearings in position on the hub with the word THRUST, or the markings stamped on the edge of the inner race, facing away from the centre of the hub (photo).
9   Using a tube of suitable diameter, a large socket or a soft metal drift, drive the bearing into the hub bore until it contacts the shoulder in the centre of the hub. Ensure that the bearing does not tip slightly and bind as it is being fitted. *If this happens the outer race may crack, so take great care to keep it square.*
10   Turn the hub over and repeat this procedure for the other bearing.
11   Dip the new oil seals in oil and carefully fit them to the hub using

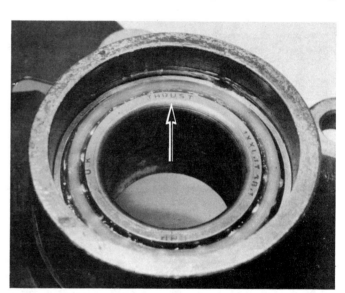

4.8 Bearing fitted with THRUST marking (arrowed) facing away from the hub centre

4.11A Fit the outer oil seal with its sealing lip facing inwards

4.11B Place the spacer in position over the bearing ...

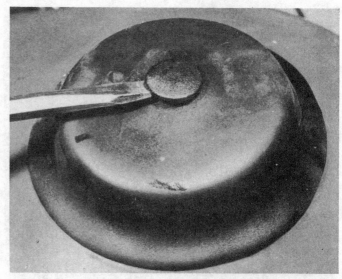

5.3 Prise off the plastic cap and remove the front strut upper mounting cover

4.11C ... then fit the inner oil seal, with its sealing lip facing inwards

5.4 Undo the strut upper mounting nut whilst holding the spindle

a tube of suitable diameter or one of the old seals to drive them in. Note that both oil seals are fitted with their sealing lips inward (photo) and that the inner seal has a second lip on its outer face. Don't forget to fit the spacer between the bearing and inner oil seal (photos).
12  The swivel hub can now be refitted to the car, as described in the previous Section.

## 5  Front suspension strut – removal and refitting

1  Securely apply the handbrake, chock the rear wheels and remove the trim from the front roadwheel.
2  Slacken the roadwheel retaining nuts, then jack up the front of the car and support it securely on axle stands. Remove the roadwheel.
3  From within the engine compartment prise off the small plastic cap in the centre of the strut upper mounting (photo).
4  Insert an Allen key of the appropriate size into the centre of the strut spindle and, while holding the key to prevent the spindle turning, unscrew the mounting nut (photo). Lift off the spring washer and the cup and washer assembly.
5  Undo and remove the two nuts, bolts and spring washers securing

5.5A Suspension strut to swivel hub upper retaining bolt and nut

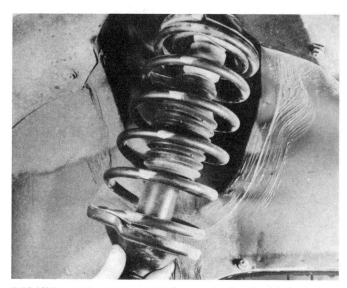

5.5B Lifting out the strut assembly from under the wheel arch

Fig. 10.4 Suspension strut bearing retainer nut removal tool
(Sec 6)

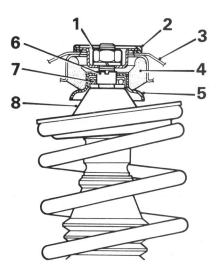

Fig. 10.5 Front suspension strut upper mounting details (Sec 6)

| | | | |
|---|---|---|---|
| 1 | Mounting nut | 5 | Bearing housing |
| 2 | Cup and washer assembly | 6 | Bearing retainer nut |
| 3 | Inner front wing valance | 7 | Upper bearing |
| 4 | Upper mounting rubber | 8 | Spring seat |

the suspension strut to the swivel hub (photo). Release the strut from the hub and withdraw it from under the wheel arch (photo).
6   Refitting is the reverse sequence to removal. Ensure that all retaining nuts and bolts are tightened to the specified torque, where applicable.

## 6   Front suspension strut – dismantling and reassembly

**Note:** *Before attempting to dismantle the front suspension strut, a suitable tool to hold the coil spring in compression must be obtained. Adjustable coil spring compressors are readily available and are recommended for this operation. Any attempt to dismantle the strut without such a tool is likely to result in damage or personal injury.*
1   Proceed by removing the front suspension strut, as described in the previous Section.
2   Position the spring compressors on either side of the spring (photo) and compress the spring evenly until there is no tension on the upper spring seat or upper mounting.
3   Lift off the upper mounting rubber (photo) and bearing housing (photo).
4   To remove the bearing retainer nut it will be necessary to make up a suitable tool which will engage in the slots of the nut enabling it to be unscrewed. A tool can be made out of a large nut with one end suitably shaped by cutting or filing so that two projections are left which will engage with the slots in the retainer nut (Fig. 10.4).
5   While holding the strut spindle with an Allen key, unscrew the

bearing retainer nut. Lift off the upper bearing, spring seat and insulator ring.
6   The coil spring can now be removed, with compressors still in position if desired, followed by the spring aid, rubber boot and lower insulator ring.
7   With the strut completely dismantled, the components can be examined as follows.
8   Examine the strut for signs of fluid leakage. Check the strut spindle for signs of wear or pitting along its entire length and check the strut

6.2 Compressors in position on the front coil spring

6.3A Lift off the upper mounting rubber ...

6.3B ... and bearing housing (arrowed)

body for signs of damage or elongation of the mounting bolt holes. Test the operation of the strut, while holding it in an upright position, by moving the spindle through a full stroke and then through short strokes of 2 to 4 in (50 to 100 mm). In both cases the resistance felt should be smooth and continuous. If the resistance is jerky or uneven, or if there is any visible sign of wear or damage to the strut, renewal is necessary.

9   If any doubt exists about the condition of the coil spring, remove the spring compressors and check the spring for distortion. Also measure the free length of the spring and compare the measurement with the figure given in the Specifications. Renew the spring if it is distorted or outside the specified length.

10  Begin assembly by fitting the rubber boot and spring aid to the strut.

11  Place the lower insulator ring in position, followed by the spring, ensuring that the end of the bottom coil locates in the step of the spring seat.

12  With the spring suitably compressed, withdraw the strut spindle as far as it will go and refit the upper insulator ring and spring seat. Make sure that the end of the spring upper coil locates in the step of the spring seat.

13  Place the bearing, with the small internal diameter downwards, over the spindle and refit the bearing retainer nut. Tighten the nut to the specified torque and remove the spring compressors.

14  Finally refit the bearing housing and the upper mounting rubber, large opening uppermost.

15  The strut can now be refitted to the car, as described in the previous Section.

## 7   Front lower suspension arm – removal and refitting

1   Apply the handbrake, chock the rear wheels and remove the front wheel trim. Slacken the wheel nuts. Jack up the front of the car, support it securely on axle stands and remove the roadwheel.

2   Undo and remove the nut and washer and then withdraw the clamp bolt securing the lower balljoint to the swivel hub.

3   Lever the lower suspension arm downwards to release the balljoint from the hub.

4   Undo and remove the three bolts securing the suspension arm rear mounting to the underbody (photo).

5   Undo and remove the nut and bolt securing the front mounting to the crossmember (photo).

6   Undo and remove the nut and washer securing the anti-roll bar mounting bush to the suspension arm.

7   Ease the front mounting out of the subframe, lift the suspension arm to disengage the anti-roll bar and manoeuvre the arm out from beneath the vehicle.

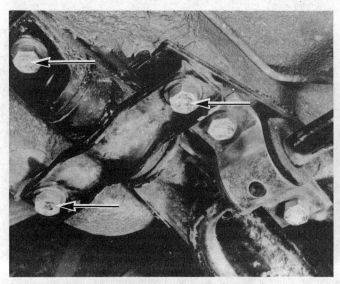

7.4 Front suspension arm rear mounting bolts (arrowed)

7.5 Front suspension arm front mounting retaining bolt (arrowed)

8   With the arm removed from the car, carefully examine the rubber mounting bushes for swelling or deterioration, the balljoint for slackness, and check for damage to the rubber boot. Renewal of the balljoint is described in Section 8. Renewal of the suspension arm mounting bushes may be carried out as follows.

9   Using a two-legged puller, or similar tool, draw the rear bush assembly off its spigot. The rear bush can be removed from its mounting bracket by pressing it out in a vice with the aid of tubes of suitable length and diameter. Removal of the front bush in the arm follows the same procedure.

10  To refit the new bushes first lubricate them with rubber grease and then press them fully into place using a vice. Slide the rear bush assembly onto its spigot on the suspension arm. Note that the rear bush housings are handed and must not be interchanged from side to side. When fitted the small hole in the bush housing must be toward the centre of the car.

11  Refitting the lower suspension arm is the reverse sequence to removal. Ensure that all retaining nuts and bolts are tightened to the specified torque.

## 8   Front lower suspension arm balljoint – removal and refitting

1   Remove the lower suspension arm, as described in the previous Section.

2   Drill out the heads of the three rivets securing the balljoint to the arm and drive out the rivets with a punch.

3   Withdraw the balljoint from the suspension arm.

4   Replacement balljoint kits are supplied with nuts and bolts to secure the joint, in place of the factory fitted rivets. Slide the new balljoint into position and fit the bolts so that the bolt heads are located above the arm. Screw on the nuts and tighten them to the specified torque.

5   The lower suspension arm can now be fitted to the car, as described in the previous Section.

## 9   Anti-roll bar – removal and refitting

**Note:** *For this operation the front suspension must be kept in a laden condition. If a vehicle lift or inspection pit are not available, it will be necessary to drive the front of the car up on ramps.*

1   Undo and remove the nuts and washers securing the anti-roll bar mounting bush, one each side, to the lower suspension arms.

2   Undo and remove the four bolts securing the anti-roll bar clamps to the subframe. Lift away the clamps, disengage the mounting bushes from the lower suspension arms and lower the anti-roll bar to the ground.

3    If the mounting blocks require renewal, slip them off the bar and place new blocks in position after lubricating liberally with rubber grease.

4    To renew the mounting bushes first measure and record the distance from the outer face of the bush to the end of the anti-roll bar. Now draw off the bushes using a two-legged puller.

5    Lubricate the new bushes with rubber grease and drive them onto the anti-roll bar using a hammer and tube of suitable diameter. Use the measurement recorded during removal as a setting dimension.

6    To refit the anti-roll bar to the car, first locate the bush assemblies in the lower suspension arms and fit the washers and retaining nuts finger tight.

7    Place the clamps over the mounting blocks and refit the retaining bolts, noting that the long bolts locate in the rear holes.

8    First tighten the clamp bolts to the specified torque, followed by the bush retaining nuts.

9    Lower the car to the ground (if applicable).

## 10  Rear hub assembly – removal and refitting

1    Chock the front wheels, remove the rear wheel trim and slacken the wheel nuts. Jack up the rear of the car and support it securely on axle stands. Remove the roadwheel and release the handbrake.

2    By judicious tapping and levering, extract the hub cap and withdraw the retaining split pin from the hub retaining nut.

3    Using a large socket and bar, undo and remove the hub retaining nut and flat washer. *Note that the left-hand nut has a left-hand thread and the right-hand nut has a conventional right-hand thread.* **Take care not to tip the car from the axle stands.** If the hub nuts are particularly tight, temporarily refit the roadwheel and lower the car to the ground. Slacken the nut in this more stable position and then raise and support the car before removing the nut.

4    Withdraw the hub and brake drum assembly from the stub axle (photo). If it is not possible to withdraw the hub due to the brake drum

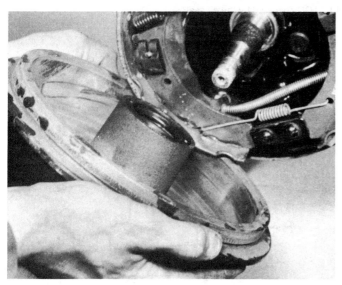

10.4 Withdraw the hub and brake drum from the stub axle

binding on the brake shoes, the following procedure should be adopted. Refer to Chapter 8, if necessary, and slacken off the handbrake cable at the cable adjuster. From the rear of the brake backplate, prise out the handbrake lever stop, which will allow the brake shoes to retract sufficiently for the hub assembly to be removed. It will, however, be necessary to remove the brake shoes and fit a new handbrake lever stop to the backplate.

5    Refitting the hub assembly is the reverse sequence to removal. Tighten the hub retaining nut to the specified torque and then align the next split pin hole. Always use a new split pin.

## 11  Rear hub bearings – removal and refitting

1    Remove the rear hub assembly from the car, as described in the previous Section.

2    With the hub on the bench, prise out the rear oil seal using a stout screwdriver or suitable flat bar.

3    Support the hub on blocks and, using a soft metal drift, tap out the two bearing inner races. Take care not to lose any of the balls which will be released from the ball cage as the inner races are removed.

4    Withdraw the spacer located between the two bearings and then drive the two outer races from the centre of the hub.

5    Wipe away any surplus grease and then thoroughly clean all the parts in paraffin or a suitable solvent. Dry with a lint-free rag.

6    Carefully examine the bearing inner and outer races, the ball cage and the balls for scoring, pitting or wear ridges. Renew both bearings as a set if any of these conditions are apparent. The hub oil seal must be renewed as it will have been damaged during removal. If the bearings are in a satisfactory condition, reassemble the balls and ball cage on the outer race, holding them in place with grease, and then press the inner race back into position (photo).

7    Before refitting the bearings, remove any burrs that may be present in the bore of the hub using a fine file or scraper.

8    Pack the bearings with a high melting-point grease and place the outer bearing in position with the word THRUST, or the markings stamped on the edge of the inner race, facing outwards. Drive the bearing into position using a tube of suitable diameter, or a drift, in contact with the bearing outer race. Take great care to keep the bearing square as it is installed, otherwise it will jam in the hub bore which could crack the outer race.

9    Dip a new oil seal in clean engine oil and position it over the outer bearing with its sealing lip facing inwards. Using a hammer and block of wood, tap the seal into the hub until it is flush with the end face of the hub.

10   Turn the hub over and lay the spacer over the installed outer bearing, with the smaller diameter of the spacer facing outwards (photo).

11   Fit the inner bearing and oil seal to the hub, using the same procedure as for the outer bearing, with the word THRUST, or the bearing markings, also facing out from the hub centre (photos).

12   The rear hub assembly can now be refitted to the car, as described in the previous Section.

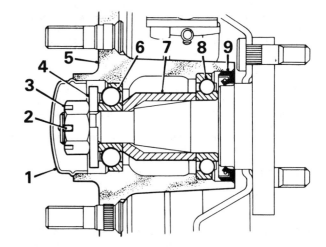

**Fig. 10.6 Cross-sectional view of the rear hub (Sec 11)**

| | |
|---|---|
| 1   Hub cap | 6   Outer bearing |
| 2   Split pin | 7   Spacer |
| 3   Hub retaining nut | 8   Inner bearing |
| 4   Flat washer | 9   Oil seal |
| 5   Hub and brake drum assembly | |

11.6 Reassemble the balls in the hub bearing cage then press the inner race into position

11.10 Fit the spacer between the bearings, smaller diameter outward

11.11A Fit the bearings with the word THRUST or the markings facing away from the hub centre

11.11B Fit the inner bearing to the hub ...

11.11C ... followed by the oil seal

## 12 Rear stub axle – removal and refitting

1   Remove the rear hub assembly, as described in Section 10.
2   Working under the car, disconnect the handbrake inner cable at the cable connector located beneath the rear axle transverse member.
3   Using circlip pliers, extract the circlip securing the handbrake outer cable to the bracket on the rear axle.
4   Using a brake hose clamp, or self-locking wrench with protected jaws, clamp the flexible brake hose located just in front of the rear axle. This will minimise brake fluid loss during subsequent operations.
5   Unscrew the brake pipe union nut at the rear of the wheel cylinder and carefully ease the pipe out of the cylinder. Plug the end of the pipe to prevent dirt entry.
6   Undo and remove the four nuts and spring washers securing the stub axle to the trailing arm and lift away the stub axle and brake backplate as an assembly.
7   Support the stub axle in a vice and drive out the four mounting studs using a soft metal drift. Take care not to damage the ends of the threads during this operation.
8   With the studs removed, separate the backplate from the stub axle.
9   Refitting the stub axle is the reverse sequence to removal. Tighten all retaining nuts and bolts to the specified torque and, on completion, bleed the brake hydraulic system, as described in Chapter 8; if suitable precautions were taken to minimise fluid loss, as described, it should only be necessary to bleed the relevant brake and not the entire system.

## 13 Rear suspension strut – removal and refitting

1   Chock the wheels, prise off the rear wheel trim and slacken the wheel nuts. Jack up the rear of the car and support it securely on axle stands. Remove the roadwheel.

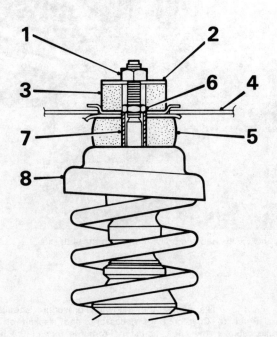

Fig. 10.7 Rear suspension strut upper mounting details (Sec 13)

| | |
|---|---|
| 1   Upper mounting nut | 5   Large mounting rubber |
| 2   Flat washer | 6   Spring retainer nut |
| 3   Small mounting rubber | 7   Spacer |
| 4   Rear wheel arch | 8   Spring seat |

2   As a safety precaution place a jack on suitable blocks beneath, and in contact with, the suspension trailing arm.

3   From inside the luggage compartment release the plastic cap from the strut upper mounting by rotating it whilst at the same time lifting upwards.

4   Engage a suitable small spanner on the flats of the strut spindle to prevent it turning and unscrew the upper mounting nut (photo). Lift off the flat washer and small mounting rubber.

5   From underneath the car, undo and remove the bolt securing the strut lower mounting to the rear suspension trailing arm (photo). Withdraw the strut from under the wheel arch and remove the large mounting rubber.

6   Refitting the suspension strut is the reverse sequence to removal. Ensure that the upper and lower mountings are tightened to the specified torque.

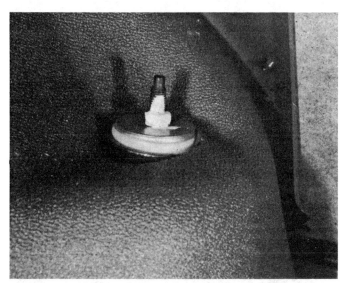

13.4 Rear suspension upper mounting, accessible from inside the luggage compartment

13.5 Rear suspension lower mounting bolt (arrowed)

## 14  Rear suspension strut – dismantling and reassembly

**Note**: *Before attempting to dismantle the rear suspension strut, a suitable tool to hold the coil spring in compression must be obtained. Adjustable coil spring compressors are readily available and are*

*recommended for this operation. Any attempt to dismantle the strut without such a tool is likely to result in damage or personal injury.*

1   Proceed by removing the rear suspension strut, as described in the previous Section.

2   Position the spring compressors on either side of the spring and compress the spring evenly until all the tension on the spring retainer nut is released.

3   With a spanner engaged with the flats of the strut spindle to stop it turning, unscrew the spring retainer nut. Lift off the spacer, spring seat and insulator ring.

4   The coil spring can now be removed, with the compressors still in position, if desired.

5   Carefully tap the spring aid retaining collar off the strut spindle then lift off the spring aid and the protective rubber boot.

6   With the strut completely dismantled, the components can be examined as follows.

7   Examine the strut body for signs of damage or corrosion, and for any trace of fluid leakage. Check the strut spindle for distortion, wear, pitting, or corrosion along its entire length. Test the operation of the strut, while holding it in an upright position, by moving the spindle through a full stroke and then through short strokes of 2 to 4 in (50 to 100 mm). In both cases the resistance felt should be smooth and continuous. If the resistance is jerky or uneven, or if there is any sign of wear or damage to the strut, renewal is necessary.

8   If the strut is in a satisfactory condition check the condition of the circlip on the strut spindle and, if it is in any way damaged or distorted, fit a new circlip.

9   If any doubt exists about the condition of the coil spring, remove the spring compressors and check the spring for distortion. Also measure the free length of the spring and compare the measurement with the figure given in the Specifications. Renew the spring if it is distorted or not of the specified length.

10  Begin reassembly by fitting the rubber boot to the strut, followed by the spring aid and retaining collar.

11  Place the spring in position, ensuring that the end of the bottom coil locates in the step of the retainer.

12  With the spring suitably compressed, withdraw the strut spindle as far as it will go and refit the insulator ring and spring seat, ensuring that the end of the coil locates in the step of the spring seat.

13  Refit the spacer, followed by the spring retainer nut. Tighten the nut to the specified torque.

14  The spring compressors can now be removed and the strut assembly refitted to the car, as described in the previous Section.

## 15  Rear axle – removal and refitting

1   Chock the wheels, prise off the rear wheel trim and slacken the wheel nuts. Jack up the rear of the car and support it securely on axle stands. Remove the roadwheel and release the handbrake.

2   Disconnect the two rear handbrake inner cables at the connectors located behind the rear axle transverse member.

3   Using two brake hose clamps, or self-locking wrenches with protected jaws, clamp the brake hydraulic flexible hoses, one located on each side of the car, adjacent to the rear axle front pivot bolts. This will minimise brake fluid loss during subsequent operations.

4   Undo and remove the two rear brake pipe unions at their connections with the flexible hoses. Plug or tape over the pipe and hose ends after disconnection to prevent dirt entry.

5   Using pliers extract the retaining clips and release the flexible hoses from the brackets on the rear axle.

6   Place a jack beneath one of the rear axle trailing arms and just take the weight of the axle.

7   Undo and remove the single bolt each side securing the rear suspension strut lower mountings to the rear axle trailing arms.

8   Undo and remove the nuts from the rear axle pivot bolts (one each side) (photo). Suitably support the axle beneath the transverse member and drift the pivot bolts from their locations.

9   Lever the axle out of its pivot mountings, lower it to the ground and withdraw it from under the car.

10  If necessary the axle can be completely dismantled by removing the rear hub assemblies and stub axles, as described in Sections 10 and 12 respectively.

11  If the axle pivot bushes require renewal, the old ones may be prised out and new ones inserted after lubricating them thoroughly in rubber grease.

15.8 Rear axle pivot bolt retaining nut (arrowed)

16.3 ... undo the nut ...

12  Refitting the axle to the car is a straightforward reversal of the removal sequence. Ensure that all nuts and bolts are tightened to the specified torque and, on completion, bleed the brake hydraulic system, as described in Chapter 8.

## 16  Steering wheel – removal and refitting

1    Set the front wheels in the straight-ahead position.
2    Ease off the steering wheel pad to provide access to the retaining nut (photo).
3    Using a socket and bar, undo and remove the retaining nut and lockwasher (photo).
4    Mark the steering wheel and inner column in relation to each other and withdraw the wheel from the inner column splines (photo). If it is tight, tap it upwards near the centre, using the palm of your hand. *Refit the steering wheel retaining nut two turns before doing this, for obvious reasons.*
5    Refitting is the reverse of removal, but align the previously made marks and tighten the retaining nut to the specified torque. Note that there is a small arrow stamped on the striker bush of the multi-function switch and this must point towards the direction indicator switch when fitting the steering wheel.

16.4 ... then pull the steering wheel off the inner column

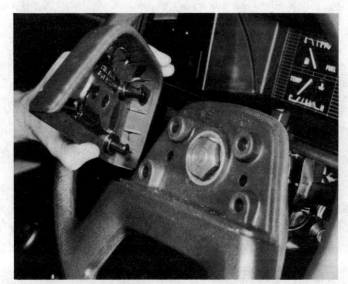

16.2 Ease off the steering wheel pad ...

## 17  Steering column assembly – removal, overhaul and refitting

1    Disconnect the battery negative terminal.
2    Remove the steering wheel, as described in the previous Section.
3    Undo and remove the screws securing the two steering column cowl halves to the column and lift off both cowls.
4    Disconnect the wiring multi-plugs from the steering column multi-function switches and from the ignition switch. Release any cable ties securing the wiring harness to the column.
5    From under the facia, remove the steering column cover and seal and then, to assist refitting, mark the inner column in relation to the intermediate shaft.
6    Undo and remove the intermediate shaft clamp bolt and nut securing the intermediate shaft to the inner column.
7    Undo and remove the upper and lower steering column mounting bolts, lift the column off the intermediate shaft and remove the assembly from the car.
8    Slacken the clamp screw and lift the multi-function switch off the steering column. Release the wiring from the clip.
9    Carefully support the outer column in a vice and withdraw the inner column from the top of the outer column.

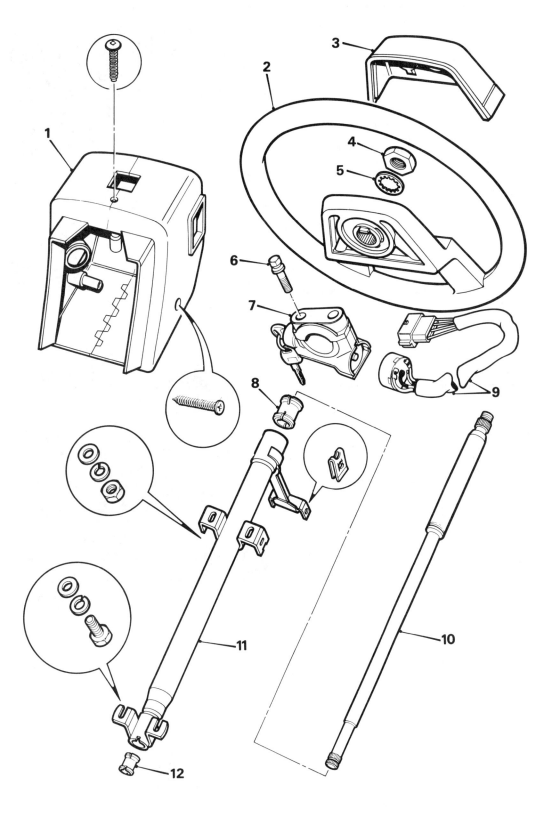

**Fig. 10.8 Exploded view of the steering column assembly (Sec 17)**

| | | |
|---|---|---|
| 1 | Steering column cowl | |
| 2 | Steering wheel | |
| 3 | Steering wheel pad | |
| 4 | Steering wheel retaining nut | |

| | |
|---|---|
| 5 | Lockwasher |
| 6 | Shear bolt |
| 7 | Steering lock/ignition switch |
| 8 | Top bush |

| | |
|---|---|
| 9 | Ignition switch |
| 10 | Inner column |
| 11 | Outer column |
| 12 | Bottom bush |

10  Lift out the top bush, bend back the retaining tag and extract the bottom bush.

11  Clean the parts in paraffin, or a suitable solvent, and wipe dry. Examine the bushes for wear and renew, if necessary.

12  Reassembly and refitting is a reverse of removal and dismantling, bearing in mind the following points:

(a)  Smear the outer surfaces and inner grooves of the bushes with graphite grease before fitting

(b)  Ensure that the marks made on the inner column and intermediate shaft are aligned when fitting the column

(c)  Adjust the position of the outer column so that it is 3.25 in (82.6 mm) below the top of the inner column before tightening the retaining bolts

(d)  Position the arrow on the striker bush of the multi-function switch so that it points towards the direction indicator switch before fitting the steering wheel

(e)  Tighten all retaining nuts and bolts to the specified torque

20.3A Using a separator tool to release the tie-rod outer balljoint

## 18  Steering column intermediate shaft – removal and refitting

1  Disconnect the battery negative terminal.

2  Apply the handbrake, chock the rear wheels, jack up the front of the car and support it securely on axle stands.

3  From inside the car, pull back the carpet and unscrew the nut and bolt securing the intermediate shaft lower universal joint to the pinion shaft.

4  From under the facia remove the steering column cover and seal.

5  Unscrew the nut and bolt securing the intermediate shaft upper universal joint to the inner column.

6  Release the intermediate shaft from the pinion and inner column, and withdraw it from the car.

7  Before refitting it is first necessary to centralize the steering gear. To do this move the rubber band on the rack housing to one side to expose the centralizing hole or, if power-assisted steering is fitted, undo the centralizing plug on the rack housing. Turn the pinion shaft or move the roadwheels until a corresponding hole in the rack is in alignment. Insert a 0.24 in (6.0 mm) diameter bolt or drill bit (0.187 in/4.7 mm diameter for power-assisted steering) into the hole to hold the steering gear in the central position.

8  From inside the car, engage the intermediate shaft upper universal joint with the inner column. Refit the clamp bolt and nut and tighten to the specified torque.

9  Position the steering wheel in the straight-ahead position and engage the intermediate shaft lower universal joint with the pinion shaft. Refit and tighten the clamp nut and bolt.

10  Refit the steering column cover and seal and place the carpet in position.

11  Remove the centralizing bolt or drill from the rack housing, refit the rubber band or centralizing plug over the hole and lower the car to the ground. Reconnect the battery.

20.3B Withdraw the balljoint from the steering arm after releasing the taper

## 19  Steering column lock/ignition switch – removal and refitting

1  Remove the steering column, as described in Section 17.

2  With the column assembly on the bench, drill out the shear bolt heads and remove the clamp plate and lock/ignition switch from the outer column.

3  Locate the new lock body centrally over the slot in the outer column. Lightly bolt the clamp plate into position, but take care not to shear the bolt heads.

4  Refit the steering column, as described in Section 17, but before fitting the cowls check that the lock and ignition switch operate correctly.

5  Tighten the shear bolts until the heads break off, and then refit the cowls.

## 20  Tie-rod outer balljoint – removal and refitting

1  Apply the handbrake, chock the rear wheels, prise off the front wheel trim and slacken the wheel nuts. Jack up the front of the car and support it securely on axle stands. Remove the roadwheel.

2  Using a suitable spanner, slacken the balljoint locknut on the tie-rod by a quarter of a turn.

3  Undo and remove the locknut securing the balljoint to the steering arm and then release the tapered ball-pin using a balljoint separator tool (photos).

4  Hold the tie-rod with a pair of grips or a self-locking wrench and unscrew the balljoint.

5  Screw the new balljoint onto the tie-rod until it contacts the locknut.

6  Insert the balljoint into the steering arm, refit the locknut and tighten it to the specified torque. If the ball-pin slips in the steering arm taper as the locknut is tightened, lever down on the top of the balljoint using a stout bar. This should lock the bolt-pin sufficiently to allow the nut to be tightened.

7  Finally, tighten the locknut on the tie-rod securely against the balljoint. After fitting check the front wheel alignment as described in Section 31.

## 21  Steering rack rubber gaiter – removal and refitting

1  Remove the tie-rod outer balljoint, as described in the previous Section.

2  Count and record the number of exposed threads from the rear

face of the locknut to the inner end of the tie-rod, then unscrew the locknut.

3    Release the retaining wire or unscrew the securing clip screws and slide the gaiter off the rack housing and tie-rod.

4    Lubricate a new rubber gaiter with rubber grease and position it over the housing and tie-rod.

5    Using new clips, or two or three turns of soft iron wire, secure the gaiter in position.

6    Screw the locknut back onto the tie-rod and position it in exactly the same place as noted during removal.

7    Refit the outer balljoint, as described in the previous Section.

## 22 Rack and pinion steering gear (manual steering) – removal and refitting

1    Disconnect the battery negative terminal.

2    Apply the handbrake, chock the rear wheels and remove the trim from both front roadwheels. Slacken the wheel nuts before jacking up the front of the car and then support it securely on axle stands. Remove the roadwheels.

3    Unscrew the nuts securing the tie-rod outer balljoints to the steering arms and release the tapered ballpins using a balljoint separator tool.

4    From inside the car pull back the carpet and unscrew the nut and bolt securing the intermediate shaft lower universal joint to the pinion shaft.

5    Using a sharp knife make diagonal cuts in the sound-deadening material around the pinion shaft to provide access to the pinion cover plate. The pinion cover plate can now be removed.

6    Undo and remove the two bolts and washers securing the pinion end of the steering gear to the bulkhead. Now remove the two bolts and washers, clamp plate and plastic seating securing the other end of the steering gear to the bulkhead.

7    Slacken the front crossmember support strut bolts on the driver's side. Remove the two short bolts, but leave the long bolt in position. Pivot the support strut outward at the top.

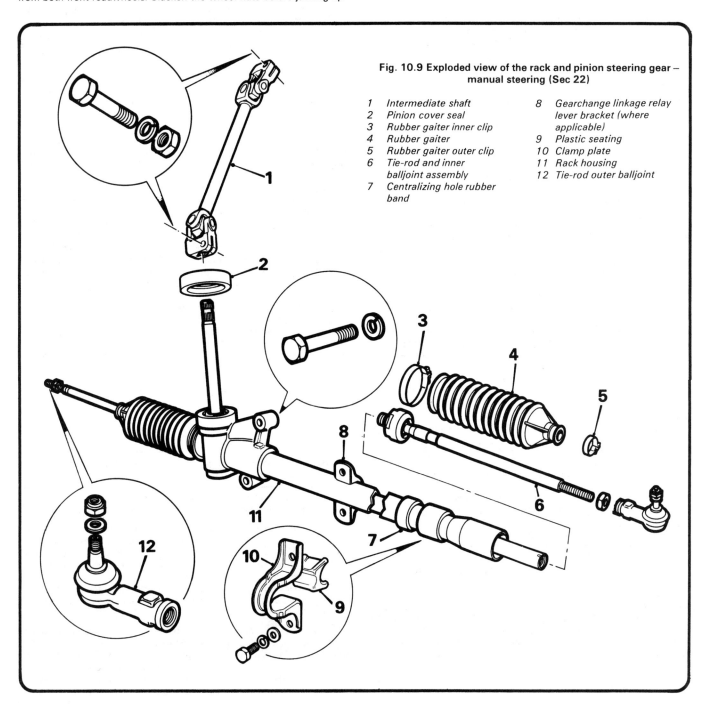

**Fig. 10.9 Exploded view of the rack and pinion steering gear – manual steering (Sec 22)**

1    Intermediate shaft
2    Pinion cover seal
3    Rubber gaiter inner clip
4    Rubber gaiter
5    Rubber gaiter outer clip
6    Tie-rod and inner balljoint assembly
7    Centralizing hole rubber band
8    Gearchange linkage relay lever bracket (where applicable)
9    Plastic seating
10    Clamp plate
11    Rack housing
12    Tie-rod outer balljoint

8  Ease the pinion shaft out of the intermediate shaft splines, manoeuvre the steering gear as necessary and withdraw the assembly out through the driver's side wheel arch.

9  To refit the steering gear, manoeuvre it into position through the wheel arch and locate the pinion in the bulkhead aperture.

10  Fit the retaining bolts, finger tight, to the pinion end of the steering gear. At the other end place the plastic seating and clamp in position, refit the retaining bolts and tighten them to the specified torque. Now tighten the pinion end retaining bolts to the specified torque.

11  Refit the two crossmember support strut bolts, but don't tighten the bolts fully until the full weight of the car is on its wheels.

12  Move the rubber band on the rack housing to one side to expose the centralizing hole. Move the rack as necessary, by turning the pinion, until a 0.24 in (6 mm) drill or bolt can be inserted through the hole in the housing and into the corresponding hole in the rack.

13  From inside the car, refit the pinion cover plate, position the steering wheel in the straight-ahead position and refit the intermediate shaft universal joint to the pinion shaft. Refit and tighten the clamp bolt and nut.

14  Reposition the sound-deadening material around the pinion shaft and tape over the diagonal cuts. Refit the carpets.

15  Remove the centralizing bolt or drill, and cover the hole with the rubber band.

16  Refit the tie-rod outer balljoints to the steering arms and secure with the retaining nuts tightened to the specified torque.

17  Refit the roadwheels and wheel nuts, lower the car to the ground and fully tighten the wheel nuts. Tighten the crossmember support strut bolts and refit the wheel trim.

18  Check and if necessary adjust the front wheel toe setting as described in Section 31.

## 23  Rack and pinion steering gear (manual steering) – dismantling and reassembly

**Note**: *The manual steering gear fitted to Montego models cannot be fully dismantled for repair or overhaul and, with one exception, individual parts are not available separately. Should repair of the steering gear be necessary, due to wear or damage, a complete assembly must be obtained. It is, however, possible to renew the tie-rods individually and this Section describes the procedure.*

1  Begin by removing the steering gear from the car, as described in the previous Section.

2  Measure the distance between the ball-pin centres of the two outer balljoints and record this figure as an aid to reassembly.

3  Slacken the tie-rod outer balljoint locknut and unscrew the balljoint, followed by the locknut.

4  Remove the wire or retaining clips securing the rubber gaiter to the rack housing. Slide the gaiter off the rack housing and tie-rod.

5  Turn the pinion to extend the rack fully and support the exposed portion of rack between protected vice jaws.

6  Unscrew the ballhousing and withdraw the tie-rod and ballhousing from the rack.

7  Repeat paragraphs 3 to 6 inclusive for the other tie-rod, if required.

8  Liberally lubricate the tie-rod and ballhousing, using the specified lubricant.

9  Refit the ballhousing and tie-rod to the rack and tighten the housing fully. Secure the ballhousing by staking its edge into the groove in the rack.

10  Liberally lubricate the rack teeth with the specified lubricant and refit the rubber gaiter. Secure the gaiter with the retaining clips or two to three turns of soft iron wire.

11  Refit the outer balljoint locknut and balljoint to the tie-rod, but do not tighten the locknut at this stage.

12  Repeat paragraphs 8 to 11 inclusive for the other tie-rod, if this was also removed.

13  Position the two balljoints so that the dimension between the ballpin centres is as recorded during dismantling with an equal number of exposed threads visible on each tie-rod. Now secure the balljoints with the locknuts

14  The steering gear can now be refitted to the car, as described in the previous Section. It will, however, be necessary to adjust the toe setting after refitting, as described in Section 31.

## 24  Rack and pinion steering gear (power-assisted steering) removal and refitting

1  Refer to Section 9 and remove the anti-roll bar.

2  Jack up the front of the car and support it on axle stands.

3  Undo the nuts securing the tie-rod outer balljoints to the steering arms and release the tapered ball-pins using a balljoint separator tool.

4  From inside the car pull back the carpet then undo the nut and bolt securing the intermediate shaft lower universal joint to the pinion shaft.

5  Undo the front crossmember support strut bolts on the driver's side. Remove the bolts and withdraw the support strut from under the car.

6  Wipe clean the area around the fluid hose unions on the pinion housing.

7  Place a container beneath the housing to catch any spilled fluid then undo the feed and return hose union nuts. Ease the pipe unions out of the pinion housing and plug or tape over their ends to prevent further loss of fluid. Plug or tape over the orifices in the pinion housing to prevent dirt entry and further fluid spillage.

8  Undo the steering gear mounting bolts, recover the mounting rubbers and manoeuvre the unit out through the driver's side wheel arch.

9  To refit the steering gear, ensure that the pinion cover seal is in place then manoeuvre the unit into position through the wheel arch.

10  Refit the steering gear mounting bolts and rubbers, but tighten the bolts finger tight only at this stage.

11  Remove the plug over the centralizing hole in the centre of the rack housing and insert a 0.187 in (4.7 mm) diameter dowel rod or drill bit in the hole. Move the rack as necessary by turning the pinion until the dowel or drill bit engages fully with the corresponding hole in the rack.

12  Tighten the steering gear mounting bolts to the specified torque.

13  Refit the fluid feed and return hose unions to the pinion housing and securely tighten the union nuts.

14  Refit the crossmember support strut, but tighten the retaining bolts finger tight only at this stage.

15  Position the steering wheel in the straight-ahead position and engage the intermediate shaft universal joint over the pinion shaft. Refit the clamp nut and bolt and tighten them to the specified torque.

16  Remove the centralizing rod or drill bit from the rack and refit the plug.

17  Refit the tie-rod outer balljoints to the steering arms and secure with the retaining nuts tightened to the specified torque.

18  Refit the roadwheels and lower the car to the ground. Tighten the wheel nuts and refit the trim.

19  Tighten the crossmember support strut retaining bolts to the specified torque.

20  Fill and bleed the steering gear as described in Section 30, then refit the anti-roll bar as described in Section 9.

21  Check and if necessary adjust the front wheel toe setting as described in Section 31.

## 25  Rack and pinion steering gear (power-assisted steering) overhaul – general precautions

1  It is not generally recommended that the power-assisted steering gear be dismantled by the DIY mechanic for the following reasons:

(a)  *The only parts available separately are the valve and pinion assembly, the internal seals and washers, the tie-rods and the rubber gaiters. If the rack, rack housing or any of the other mechanical components are worn or damaged a complete steering gear assembly will be required*

(b)  *Special BL tools are needed for certain operations and unless these can be obtained no attempt should be made to overhaul the unit*

2  If overhaul is to be attempted a perfectly clean working area must be available as cleanliness during dismantling and reassembly is of the utmost importance. Read through the following Section fully to familiarize yourself with the procedure, the special tools required, and their use. Make sure that all the tools are available and that all components likely to be required can be obtained from your BL dealer.

## 26 Rack and pinion steering gear (power-assisted steering) — dismantling and reassembly

*Refer to Section 25 before proceeding*

1  Remove the steering gear from the car as described in Section 24.
2  Clean the unit externally using a rag moistened with paraffin or a suitable solvent. Dry with a lint-free cloth.
3  Remove the seal and tube assembly from the top of the pinion housing then support the steering gear in a vice by means of the pinion end mounting lug. Do not grip the rack housing with the vice.
4  Slacken the locknuts then unscrew the tie-rod outer balljoints from the tie-rods, followed by the locknuts.
5  Release the wire retaining clips and slide off the two rubber gaiters. Have a container handy to catch any spilled fluid as the gaiters are released.
6  Using tool 18G 1440, slacken the tie-rod inner balljoints, but do not remove them from the rack at this stage.
7  Extract the end cap located at the base of the pinion housing then undo the locknut from the end of the pinion.
8  Now remove the two tie-rod inner balljoints complete with tie-rods.
9  Prise out the dust shield from the top of the pinion housing then extract the circlip located just below it.
10  Using a suitable box spanner undo the yoke cover at the rear of the pinion housing and withdraw the spring and yoke.
11  Remove the steering gear from the vice and support the pinion housing. Using a soft metal drift inserted through the base of the pinion housing, tap out the pinion assembly. Recover the fluid seal and bearing from the pinion.
12  Extract the circlip securing the pinion lower bearing in the housing. Using a drift inserted through the top of the housing, drive out the bearing.
13  Secure the pinion end mounting lug in the vice once more and extract the rack travel stop locking wire.
14  Withdraw the rack and support bush assembly from the passenger's side end of the rack housing. Have a container handy to catch the spilled fluid as the rack is removed.
15  Remove the travel stop and support bush from the rack, and the piston ring and O-ring from the rack piston.
16  Using a soft metal drift inserted through the pinion end of the rack housing, tap out the rack fluid seal and support ring.
17  Remove the steering gear from the vice, support the pinion housing and drift out the pinion bush and fluid seal.
18  Inspect all the parts for scoring, wear, or damage, paying particular attention to the rack and pinion teeth, the tie-rod inner balljoints and the pinion bearings. Renew all parts as necessary or, if the rack or pinion are worn, obtain a complete steering gear assembly. If the original unit is to be reassembled, obtain a new set of seals as a matter of course.
19  Using tool 18G 1442 drift the pinion bush, followed by the fluid seal into the pinion housing.
20  Fit the pinion lower bearing using tool 18G 1443 then secure the bearing with the circlip.
21  Using tool 18G 1448 carefully fit the O-ring and piston ring to the rack piston.
22  Fit the protector sleeve, tool 18G 1446 to the rack housing fluid seal and support ring then slide this assembly over the rack teeth and onto the plain portion of the rack. Note that the fluid seal faces the rack piston. Hold these parts in position on the rack and remove the protector sleeve.
23  Carefully slide the rack into the rack housing and push it towards the pinion end to locate the support ring and seal against the housing end.
24  Turn the rack so that the teeth are in position to accept the pinion. Insert the pinion from the top of the housing and then refit the locknut, finger tight only at this stage.
25  Using tool 18G 1444 fit the pinion upper bearing with the markings on the bearing uppermost.
26  Place the protector sleeve, tool 18G 1281, over the pinion splines and locate the fluid seal in the pinion housing. Remove the protector sleeve then use tool 18G 1444 to fit the seal fully into the housing.
27  Secure the bearing and fluid seal with the circlip then using the protector sleeve again, fit the dust shield to the top of the pinion housing. Remove the protector sleeve.
28  Locate the protector sleeve, 18G 1446 in the rack support bush

assembly and fit the bush to the rack housing with the fluid seal facing away from the pinion. Remove the protector sleeve.
29  Insert the rack travel stop into the rack housing with the groove in the stop aligned with the slot in the housing.
30  Insert the hooked end of the locking wire into the hole in the rack and turn the travel stop until the complete length of locking wire is drawn into the rack housing.
31  Using tool 18G 1440 refit the tie-rods and inner balljoints to the ends of the rack and tighten them securely. Using a small punch, stake the edge of the inner balljoint housings into the grooves in the rack.
32  Hold the pinion splines and tighten the locknut at the base of the pinion to the specified torque. Refit the end cap to the pinion housing.
33  Refit the yoke and spring then apply a thread locking compound to the threads of the yoke cover. Fit the cover to the pinion housing and, using a suitable box spanner tighten the cover to between 30 and 40 lbf in (3.4 and 4.5 Nm). Using tools 18G 1395 and MS103, check that the minimum torque needed to turn the pinion is 12 lbf in (1.3 Nm). Now back off the yoke cover by 45° to 50° and check that the maximum torque required to turn the pinion is 15 lbf in (1.7 Nm). If the pinion torque is excessive the yoke cover may be backed off by a further 15°. When the correct turning torque is achieved lock the yoke cover by peening it in three places.
34  Inject 50 cc of the specified grease into the pinion end of the rack housing, spreading it liberally over the rack teeth. Smear a coating of the grease on the portion of the rack extending beyond the rack travel stop, and also on the tie-rods where the small end of the rubber gaiters contact.
35  Slide the two rubber gaiters into place over the tie-rods and rack housing, and secure the gaiters with the wire retaining clips.
36  Screw the tie-rod outer balljoint locknuts onto the tie-rods, followed by the two outer balljoints. Position the balljoints so that their ballpin centres are 50.47 in (1281 mm) apart with an equal number of exposed threads visible on each tie-rod. Secure the balljoints by tightening the locknuts.
37  Fit the seal and tube assembly to the top of the pinion housing.
38  The steering gear can now be refitted to the car as described in Section 24. Adjust the toe-setting after refitting as described in Section 31.

## 27 Power-assisted steering pump drivebelt — removal, refitting and adjustment

1  Slacken the two pivot mounting bolts securing the pump mounting bracket to the engine bracket.
2  Slacken the two pump bracket adjustment bolts, push the pump in towards the engine and slip the drivebelt off the two pulleys.
3  Inspect the belt for cracks or fraying and renew it if there is any sign of deterioration.
4  Slip the belt into position ensuring that it seats fully into the vee of the pulleys. Move the pump away from the engine until, using moderate finger pressure at the mid point of the belt, it is possible to deflect it by approximately 0.5 in (13.0 mm).
5  Hold the pump in this position and tighten the adjustment and pivot mounting bolts. **Note**: *do not lever or apply any force to the pump reservoir when adjusting the drivebelt tension.*

## 28 Power-assisted steering pump — removal and refitting

1  Remove the drivebelt as described in the previous Section.
2  Unscrew the filler cap and remove it from the pump reservoir.
3  Hold the high pressure pipe adaptor at the rear of the pump with a spanner then using a second spanner, undo the high pressure pipe union nut.
4  With a suitable container handy, slacken the low pressure hose clip, disconnect the hose and allow the fluid to drain into the container. After draining, plug or tape over the pipe, hose and pump connections to prevent dirt entry. Refit the filler cap.
5  Undo and remove the two pump bracket adjustment bolts and the two pivot mounting bolts. Withdraw the pump complete with mounting bracket off the engine bracket. Note that the pump is a sealed unit and cannot be dismantled for overhaul or repair. In the event of pump failure a new unit must be obtained. Note also that the mounting bracket is an integral part of the pump assembly and must not be removed.

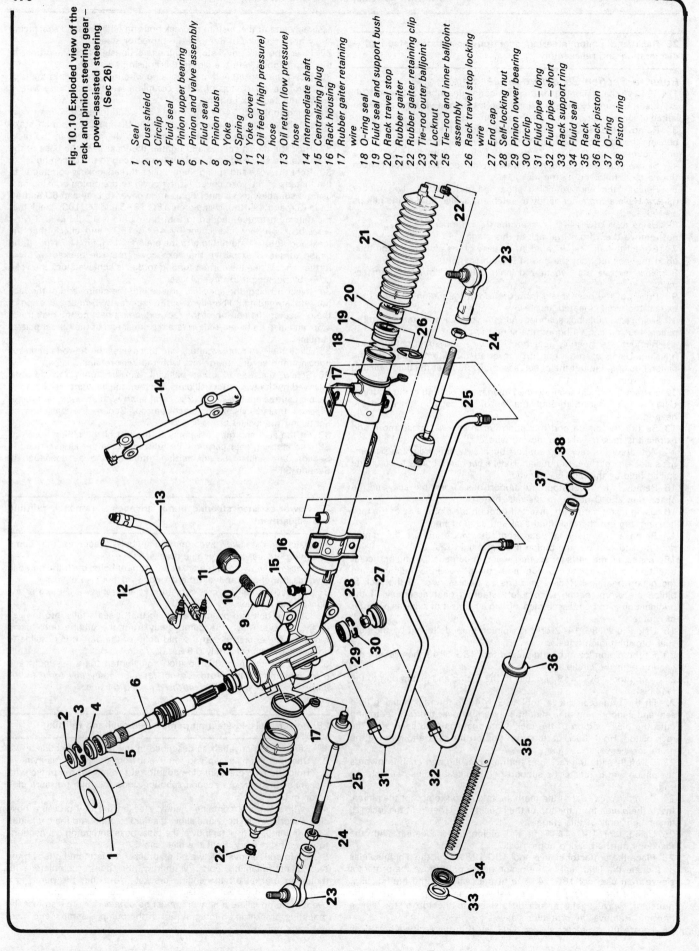

**Fig. 10.10 Exploded view of the rack and pinion steering gear – power-assisted steering (Sec 26)**

1 Seal
2 Dust shield
3 Circlip
4 Fluid seal
5 Pinion upper bearing
6 Pinion and valve assembly
7 Fluid seal
8 Pinion bush
9 Yoke
10 Spring
11 Yoke cover
12 Oil feed (high pressure) hose
13 Oil return (low pressure) hose
14 Intermediate shaft
15 Centralizing plug
16 Rack housing
17 Rubber gaiter retaining wire
18 O-ring seal
19 Fluid seal and support bush
20 Rack travel stop
21 Rubber gaiter
22 Rubber gaiter retaining clip
23 Tie-rod outer balljoint
24 Locknut
25 Tie-rod and inner balljoint assembly
26 Rack travel stop locking wire
27 End cap
28 Self-locking nut
29 Pinion lower bearing
30 Circlip
31 Fluid pipe – long
32 Fluid pipe – short
33 Rack support ring
34 Fluid seal
35 Rack
36 Rack piston
37 O-ring
38 Piston ring

**Fig. 10.11 Power-assisted steering pump mounting and adjustment points (Secs 27 and 28)**

2  *High pressure pipe adaptor*
3  *High pressure pipe union nut*
4  *Low pressure hose clip*
5  *Pump bracket pivot mounting bolts*
6  *Pump bracket adjustment bolts*

6   To refit the pump place it in position and fit the pivot mounting and adjustment bolts moderately tight only at this stage.
7   Refit the low pressure hose and high pressure pipe ensuring that the unions are tight.
8   Refit and adjust the drivebelt as described in Section 27, then refill and bleed the steering gear as described in Section 30.

## 29 Power-assisted steering fluid – level checking

1   The fluid level in the steering pump reservoir should be checked when the system is cold and with the engine stopped.
2   Wipe clean the filler cap and the filler neck of the pump reservoir. Unscrew the cap, wipe the dipstick then refit the cap fully. Unscrew the cap once more and note the fluid level on the dipstick which should be between the MAX and MIN marks.
3   If topping up is necessary, add the specified type and grade of fluid through the filler neck to bring the level up to the MAX mark on the dipstick. Take care not to overfill the system. Refit the cap after checking or topping up the fluid.
4   If frequent topping up is necessary this would indicate a leak somewhere in the system and a careful inspection should be made of all pipes, hoses and unions. If no leaks can be found, check the rubber gaiters on the steering gear for any sign of wetness or obvious signs of fluid inside them which would indicate failed seals within the steering gear. Renew any pipe or hose unions as necessary or rectify any internal leakage with reference to Section 26.

## 30 Power-assisted steering fluid – bleeding

1   If any of the components of the power-assisted steering system have been renewed, or if any of the fluid hoses or pipes have been disconnected, bleeding should be carried out as follows to remove all air from the system.
2   Before starting ensure that all hoses and pipe unions are tight and that the pump reservoir is filled to the MAX mark on the dipstick with

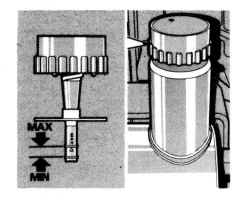

**Fig. 10.12 Fluid level markings on power-assisted steering pump filler cap (Sec 29)**

fresh fluid. Never re-use fluid that has been previously drained from the system.
3   Position the roadwheels in the straight-ahead position then refer to Section 27 and remove the pump drivebelt.
4   Turn the pump pulley by hand in the normal direction of rotation through at least ten revolutions to prime the system.
5   Refit and tension the drivebelt as described in Section 27.
6   Check and if necessary top up the fluid to the MAX mark on the dipstick and then start the engine.
7   Turn the steering wheel to full lock then return it to the straight-ahead position. Switch off the engine, check and if necessary top up the fluid.
8   Start the engine again and turn the steering wheel to the opposite lock, returning it once more to the straight-ahead position. Switch off and check the fluid, topping-up if necessary.
9   Start the engine again and turn the steering wheel from lock to lock several times to exhaust any air remaining in the system.
10  Switch off and finally top up the fluid level if necessary.

## 31 Front wheel alignment and steering angles

1   Accurate front wheel alignment is essential to provide positive steering and prevent excessive tyre wear. Before considering the steering/suspension geometry, check that the tyres are correctly inflated, the front wheels are not buckled and the steering linkage and suspension joints are in good order, without slackness or wear.
2   Wheel alignment consists of four factors: Camber is the angle at which the front wheels are set from the vertical when viewed from the front of the car. 'Positive camber' is the amount (in degrees) that the wheels are tilted outward at the top from the vertical. Castor is the angle between the steering axis and a vertical line when viewed from each side of the car. 'Positive castor' is when the steering axis is inclined rearward. Steering axis inclination is the angle (when viewed from the front of the car) between the vertical and an imaginary line drawn between the suspension strut upper mounting and the lower suspension arm balljoint. Toe setting is the amount by which the distance between the front inside edges of the roadwheels (measured at hub height) differs from the diametrically opposite distance measured between the rear inside edges of the front roadwheels.
3   With the exception of the toe setting, all other steering angles on Montego models are set during manufacture and no adjustment is possible. It can be assumed, therefore, that unless the car has suffered accident damage all the preset steering angles will be correct. Should there be some doubt about their accuracy it will be necessary to seek the help of a BL dealer, as special gauges are needed to check the steering angles.
4   Two methods are available to the home mechanic for checking the toe setting. One method is to use a gauge to measure the distance between the front and rear inside edges of the roadwheels. The other method is to use a scuff plate in which each front wheel is rolled across a movable plate which records any deviation, or scuff, of the tyre from the straight-ahead position as it moves across the plate.

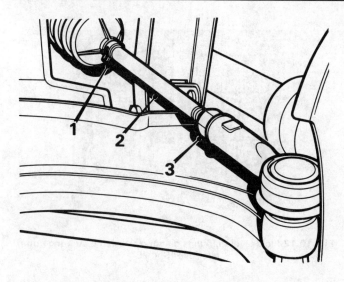

**Fig. 10.13 Toe setting adjustment (Sec 31)**

*1   Rubber gaiter clip*          *3   Outer balljoint locknut*
*2   Tie-rod*

Relatively inexpensive equipment of both types is available from accessory outlets to enable these checks, and subsequent adjustments, to be carried out at home.

5    If, after checking the toe setting using whichever method is preferable, it is found that adjustment is necessary, proceed as follows.

6    Turn the steering wheel onto full left lock and record the number of exposed threads on the right-hand steering tie-rod. Now turn the steering onto full right lock and record the number of threads on the left-hand tie-rod. If there are the same number of threads visible on both tie-rods then subsequent adjustments can be made equally on both sides. If there are more threads visible on one side than the other it will be necessary to compensate for this during adjustment. *After adjustment there must be the same number of threads visible on each tie-rod. This is most important.*

7    To alter the toe setting slacken the locknut on the tie-rod and turn the rod using a self-grip wrench to achieve the desired setting. When viewed from the side of the car, turning the tie-rod clockwise will increase the toe-in, turning it anti-clockwise will increase the toe-out.

Only turn the tie-rods by a quarter of a turn each time and then recheck the setting using the gauges, or scuff plate.

8    After adjustment tighten the locknuts and reposition the steering gear rubber gaiters, if necessary, to remove any twist caused by turning the tie-rods.

## 32  Wheels and tyres

1    Check the tyre pressures regularly (see Routine Maintenance) when the tyres are cold, and periodically check the tread depth using a depth gauge.

2    Frequently inspect the tyre walls and treads for damage and pick out any stones which have become trapped in the tread pattern.

3    In the interests of extending tread life, the wheels and tyres can be moved between front and rear on the same side of the car and the spare incorporated in the rotational pattern. If the wheels have previously been balanced on the car it will be necessary to have them rebalanced after rotation.

4    Never mix tyres of different construction, or very dissimilar tread patterns. This is particularly important on models equipped with TD (Total Deflation) wheels and tyres in which two grooves running circumferentially around the wheel retain the tyre in place in the event of a puncture. TD tyres and wheels are not interchangeable with wheels and tyres of different design and construction, and replacements must be of identical specification to those originally fitted to the car.

5    Always keep the roadwheel nuts tightened to the specified torque and if the wheel stud holes become elongated or flattened, renew the wheel.

6    Occasionally clean the inner faces of the roadwheels and, if there is any sign of rust or corrosion, paint them with metal preservative paint.

7    Before removing a roadwheel which has been balanced on the car, always mark one wheel stud hole and the hub, so that the roadwheel may be refitted in the same position, to maintain the balance.

8    Should unexpected excessive wear be noticed on any of the tyres its cause must be identified and rectified immediately. Generally speaking the wear pattern can be used as a guide to the cause. If a tyre is worn excessively in the centre of the tread face, but not on the edges, over inflation is indicated. Similarly if the edges are worn, but not the centre, this may be due to under inflation. If both the front or rear tyres are wearing on their inside or outside edges, this is likely to be due to incorrect toe setting. If only one tyre is exhibiting this tendency then there may be a problem with the steering geometry, a worn steering or suspension component, or a faulty tyre. Wheel and tyre imbalance is indicated by irregular and uneven wear patches appearing periodically around the tread face.

## 33 Fault diagnosis – suspension and steering

**Note**: *Before diagnosing steering or suspension faults, be sure that the trouble is not due to incorrect tyre pressures, a mixture of tyre types or binding brakes*

| Symptom | Reason(s) |
|---|---|
| Vehicle pulls to one side | Incorrect wheel alignment<br>Wear in suspension or steering components<br>Accident damage to steering or suspension components<br>Faulty tyre |
| Steering stiff or heavy | Lack of steering gear lubricant<br>Seized balljoint<br>Incorrect wheel alignment<br>Steering rack or column bent<br>Broken or incorrectly adjusted pump drivebelt (power-assisted steering)<br>Power-assisted steering pump faulty |
| Excessive play in steering | Wear in steering or suspension components<br>Wear in intermediate shaft universal joints<br>Worn rack and pinion assembly |
| Wheel wobble and vibration | Roadwheels out of balance<br>Roadwheels buckled or distorted<br>Faulty or damaged tyre<br>Worn shock absorber (suspension struts)<br>Worn steering or suspension joints<br>Wheel nuts loose<br>See also Fault diagnosis – driveshafts (Chapter 7) |
| Tyre wear uneven | Incorrect wheel alignment<br>Worn steering or suspension components<br>Roadwheels out of balance<br>Accident damage |
| Excessive pitching and rolling on corners and during braking | Worn or faulty shock absorbers (suspension struts)<br>Weak or broken coil spring<br>Worn or broken anti-roll bar mountings |
| Excessive noise from steering gear (power-assisted steering) | Drivebelt slipping<br>Insufficient fluid in pump reservoir<br>Worn or faulty power-assisted steering pump |
| Jerky steering, intermittent loss of power assistance (power-assisted steering) | Drivebelt slipping<br>Faulty power-assisted steering pump<br>Kinked or restricted fluid hose<br>Insufficient fluid in pump reservoir<br>Fluid leakage<br>Faulty rack and pinion assembly |

**Fault diagnosis appears overleaf**

# Chapter 11 Bodywork

*For modifications, and information applicable to later models, see Supplement at end of manual*

## Contents

## Specifications

### Torque wrench settings

| | lbf ft | Nm |
|---|---|---|
| Seat belt mountings | 24 | 32 |
| Bumper bracket retaining bolts | 18 | 25 |
| Front body panel retaining bolts | 6 | 8 |
| Door lock retaining bolts | 6 | 8 |
| Door striker pins | 33 | 45 |
| Bonnet lock retaining bolts | 6 | 8 |
| Boot lock retaining bolts | 6 | 8 |
| Boot lock striker retaining bolts | 18 | 25 |

## 1 General description

The bodyshell and underframe is of all-steel welded construction and is of computer-based design. The assembly and welding of the main body unit is completed entirely by computer-controlled robots, and the finished unit is checked for dimensional accuracy using modern computer and laser technology.

The front wings are bolted in position and are detachable should renewal be necessary after a front end collision.

## 2 Maintenance – bodywork and underframe

1 The general condition of a vehicle's bodywork is the one thing that significantly affects its value. Maintenance is easy but needs to be regular. Neglect, particularly after minor damage, can lead quickly to further deterioration and costly repair bills. It is important also to keep watch on those parts of the vehicle not immediately visible, for instance the underside, inside all the wheel arches and the lower part of the engine compartment.

2 The basic maintenance routine for the bodywork is washing – preferably with a lot of water, from a hose. This will remove all the loose solids which may have stuck to the vehicle. It is important to flush these off in such a way as to prevent grit from scratching the finish. The wheel arches and underframe need washing in the same way to remove any accumulated mud which will retain moisture and tend to encourage rust. Paradoxically enough, the best time to clean the underframe and wheel arches is in wet weather when the mud is thoroughly wet and soft. In very wet weather the underframe is usually cleaned of large accumulations automatically and this is a good time for inspection.

3 Periodically, except on vehicles with a wax-based underbody protective coating, it is a good idea to have the whole of the underframe of the vehicle steam cleaned, engine compartment included, so that a thorough inspection can be carried out to see what minor repairs and renovations are necessary. Steam cleaning is available at many garages and is necessary for removal of the accumulation of oily grime which sometimes is allowed to become thick in certain areas. If steam cleaning facilities are not available, there are one or two excellent grease solvents available such as Holts Engine Cleaner or Holts Foambrite which can be brush applied. The dirt can then be simply hosed off. Note that these methods should not be used on vehicles with wax-based underbody protective coating or the coating will be removed. Such vehicles should be inspected annually, preferably just prior to winter, when the underbody should be washed down and any damage to the wax coating repaired using Holts

Undershield. Ideally, a completely fresh coat should be applied. It would also be worth considering the use of such wax-based protection for injection into door panels, sills, box sections, etc, as an additional safeguard against rust damage where such protection is not provided by the vehicle manufacturer.

4 After washing paintwork, wipe off with a chamois leather to give an unspotted clear finish. A coat of clear protective wax polish, like the many excellent Turtle Wax polishes, will give added protection against chemical pollutants in the air. If the paintwork sheen has dulled or oxidised, use a cleaner/polisher combination such as Turtle Extra to restore the brilliance of the shine. This requires a little effort, but such dulling is usually caused because regular washing has been neglected. Care needs to be taken with metallic paintwork, as special non-abrasive cleaner/polisher is required to avoid damage to the finish. Always check that the door and ventilator opening drain holes and pipes are completely clear so that water can be drained out (photos). Bright work should be treated in the same way as paint work. Windscreens and windows can be kept clear of the smeary film which often appears, by the use of a proprietary glass cleaner like Holts Mixra. Never use any form of wax or other body or chromium polish on glass.

## 3  Maintenance – upholstery and carpets

Mats and carpets should be brushed or vacuum cleaned regularly to keep them free of grit. If they are badly stained remove them from the vehicle for scrubbing or sponging and make quite sure they are dry before refitting. Seats and interior trim panels can be kept clean by wiping with a damp cloth and Turtle Wax Carisma. If they do become stained (which can be more apparent on light coloured upholstery) use a little liquid detergent and a soft nail brush to scour the grime out of the grain of the material. Do not forget to keep the headlining clean in the same way as the upholstery. When using liquid cleaners inside the vehicle do not over-wet the surfaces being cleaned. Excessive damp could get into the seams and padded interior causing stains, offensive odours or even rot. If the inside of the vehicle gets wet accidentally it is worthwhile taking some trouble to dry it out properly, particularly where carpets are involved. *Do not leave oil or electric heaters inside the vehicle for this purpose.*

## 4  Minor body damage – repair

*The colour bodywork repair photographic sequences between pages 32 and 33 illustrate the operations detailed in the following sub-sections.*
**Note**: *For more detailed information about bodywork repair, the Haynes Publishing Group publish a book by Lindsay Porter called The Car Bodywork Repair Manual. This incorporates information on such aspects as rust treatment, painting and glass fibre repairs, as well as details on more ambitious repairs involving welding and panel beating.*

### Repair of minor scratches in bodywork

If the scratch is very superficial, and does not penetrate to the metal of the bodywork, repair is very simple. Lightly rub the area of the scratch with a paintwork renovator like Turtle Wax New Color Back, or a very fine cutting paste like Holts Body + Plus Rubbing Compound, to remove loose paint from the scratch and to clear the surrounding bodywork of wax polish. Rinse the area with clean water.

Apply touch-up paint, such as Holts Dupli-Color Color Touch or a paint film like Holts Autofilm, to the scratch using a fine paint brush; continue to apply fine layers of paint until the surface of the paint in the scratch is level with the surrounding paintwork. Allow the new paint at least two weeks to harden: then blend it into the surrounding paintwork by rubbing the scratch area with a paintwork renovator or a very fine cutting paste, such as Holts Body + Plus Rubbing Compound or Turtle Wax New Color Back. Finally, apply wax polish from one of the Turtle Wax range of wax polishes.

Where the scratch has penetrated right through to the metal of the bodywork, causing the metal to rust, a different repair technique is required. Remove any loose rust from the bottom of the scratch with a penknife, then apply rust inhibiting paint, such as Turtle Wax Rust Master, to prevent the formation of rust in the future. Using a rubber or nylon applicator fill the scratch with bodystopper paste like Holts Body + Plus Knifing Putty. If required, this paste can be mixed with cellulose thinners, such as Holts Body + Plus Cellulose Thinners, to provide a very thin paste which is ideal for filling narrow scratches.

Before the stopper-paste in the scratch hardens, wrap a piece of smooth cotton rag around the top of a finger. Dip the finger in cellulose thinners, such as Holts Body + Plus Cellulose Thinners, and then quickly sweep it across the surface of the stopper-paste in the scratch; this will ensure that the surface of the stopper-paste is slightly hollowed. The scratch can now be painted over as described earlier in this Section.

### Repair of dents in bodywork

When deep denting of the vehicle's bodywork has taken place, the first task is to pull the dent out, until the affected bodywork almost attains its original shape. There is little point in trying to restore the original shape completely, as the metal in the damaged area will have stretched on impact and cannot be reshaped fully to its original contour. It is better to bring the level of the dent up to a point which is about ⅛ in (3 mm) below the level of the surrounding bodywork. In cases where the dent is very shallow anyway, it is not worth trying to pull it out at all. If the underside of the dent is accessible, it can be hammered out gently from behind, using a mallet with a wooden or plastic head. Whilst doing this, hold a suitable block of wood firmly against the outside of the panel to absorb the impact from the hammer blows and thus prevent a large area of the bodywork from being 'belled-out'.

Should the dent be in a section of the bodywork which has a double skin or some other factor making it inaccessible from behind, a different technique is called for. Drill several small holes through the metal inside the area – particularly in the deeper section. Then screw long self-tapping screws into the holes just sufficiently for them to gain a good purchase in the metal. Now the dent can be pulled out by pulling on the protruding heads of the screws with a pair of pliers.

The next stage of the repair is the removal of the paint from the damaged area, and from an inch or so of the surrounding 'sound' bodywork. This is accomplished most easily by using a wire brush or abrasive pad on a power drill, although it can be done just as effectively by hand using sheets of abrasive paper. To complete the preparation for filling, score the surface of the bare metal with a screwdriver or the tang of a file, or alternatively, drill small holes in the affected area. This will provide a really good 'key' for the filler paste.

To complete the repair see the Section on filling and re-spraying.

### Repair of rust holes or gashes in bodywork

Remove all paint from the affected area and from an inch or so of the surrounding 'sound' bodywork, using an abrasive pad or a wire brush on a power drill. If these are not available a few sheets of abrasive paper will do the job just as effectively. With the paint removed you will be able to gauge the severity of the corrosion and therefore decide whether to renew the whole panel (if this is possible) or to repair the affected area. New body panels are not as expensive as most people think and it is often quicker and more satisfactory to fit a new panel than to attempt to repair large areas of corrosion.

Remove all fittings from the affected area except those which will act as a guide to the original shape of the damaged bodywork (eg headlamp shells etc). Then, using tin snips or a hacksaw blade, remove all loose metal and any other metal badly affected by corrosion. Hammer the edges of the hole inwards in order to create a slight depression for the filler paste.

Wire brush the affected area to remove the powdery rust from the surface of the remaining metal. Paint the affected area with rust inhibiting paint like Turtle Wax Rust Master; if the back of the rusted area is accessible treat this also.

Before filling can take place it will be necessary to block the hole in some way. This can be achieved by the use of aluminium or plastic mesh, or aluminium tape.

Aluminium or plastic mesh or glass fibre matting, such as the Holts Body + Plus Glass Fibre Matting, is probably the best material to use for a large hole. Cut a piece to the approximate size and shape of the hole to be filled, then position it in the hole so that its edges are below the level of the surrounding bodywork. It can be retained in position by several blobs of filler paste around its periphery.

Aluminium tape should be used for small or very narrow holes. Pull a piece off the roll and trim it to the approximate size and shape required, then pull off the backing paper (if used) and stick the tape over the hole; it can be overlapped if the thickness of one piece is insufficient. Burnish down the edges of the tape with the handle of a screwdriver or similar, to ensure that the tape is securely attached to the metal underneath.

## Bodywork repairs – filling and re-spraying

Before using this Section, see the Sections on dent, deep scratch, rust holes and gash repairs.

Many types of bodyfiller are available, but generally speaking those proprietary kits which contain a tin of filler paste and a tube of resin hardener are best for this type of repair, like Holts Body+Plus or Holts No Mix which can be used directly from the tube. A wide, flexible plastic or nylon applicator will be found invaluable for imparting a smooth and well contoured finish to the surface of the filler.

Mix up a little filler on a clean piece of card or board – measure the hardener carefully (follow the maker's instructions on the pack) otherwise the filler will set too rapidly or too slowly. Alternatively, Holts No Mix can be used straight from the tube without mixing, but daylight is required to cure it. Using the applicator apply the filler paste to the prepared area; draw the applicator across the surface of the filler to achieve the correct contour and to level the filler surface. As soon as a contour that approximates to the correct one is achieved, stop working the paste – if you carry on too long the paste will become sticky and begin to 'pick up' on the applicator. Continue to add thin layers of filler paste at twenty-minute intervals until the level of the filler is just proud of the surrounding bodywork.

Once the filler has hardened, excess can be removed using a metal plane or file. From then on, progressively finer grades of abrasive paper should be used, starting with a 40 grade production paper and finishing with 400 grade wet-and-dry paper. Always wrap the abrasive paper around a flat rubber, cork, or wooden block – otherwise the surface of the filler will not be completely flat. During the smoothing of the filler surface the wet-and-dry paper should be periodically rinsed in water. This will ensure that a very smooth finish is imparted to the filler at the final stage.

At this stage the 'dent' should be surrounded by a ring of bare metal, which in turn should be encircled by the finely 'feathered' edge of the good paintwork. Rinse the repair area with clean water, until all of the dust produced by the rubbing-down operation has gone.

Spray the whole repair area with a light coat of primer, either Holts Body+Plus Grey or Red Oxide Primer – this will show up any imperfections in the surface of the filler. Repair these imperfections with fresh filler paste or bodystopper, and once more smooth the surface with abrasive paper. If bodystopper is used, it can be mixed with cellulose thinners to form a really thin paste which is ideal for filling small holes. Repeat this spray and repair procedure until you are satisfied that the surface of the filler, and the feathered edge of the paintwork are perfect. Clean the repair area with clean water and allow to dry fully.

The repair area is now ready for final spraying. Paint spraying must be carried out in a warm, dry, windless and dust free atmosphere. This condition can be created artificially if you have access to a large indoor working area, but if you are forced to work in the open, you will have to pick your day very carefully. If you are working indoors, dousing the floor in the work area with water will help to settle the dust which would otherwise be in the atmosphere. If the repair area is confined to one body panel, mask off the surrounding panels; this will help to minimise the effects of a slight mis-match in paint colours. Bodywork fittings (eg chrome strips, door handles etc) will also need to be masked off. Use genuine masking tape and several thicknesses of newspaper for the masking operations.

Before commencing to spray, agitate the aerosol can thoroughly, then spray a test area (an old tin, or similar) until the technique is mastered. Cover the repair area with a thick coat of primer; the thickness should be built up using several thin layers of paint rather than one thick one. Using 400 grade wet-and-dry paper, rub down the surface of the primer until it is really smooth. While doing this, the work area should be thoroughly doused with water, and the wet-and-dry paper periodically rinsed in water. Allow to dry before spraying on more paint.

Spray on the top coat using Holts Dupli-Color Autospray, again building up the thickness by using several thin layers of paint. Start spraying in the centre of the repair area and then work outwards, with a side-to-side motion, until the whole repair area and about 2 inches of the surrounding original paintwork is covered. Remove all masking material 10 to 15 minutes after spraying on the last coat of paint.

Allow the new paint at least two weeks to harden, then, using a paintwork renovator or a very fine cutting paste such as Turtle Wax New Color Back or Holts Body+Plus Rubbing Compound, blend the edges of the paint into the existing paintwork. Finally, apply wax polish.

## 5  Major body damage – repair

Where serious damage has occurred or large areas need renewal due to neglect, it means that completely new sections or panels will need welding in, and this is best left to professionals. If the damage is due to impact, it will also be necessary to completely check the alignment of the bodyshell structure. Due to the principle of construction, the strength and shape of the whole car can be affected by damage to one part. In such instances the services of a BL agent with specialist checking jigs are essential. If a body is left misaligned, it is first of all dangerous, as the car will not handle properly, and secondly uneven stresses will be imposed on the steering, engine and transmission, causing abnormal wear or complete failure. Tyre wear may also be excessive.

## 6  Maintenance – hinges and locks

1   Oil the hinges of the bonnet, boot lid and doors with a drop or two of light oil at regular intervals.
2   At the same time, lightly oil the bonnet release mechanism and all door locks.
3   Do not attempt to lubricate the steering lock.

## 7  Door rattles – tracing and rectification

1   Check first that the door is not loose at the hinges, and that the latch is holding the door firmly in position. Check also that the door lines up with the aperture in the body. If the door is out of alignment, adjust it, as described in Sections 22 and 28.
2   If the latch is holding the door in the correct position, but the latch still rattles, the lock mechanism is worn and should be renewed.
3   Other rattles from the door could be caused by wear in the window operating mechanism, interior lock mechanism, or loose glass channels.

## 8  Bonnet – removal, refitting and adjustment

1   Support the bonnet in its open position and place some cardboard, or rags, beneath the corners by the hinges.
2   Mark the location of the hinges with a soft pencil, then loosen the four retaining nuts and bolts (photo).
3   With the help of an assistant, release the stay, unscrew and remove the retaining bolts and withdraw the bonnet from the car.
4   Refitting is a reversal of removal, but adjust the hinges to their original positions. The bonnet rear edge should be flush with the scuttle, and the gaps at either side equal.

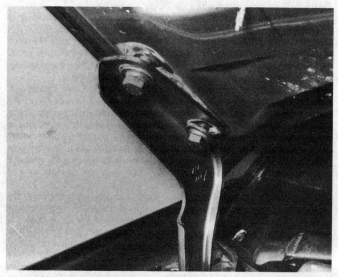

8.2 Slacken the bonnet hinge retaining bolts

## 9 Bonnet lock – adjustment

1   Adjustment is only possible at the lockpin mounted on the bonnet (photo).
2   Slacken the locknut and use a screwdriver to adjust the length of the lockpin so that the bonnet closes easily and is held firmly in place.
3   Tighten the locknut.

9.1 Bonnet lock adjustment points – locknut (A) and lockpin (B)

## 10 Bonnet lock and release cable – removal and refitting

1   Working inside the car, remove the bonnet release lever and disconnect the cable.
2   Working in the engine compartment, release the outer cable from the support bracket on the lock, and detach the inner cable from the lock lever.
3   Release the cable from its retaining clips, pull the cable through the bulkhead grommet and remove it from the car.
4   To remove the lock, undo and remove the retaining bolts and remove the lock assembly through the aperture in the front body panel.
5   Refitting is the reverse sequence to removal. Adjust the bonnet lock as described in the previous Section, if necessary.

## 11 Radiator grille – removal and refitting

1   Open the bonnet and support it in the raised position.
2   Very carefully lift the three plastic retaining tags, one at a time, and tip the grille outward at the top.
3   Lift the grille up to release the lower legs and remove the grille from the car.
4   Refitting is the reverse sequence to removal.

## 12 Front body panel – removal and refitting

1   Remove the bonnet lock and cable from the panel, as described in Section 10, and the radiator grille, as described in the previous Section.
2   Detach the air cleaner cold air intake hose from the panel.
3   Undo and remove the three screws each side securing the panel to the body side-members.
4   Release the bonnet stay, lift up the panel and remove it from the car.
5   Refitting is the reverse sequence to removal. Ensure that the lugs on the radiator and cooling fan cowl are correctly located in the body panel as it is fitted.

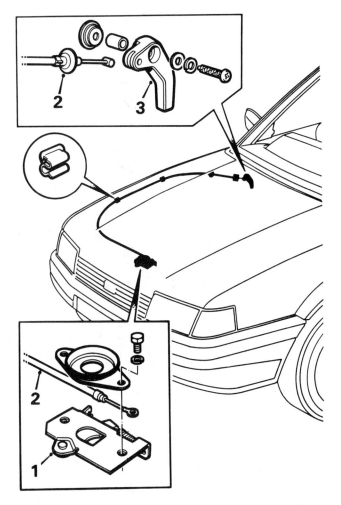

Fig. 11.1 Bonnet lock and release cable details (Sec 10)

1   Bonnet lock          3   Release lever
2   Release cable

## 13 Front wing – removal and refitting

1   Open the bonnet and support it in the raised position.
2   Remove the headlamp and front direction indicator lamp, as described in Chapter 9.
3   Remove the front bumper, as described in Section 32 of this Chapter.
4   Undo the screws and remove the wheel arch liner.
5   Refer to Fig. 11.2 and remove the wing retaining bolts from the locations shown.
6   Carefully ease the front wing off its mating flanges and remove it from the car.
7   Refitting is the reverse sequence to removal, but use a mastic sealing compound to seal the wing to the body.

## 14 Boot lid – removal and refitting

1   Release the retaining clips and plastic rivets securing the hinge covers in position and lift away the covers.
2   Prise off the private lock operating rod retaining clip and slide the rod out of the private lock lever.
3   Undo the three bolts securing the lock mechanism to the boot lid and slide the lock assembly clear of the boot lid.
4   Ease the boot release cable outer sheath out of its support plate and disengage the inner cable end from the lock lever.

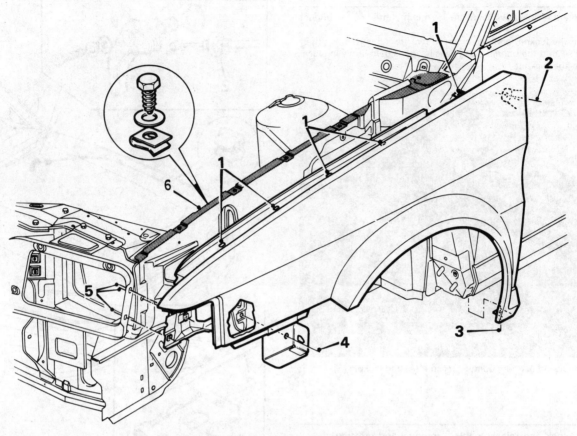

**Fig. 11.2 Front wing attachment points (Sec 13)**

1   Wing to inner valance attachments
2   Wing to front pillar attachment
3   Wing to sill panel attachment
4   Wing to lower valance attachment
5   Wing to headlamp surround attachment
6   Mastic sealer application area

5   Disconnect the wiring harness connectors then withdraw the harness and release cable from the boot lid.
6   Using a soft pencil, mark the outline of the hinges on the boot lid.
7   Have an assistant support the boot lid, then undo the hinge retaining bolts and lift the boot lid away (photo).
8   Refitting is the reverse sequence to removal. Align the hinges with the outline marks made during removal, but adjust the hinge position slightly if necessary so that the lid closes easily and is in correct alignment. Further adjustment is possible by slackening the striker plate bolts and repositioning the plate (photo).

## 15 Boot lock – removal and refitting

**Note:** *On vehicles equipped with central locking, further information will be found in Chapter 9.*
1   Prise off the private lock operating rod retaining clip and slide the rod out of the private lock lever.
2   Undo the three bolts securing the lock mechanism to the boot lid and slide the lock clear (photo).
3   Ease the boot release cable outer sheath out of its support plate

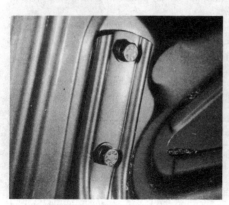

14.7 Boot lid hinge retaining bolts

14.8 Boot lock striker plate and retaining bolts

15.2 Boot lock retaining bolts (arrowed)

and disengage the inner cable end from the lock lever. Where applicable disconnect the wiring connectors then remove the lock from the boot lid.

4    Refitting is the reverse sequence to removal. If necessary adjust the striker plate position after refitting the lock so that the boot lid shuts and locks without slamming.

## 16  Boot private lock – removal and refitting

**Note**: *On vehicles equipped with central locking, further information will be found in Chapter 9.*

1    Prise off the private lock operating rod retaining clip and slide the rod out of the private lock lever.

2    Undo the large nut securing the private lock to the lock retainer (photo). Withdraw the private lock, washer and lock retainer from the boot lid.

3    Refitting is the reverse sequence to removal.

## 17  Boot release cable and lever – removal and refitting

1    Refer to Section 14 and carry out the operations described in paragraphs 1 to 5.

2    Refer to Section 30 and remove the rear seat cushion.

16.2 Boot private lock retaining nut (arrowed)

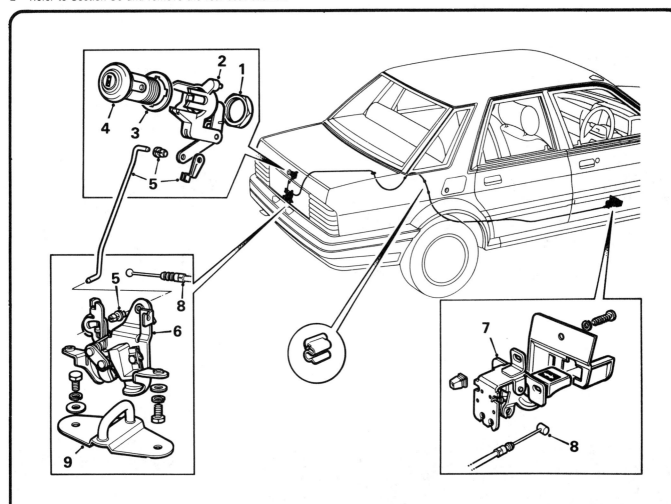

**Fig. 11.3 Boot lock and release cable details (Secs 15, 16 and 17)**

| | | |
|---|---|---|
| *1    Private lock retaining nut* | *4    Private lock barrel* | *7    Release lever* |
| *2    Lock retainer* | *5    Operating rod and clips* | *8    Release cable* |
| *3    Washer* | *6    Boot lock* | *9    Striker plate* |

3   Undo the front seat belt lower anchorage attachment and remove the sill interior trim panels.
4   Undo the driver's seat outer rail retaining bolts and withdraw the oddments tray.
5   Undo the screw securing the release lever shield to the body and remove the shield.
6   Undo the two release lever bracket retaining screws, remove the bracket assembly and withdraw the cable.
7   Refitting is the reverse sequence to removal.

## 18  Windscreen – removal and refitting

All Montego models are equipped with a flush-glazed, laminated windscreen, secured to the bodyshell by direct bonding. Due to this method of retention, special tools and equipment are required to remove and refit the screen, and this task is definitely beyond the scope of the home mechanic. If it is necessary to have the windscreen removed, this job should be left to a suitably equipped specialist or BL dealer.

## 19  Rear window and rear quarter windows – removal and refitting

The rear windows, like the windscreen, also feature direct bonding retention and their removal and refitting should be entrusted to a BL dealer.

## 20  Front door inner trim panel – removal and refitting

1   If manually operated windows are fitted, prise out the finisher cap then undo the window regulator handle retaining screw (photos). Withdraw the handle and backing plate.
2   Prise out the finisher cap and undo the door internal release lever surround retaining screw (photos). Slide the surround off the lever.

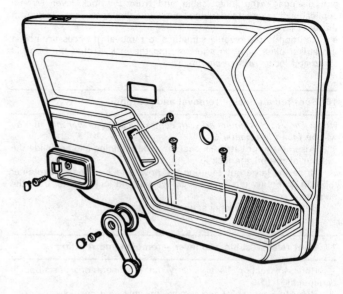

**Fig. 11.4 Front door inner trim panel removal (Sec 20)**

3   Undo the three screws securing the trim panel to the door; one located in the well of the door pull, and two located at the bottom of the door bin (photo).
4   Using a screwdriver or suitable flat bar carefully lever between the panel and the door, at the front and rear sides, to release the panel retaining buttons.
5   Pull out the bottom edge and lift the panel upwards over the lock button. Where applicable disconnect the wiring at the electric window switches and radio speakers (photo) then remove the panel.

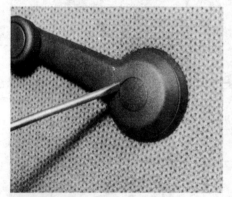

20.1A Prise out the finisher cap ...

20.1B ... then remove the regulator handle retaining screw

20.2A Prise out the finisher cap ...

20.2B ... and undo the release lever surround retaining screw

20.3 Two trim panel retaining screws (arrowed) located in the door bin

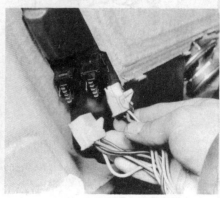

20.5 Disconnect the electrical wiring from the rear of the trim panel

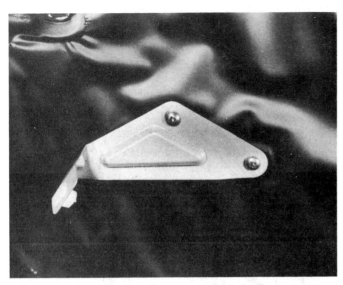

20.6 Remove the trim panel support bracket and polythene sheet for access inside the door

6   To gain access to the door internal components, undo the screws and remove the trim panel support bracket (photo) then carefully peel back the polythene condensation barrier as necessary.
7   Refitting is the reverse sequence to removal.

## 21 Front door – removal and refitting

1   Remove the inner trim panel, as described in Section 20.
2   Disconnect the wiring multi-plugs and connectors and, where fitted, at the central locking and electrically-operated window components. Remove the wiring harness from the door.
3   Have an assistant securely support the door then, using a suitable punch, drift out the hinge retaining pins (photo). Withdraw the door from the car.
4   Refitting is the reverse sequence to removal.

## 22 Front door lock – removal, refitting and adjustment

**Note**: *On vehicles equipped with central locking, further information will be found in Chapter 9.*
1   Remove the door inner trim panel, as described in Section 20.
2   Unscrew the interior lock button, temporarily refit the window regulator handle or reconnect the switch wiring and raise the window fully.
3   Prise the operating rod out of its retaining bush on the exterior door handle operating lever (photo).
4   Release the retaining clip and withdraw the operating rod from the private lock lever.
5   Undo and remove the screw securing the internal release lever to the door. Release the operating rod from its steady clip and slide the lever rearwards to release it from its location (photo).
6   Undo and remove the screw securing the rear window channel to the door. Release the channel from the felt guide and remove the channel from the door.
7   Undo and remove the three retaining screws (photo) and withdraw the lock assembly, complete with operating rods, from the door aperture.
8   The external handle can also be removed after unscrewing the two nuts and washers.
9   Refitting is the reverse sequence to removal, but adjust the lock striker pin as follows.
10  Check that the latch disc is in the open position then loosen the striker pin nut (photo) and position the pin so that the door can be closed easily and is held firmly. Close the door gently, but firmly, when making this adjustment.
11  Tighten the striker pin nut.

21.3 Front door lower hinge pin (arrowed)

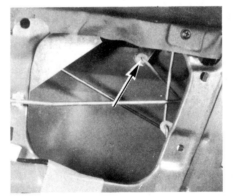

22.3 Prise the operating rod out of the lever bush (arrowed)

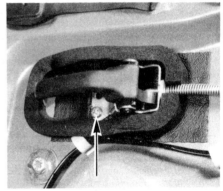

22.5 Internal release lever retaining screw (arrowed)

22.7 Door lock retaining screws

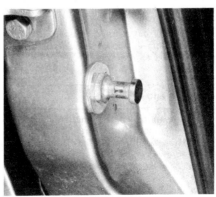

22.10 Front door striker pin

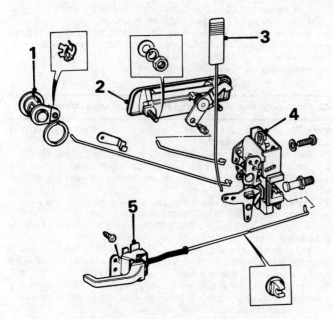

**Fig. 11.5 Front door lock components (Secs 22 and 23)**

1   Private lock
2   External handle
3   Lock button

4   Lock assembly
5   Internal release lever

## 23 Front door private lock – removal and refitting

**Note:** *On vehicles equipped with central locking, further information will be found in Chapter 9.*
1   Remove the door interior trim panel, as described in Section 20, and raise the window fully.
2   Release the retaining clip and withdraw the operating rod from the private lock lever.
3   Prise out the retaining ring and withdraw the lock and seating washer from the door.
4   To remove the lock barrel, extract the circlip and remove the spring washer and lock lever. Withdraw the lock barrel from the lock body.
5   Refitting is the reverse sequence to removal.

## 24 Front door glass and regulator – removal and refitting

**Note:** *On vehicles equipped with electrically-operated windows, further information will be found in Chapter 9.*
1   Refer to Section 20 and remove the inner trim panel.
2   Refer to Section 25 and remove the door mirror.
3   Lower the window fully and insert a small block of wood in the base of the door to support the window after removal of the regulator.
4   Undo the three bolts securing the regulator to the door panel. Release the three nylon wheels of the regulator arms from their runners and manipulate the regulator out of the door aperture.
5   Tip the window forward and remove it upwards from the door.
6   Refitting is the reverse sequence to removal.

## 25 Front door mirror – removal and refitting

1   Open the front door and, if the mirror is of the remotely-controlled type, remove the adjusting lever by moving it rearwards.
2   Carefully prise off the cheater panel internal finisher to provide access to the mirror retaining screws.
3   Undo and remove the three screws, lift off the retaining plate and withdraw the mirror from the door.
4   Refitting is the reverse sequence to removal.
5   The design of the mirror is such that it will give slightly under

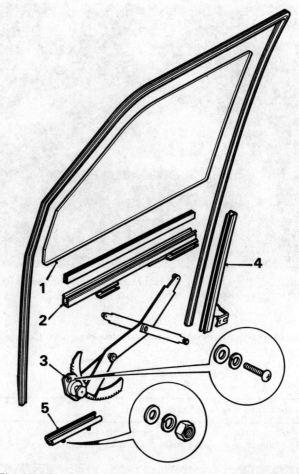

**Fig. 11.6 Front door glass and regulator components (Sec 24)**

1   Door window
2   Lifting channel
3   Window regulator

4   Rear window channel
5   Auxiliary slide

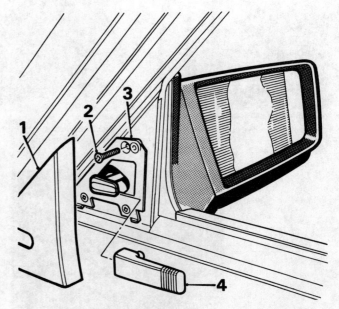

**Fig. 11.7 Remote control door mirror attachments (Sec 25)**

1   Cheater panel internal
    finisher
2   Retaining screw

3   Retaining plate
4   Adjusting lever

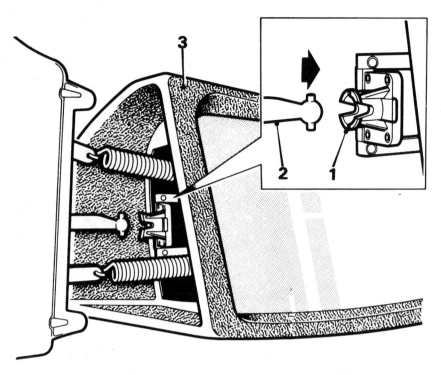

**Fig. 11.8 Mirror adjusting lever attachments (Sec 25)**

| 1    Ball plate | 2    Adjusting lever | 3    Mirror head |
| --- | --- | --- |

impact, against spring tension, thus minimising the risk of damage or injury. On remotely-controlled mirrors this will cause the adjusting lever to disengage from the internal mechanism and become inoperative. The lever can be refitted, without removing the mirror, as follows.

6    Open the door and lower the window.

7    Gently move the mirror head toward the front of the car to give a working clearance between the mirror head and the base.

8    With the mirror held in this position, grasp the ball plate with one hand and the ball-head of the adjusting lever with the other. Firmly push the ball-head into the socket on the ball plate until it fully engages. Release the mirror head and check the operation.

## 26  Rear door inner trim panel – removal and refitting

1    Removal and refitting of the rear inner trim panel is basically the same as that for the front doors, as described in Section 20. The only difference is the retaining screw locations and this will be obvious after inspection.

## 27  Rear door – removal and refitting

1    If central locking is fitted, remove the inner trim panel, as described in Section 26, and disconnect the wiring harness connectors. Remove the harness from the door.

2    Carefully mark the outline of the door hinges on the pillar using a soft pencil.

3    With the help of an assistant to support the door, undo and remove the bolts securing the upper and lower hinges to the door (photo). Carefully lift away the door.

4    Refitting is the reverse sequence to removal. Align the hinges with the pencil marks made during removal.

27.3 Rear door upper hinge

## 28  Rear door lock – removal, refitting and adjustment

**Note:** *On vehicles equipped with central locking, further information will be found in Chapter 9.*

1    Remove the door inner trim panel and capping, as described in Section 26.

2    Temporarily refit the window regulator handle and wind the window fully up. Unscrew the interior lock button.

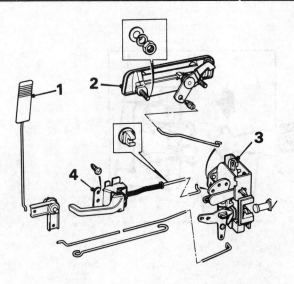

**Fig. 11.9 Rear door lock components (Sec 28)**

1  *Lock button*              3  *Lock assembly*
2  *Door exterior handle*     4  *Internal release lever*

3  Prise the operating rod out of its retaining bush on the door exterior handle operating lever.

4  Undo and remove the screw securing the internal release lever to the door. Release the operating rod from its steady clip and slide the lever rearwards to release it from its location.

5  Using a small screwdriver or punch, press out the locking pin from the centre of the bellcrank lever (photo), and withdraw the bellcrank and pivot (photo). Detach the lock operating rod from the bellcrank and from its steady clip.

6  Undo and remove the three screws securing the lock to the door and withdraw the unit, complete with operating rods, from the door aperture.

7  Refitting is the reverse sequence to removal, but adjust the lock striker pin as follows.

8  Check that the latch disc is in the open position, then loosen the striker pin nut and position the pin so that the door can be closed easily, and is held firmly. Close the door gently, but firmly, when making this adjustment.

9  Tighten the striker pin nut.

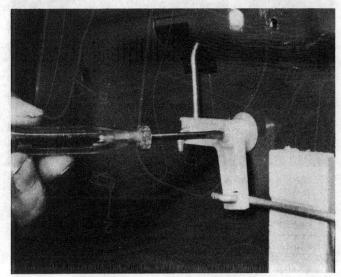

28.5A Press out the locking pin from the bellcrank lever ...

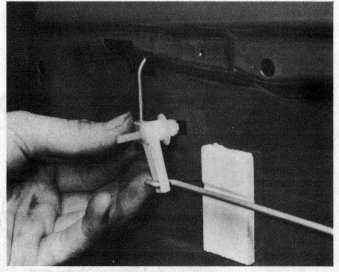

28.5B ... then withdraw the bellcrank and pivot

## 29  Rear door glass and regulator – removal and refitting

1  Wind the window fully up, and then remove the inner trim panel and capping, as described in Section 26.

2  Undo and remove the two bolts securing the window regulator to the door panel. Support the glass, disengage the regulator arm from the door glass lifting channel, and withdraw the regulator through the door aperture.

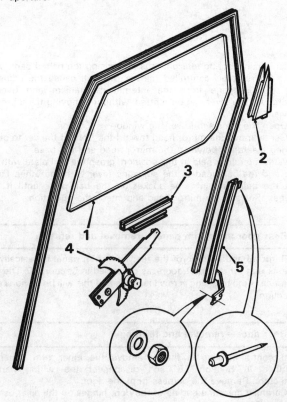

**Fig. 11.10 Rear door glass and regulator components (Sec 29)**

1  *Door window*        4  *Window regulator*
2  *Cheater panel*      5  *Rear guide channel*
3  *Lifting channel*

3   Lower the glass to the bottom of the door and then remove the rear guide channel retaining bolt. Release the channel from the glass and remove the felt from the channel.

4   Drill out the rivets securing the cheater panel to the door frame and lift off the panel. Drill an access hole in the bottom of the cheater panel and shake out the rivet debris.

5   Drill out the rivets securing the rear guide channel to the door, withdraw the channel and remove the rivet studs.

6   Carefully lift the glass upwards and out of the door. The lifting channel can now be removed from the glass, if necessary.

7   Refitting is the reverse sequence to removal.

### 30 Seats – removal and refitting

*Front seat*

1   Adjust the seat for access to the seat rail retaining bolts and undo the Torx type retaining bolts. Withdraw the seat from the car.

2   Refitting is the reverse sequence of removal.

*Rear seat*

3   Push down on the base of the rear seat cushion to release the two retaining clips and remove the cushion.

4   Release the rear seat squab latches from inside the luggage compartment and fold down the squabs.

5   Undo the squab centre hinge bracket retaining screws and lift out the squabs.

6   Undo the bolt securing the seat squab extension lower edge to the wheel arch.

7   Slide the squab extension up and remove it from the car.

8   Refitting the squab extensions, seat guards and cushion is the reverse sequence to removal.

### 31 Seat belts – removal and refitting

**Caution**: *If the vehicle has been involved in an accident in which structural damage was sustained, all the seat belt components must be renewed.*

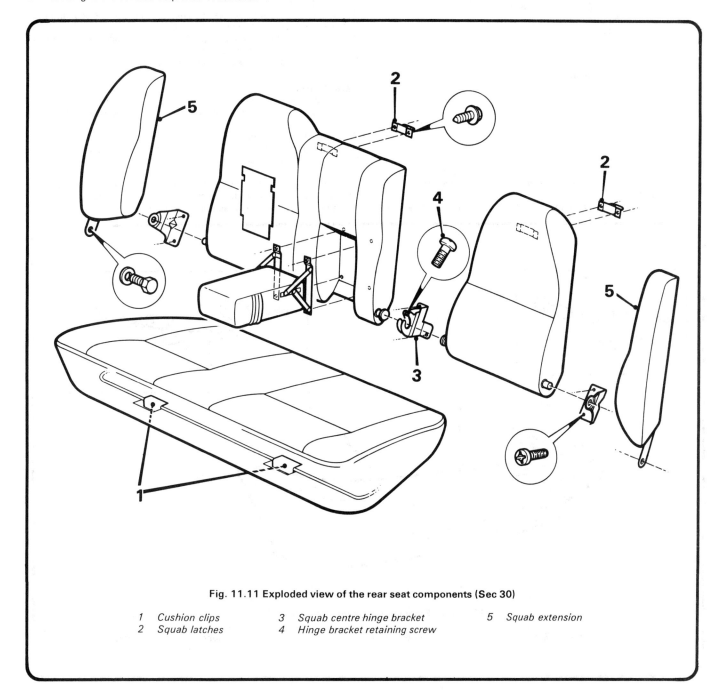

**Fig. 11.11 Exploded view of the rear seat components (Sec 30)**

| | | |
|---|---|---|
| 1   *Cushion clips* | 3   *Squab centre hinge bracket* | 5   *Squab extension* |
| 2   *Squab latches* | 4   *Hinge bracket retaining screw* | |

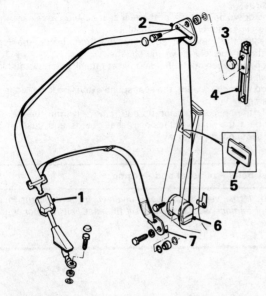

**Fig. 11.12 Front seat belts and mountings (Sec 31)**

| | | | |
|---|---|---|---|
| 1 | Flexible stalk | 5 | Belt guide |
| 2 | Upper mounting | 6 | Inertia reel |
| 3 | Adjustment slide knob | 7 | Bottom mounting |
| 4 | Adjustment slide | | |

1    Move the front seats fully forward, ease the door seal from around the central door pillar, and remove the interior trim covering the seat belt mechanism.

2    Prise off the trim capping from the top mounting. Undo and

remove the top, bottom and inertia reel mounting bolts, noting the position of the spacers and washers, and remove the assembly from the car.

3    Press out the belt guide from the interior trim and remove the belt from the trim.

4    Pull off the trim cover then undo and remove the bolt securing the flexible stalk to the floor. Note the position of the washers and spacer, and remove the stalk.

5    Refitting is the reverse sequence to removal, *but ensure that the mounting bolts are tightened to the specified torque.*

## 32 Bumpers – removal and refitting

### Front bumper

1    Refer to Chapter 9 and remove the battery.

2    Undo the two bolts each side securing the bumper front support brackets to the body.

3    Undo the nuts and remove the washers and spacers securing the bumper and brackets to the front wings.

4    Detach the wheel arch liners from the bumpers and carefully withdraw the bumper from the front of the car.

5    Refitting is the reverse sequence of removal.

### Rear bumper

6    Disconnect the battery negative terminal, remove the number plate lamps and detach the bulbholders. If necessary refer to Chapter 9.

7    Undo the two bolts each side, accessible from inside the luggage compartment, securing the bumper support brackets to the body.

8    Undo the two screws each side and remove the rear wheel arch liners.

9    Undo the nut securing each bumper front bracket to the rear valance then carefully withdraw the bumper from the car.

10   Refitting is the reverse sequence to removal.

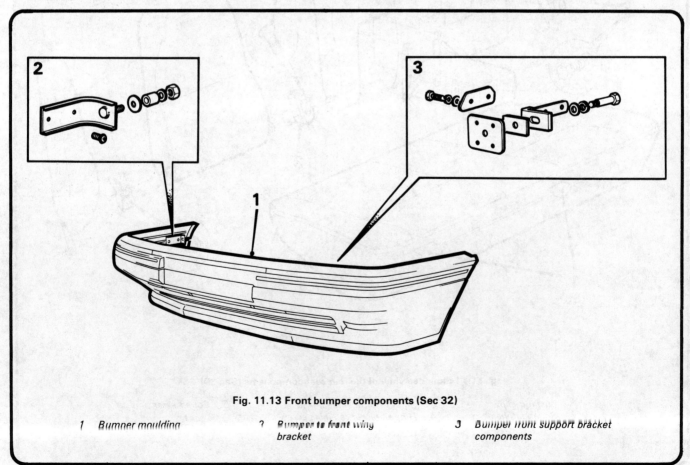

**Fig. 11.13 Front bumper components (Sec 32)**

| | | |
|---|---|---|
| 1    Bumper moulding | 2    Bumper to front wing bracket | 3    Bumper front support bracket components |

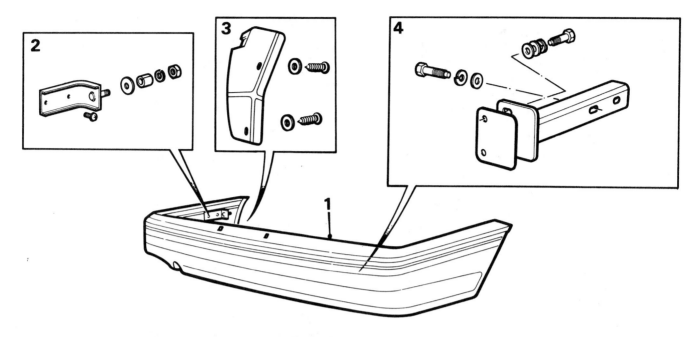

**Fig. 11.14 Rear bumper components (Sec 32)**

| | | | |
|---|---|---|---|
| *1 Bumper moulding* | *2 Bumper to rear valance support bracket* | *3 Wheel arch liner* | *4 Bumper support bracket* |

## 33 Sunroof – removal and refitting

### Sunroof panel

1  Open the trailing edge of the sunroof panel, ease each end of the panel liner down and release the trim panel clips from the legs on the liner frame.

2  Close the sunroof and disengage the liner front clips from the deflector slides. Move the liner back between the panels and detach the panel liner tensioning springs.

3  Undo and remove the four screws on each side securing the sunroof panel to the slide and tilt mechanism. *Do not remove the tilt adjustment screws.*

4  Lift out the sunroof panel and, if required, remove the weatherstrip and trim panel clips.

5  If removed, refit the weatherstrip and trim panel clips, locate the sunroof panel in position and loosely secure.

6  Close the sunroof and check its alignment. Tighten the four securing screws in sequence and adjust the tilt arm as necessary.

7  Refit the tensioning springs, open the trailing edge of the sunroof and pull the liner forward, ensuring that the clips on the liner frame engage with the deflector slides.

8  Engage the trim panel clips on the legs of the liner frame and close the sunroof.

### Sunroof liner

9  Remove the sunroof panel, as previously described, and then remove the two deflector slides.

10  Slide the liner forward, disengage it from the frame and lift it upwards and out.

11  To refit the liner, locate it on the frame with the slides under the spigots. Slide the liner rearwards, refit the deflector slides and then refit the sunroof panel.

### Sunroof frame

12  Remove the sunroof panel, sunroof liner and the wind deflector, as previously described.

13  Remove the operating handle.

14  Release the aperture trim along the front edge and ease away the headliner. Undo and remove the two screws securing the winder mechanism to the frame and front bracket. Take care to protect the roof panel in front of the aperture.

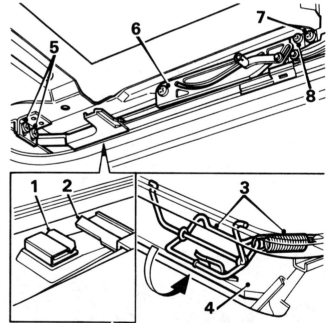

**Fig. 11.15 Sunroof panel attachments (Sec 33)**

| | | | |
|---|---|---|---|
| *1* | *Trim panel front clip* | *6* | *Tilt arm front retaining screw* |
| *2* | *Deflector slide* | *7* | *Tilt arm rear retaining screw* |
| *3* | *Trim panel clip and tensioning spring* | *8* | *Tilt arm adjustment screw* |
| *4* | *Trim panel* | | |
| *5* | *Deflector slide retaining screw* | | |

15  Undo and remove the remaining screws securing the frame to the mounting assembly, move the frame forwards and lift it out, taking care to avoid scratching the surrounding paintwork.

16  To refit the frame, position it in the aperture and engage the guide pins in the rear support brackets. Refit the retaining screws loosely, locate the wind deflector and tighten all the retaining screws.

17  Refit the sunroof liner and sunroof panel.

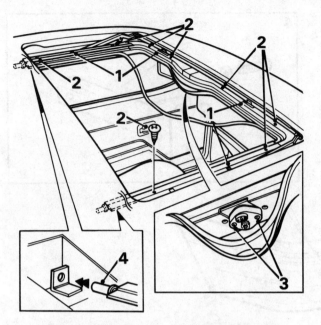

**Fig. 11.16 Sunroof frame attachments (Sec 33)**

| | | | |
|---|---|---|---|
| 1 | Wind deflector retaining screws | 3 | Winder to front bracket retaining screws |
| 2 | Frame retaining screws | 4 | Guide pin and rear support bracket |

## 34  Centre console – removal and refitting

1   Move the front seats fully forward and undo the two console rear fixing screws.

2   Move the seats fully rearward, prise out the seat belt stalk access panel and then unbolt and remove the stalks.

3   Remove both front seat backrest control handles.

4   Remove the oddments tray from the front of the console to expose the console front fixing screws. Extract the screws.

5   Prise out the panel, forward of the handbrake lever.

6   On manual transmission models, lift the console upwards; the gear lever and gaiter will remain secured by a retaining plate and screws.

7   On automatic transmission models, move the console to the rear and then lift it upwards over the handbrake and selector lever. The selector lever, index plate and bulbholder will remain attached to the floor.

34.4 Centre console front retaining screws (arrowed)

8   Refitting is a reversal of removal; tighten the seat belt stalk bolts to specified torque.

## 35  Facia panel – removal and refitting

*Note: Removal of the facia is considerably involved, entailing extensive dismantling and the disconnection of many wiring harness plugs and connectors. Read through the entire Section before starting, and familiarize yourself with the procedure by referring to the photographs and illustrations. Identify all electrical connections with a label before removal and, if necessary, make notes during dismantling.*

1   Disconnect the battery negative terminal.

2   Remove the steering column, as described in Chapter 10.

3   Remove the instrument panel, radio and facia switches, as described in Chapter 9.

4   From within the glovebox, release the two turnbuckles using a coin and lower the control unit panel from the glovebox roof. Disconnect the wiring multi-plugs and remove the panel (photo).

5   Remove the courtesy lamp switch and disconnect its wiring plug.

6   Prise off the switch blanking plate on the right-hand side of the heater control knobs.

7   Undo the screw located behind the blanking plate (photo).

8   Centralize the heater control knobs and ease the control panel,

35.4 Disconnect the wiring multi-plugs from the control unit panel

35.7 Undo the screw after prising out the blanking plate

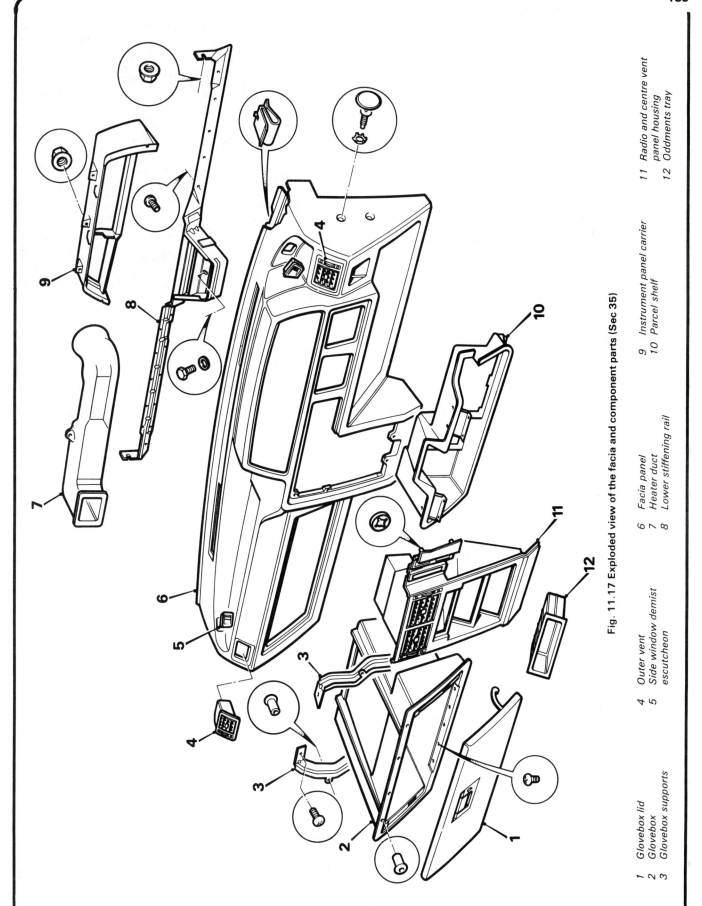

Fig. 11.17 Exploded view of the facia and component parts (Sec 35)

1  Glovebox lid
2  Glovebox
3  Glovebox supports
4  Outer vent
5  Side window demist
   escutcheon
6  Facia panel
7  Heater duct
8  Lower stiffening rail
9  Instrument panel carrier
10 Parcel shelf
11 Radio and centre vent
   panel housing
12 Oddments tray

complete with knobs, off the facia. Disconnect the wiring multi-plugs and remove the control panel.

9    Release the cigarette lighter panel from the facia, disconnect the cigarette lighter wiring plug and, where fitted, the balance control wiring plug. Remove the panel.

10   Remove the fusebox cover from the right-hand side of the facia.

11   Prise off the trim caps from the two retaining screws at each end of the facia (photo). Undo the four screws.

12   Undo the facia centre retaining screw situated at the base of the facia panel.

13   Pull the facia rearwards and disconnect any remaining wiring multi-plugs or connectors. Remove the facia from the car.

14   Refitting is a straightforward reverse of the removal sequence.

## 36  Heater – adjustments

1    The only adjustment possible is that of the heater flap linkage and is carried out as follows, with reference to Fig. 11.18.

2    Disconnect the battery negative terminal then remove the instrument panel, as described in Chapter 9.

3    Slacken the flap link rod trunnion screw situated behind the heater blend control lever.

4    Position the blend control in the fully raised position and move the heater flap lever as far as it will go towards the vehicle interior. Hold the linkage and control in this position and tighten the trunnion screw.

5    Check that the blend control lever moves freely from the fully raised to fully lowered position.

6    Refit the instrument panel and reconnect the battery.

35.11 Prise off the trim caps and undo the two bolts at each end of the facia

## 37  Heater – removal and refitting

1    Drain the cooling system, as described in Chapter 2, and then remove the facia, as described in Section 35 of this Chapter.

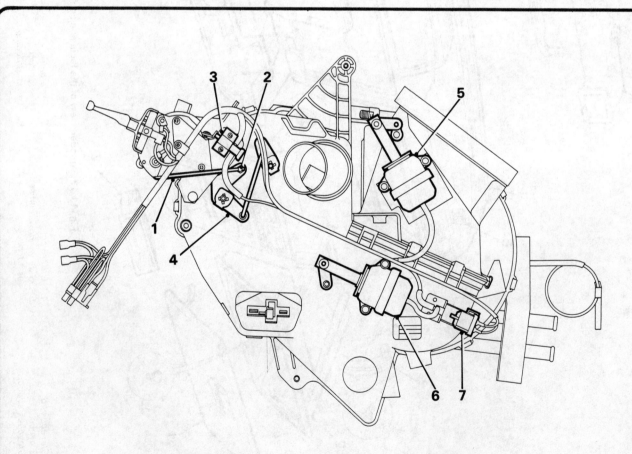

**Fig. 11.18 Heater control and adjustment details (Sec 36)**

| | | | |
|---|---|---|---|
| 1   Flap link rod | 3   Vacuum switch | 5   Upper vacuum actuator | 7   Vacuum switch solenoid |
| 2   Flap link rod trunnion screw | 4   Heater flap lever | 6   Lower vacuum actuator | |

2   From within the engine compartment slacken the clips and disconnect the two heater hoses at the outlets adjacent to the engine compartment bulkhead.

3   Disconnect the heater vacuum supply hose from the T-piece connector in the engine main vacuum line.

4   Undo the two screws securing the ventilator duct at the base of the heater.

5   Detach the windscreen demister ducts and the side vent hoses from the heater casing.

6   Disconnect the heater blower motor wiring multi-plug.

7   Undo the bolt securing the base of the heater to the support bracket and the two bolts and clamp plate securing the top of the heater to the facia rail.

8   Detach the heater drain hose from the engine compartment bulkhead, withdraw the heater and remove it from inside the car.

9   Refitting is the reverse sequence of removal. but refill the cooling system, as described in Chapter 2, after installation.

### 38  Heater – dismantling and reassembly

1   With the heater removed from the car, remove the drain hose then release the retaining spring clips and remove the heater upper support pins and carrier strap.

2   Remove the side vent elbows.

3   Remove the Rokut rivets securing the heater control panel, remove the flap link rod from the trunnion then move the control levers and panel to one side.

4   Undo the screws securing the matrix pipe bracket, pipe support clip and matrix flange. Lift out the matrix and pipes. Undo the retaining screws and remove the pipes and gasket from the matrix.

5   Undo the eight screws and two clips securing the heater motor terminal cover. Lift off the cover and disconnect the motor wiring harness plugs. Remove the harness clips.

6   Undo the two screws and washers securing the demist flaps.

7   Drill out the pop rivet adjacent to the heater lower mounting. Remove the sealing tape from inside the heater control panel aperture.

8   Release the spring retaining clips and separate the two halves of the heater body. Withdraw the resistor unit and place the wiring harness to one side. Remove the two plastic hinge bushes from the flaps.

9   Using a flat screwdriver inserted between the heater body and the rear of the blower motor casing, carefully ease out the motor.

10  With the heater dismantled, inspect the matrix for signs of leaks and, if any are apparent, renew the matrix. If the matrix appears serviceable, brush off any accumulation of dirt or debris from the fins and then reverse flush the core. Inspect the remaining heater components for any signs of damage or distortion and renew as necessary.

11  Reassembly is the reverse sequence of dismantling, but adjust the heater flap linkage, as described in Section 36, on completion.

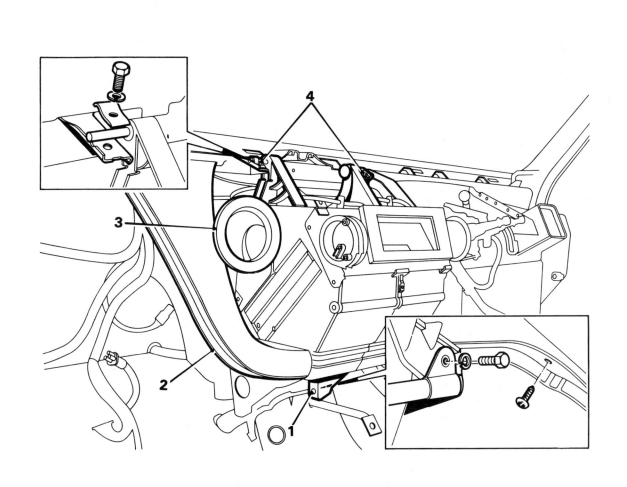

**Fig. 11.19 Heater assembly attachment details (Sec 37)**

| | | | |
|---|---|---|---|
| 1   Heater base retaining bolt | 2   Demister duct | 3   Side vent hose | 4   Upper retaining bolts and clamp plates |

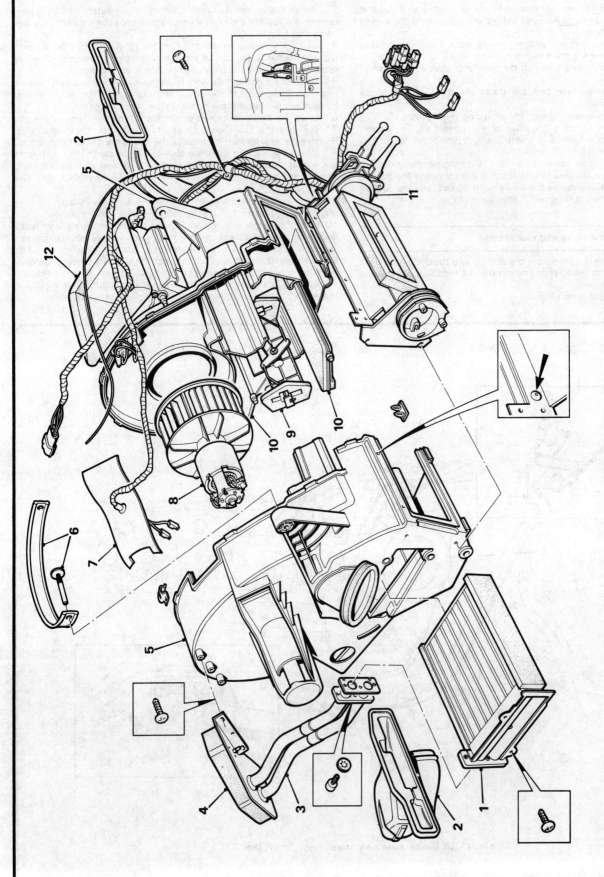

**Fig. 11.20 Exploded view of the heater assembly (Sec 38)**

1 Heater matrix
2 Demister duct
3 Matrix pipe, bracket and flange assembly
4 Bulkhead sealing pad
5 Heater body halves
6 Carrier strap and upper support pin
7 Motor cover
8 Motor and blower assembly
9 Demister flap
10 Blender flaps
11 Control panel and levers
12 Air inlet port

# Chapter 12 Supplement:
# Revisions and information on later models

## Contents

## 1  Introduction

This Supplement contains information relating to 1986 and later
models, and material which is additional to, or a revision of, that
contained in the main Chapters of the Manual.

The Sections in this Supplement follow the same order as the
Chapters to which they relate. The Specifications are all grouped
together for convenience, but follow Chapter order.

It is recommended that before any particular operation is undertaken, reference be made to the appropriate Section(s) of the Supplement. In this way, any changes to procedure or components can be
noted before referring to the main Chapters.

Austin Montego Mayfair 2.0 Estate

MG Montego Turbo

**2 Specifications**

*The specifications below are supplementary to, or revisions of, those at the beginning of the preceding Chapters.*

*Engine*

## Engine (general)
Compression ratio – Turbo models:
    Engine type 20HF50 ............................................................................. 8.0:1
    Engine type 20H and 20HE15 (MG Turbo) ............................................. 8.5:1

## Pistons
Clearance in cylinder:
    Below oil control groove:
        All models except Turbo .......................................................... 0.007 to 0.009 in (0.18 to 0.23 mm)
        Turbo models ........................................................................... 0.003 to 0.004 in (0.08 to 0.10 mm)
    Bottom of skirt:
        All models except Turbo .......................................................... 0.0004 to 0.0014 in (0.01 to 0.04 mm)
        Turbo models ........................................................................... 0.0016 to 0.002 in (0.004 to 0.005 mm)

## Piston rings
Ring gap:
    Compression rings (all models) .................................................... 0.012 to 0.020 in (0.3 to 0.5 mm)
    Oil control ring:
        All models except Turbo .......................................................... 0.013 to 0.055 in (0.33 to 1.4 mm)
        Turbo models ........................................................................... 0.010 to 0.020 in (0.25 to 0.50 mm)

## Gudgeon pins
Type:
    All models except Turbo ............................................................... Press fit in connecting rod
Turbo models ....................................................................................... Fully floating
Diameter:
    All models except Turbo ............................................................... 0.8125 to 0.8127 in (20.638 to 20.643 mm)
    Turbo models ............................................................................... 0.9374 to 0.9376 in (23.810 to 23.815 mm)
Clearance in connecting rod (Turbo models) ...................................... 0.0001 to 0.0005 in (0.003 to 0.025 mm) at 20°C (68°F)
Gudgeon pin bore offset from centre:
    Carburettor models ...................................................................... 0.059 in (1.5 mm)
    EFi models .................................................................................... 0.031 in (0.80 mm)
    Turbo models ............................................................................... 0.063 in (1.6 mm)

## Camshaft
Diametrical clearance in bearings (Turbo models) ............................ 0.0019 to 0.0038 in (0.048 to 0.097 mm)

## Valves
Seat angle:
    Exhaust (Turbo models) ............................................................... 46°15′
Stem diameter:
    Exhaust (Turbo models) ............................................................... 0.3137 to 0.3144 in (7.97 to 7.99 mm)
Valve head protrusion ......................................................................... 0.040 in (1.0 mm)
Stem-to-guide clearance:
    Inlet ............................................................................................. 0.0012 to 0.0019 in (0.03 to 0.05 mm)
    Exhaust:
        Except Turbo ............................................................................ 0.0011 to 0.0026 in (0.03 to 0.07 mm)
        Turbo ....................................................................................... 0.0019 to 0.0035 in (0.05 to 0.09 mm)

## Valve guides
Inside diameter (exhaust):
    Except Turbo ................................................................................ 0.2933 to 0.2937 in (7.450 to 7.460 mm)
    Turbo ........................................................................................... 0.3164 to 0.3172 in (8.037 to 8.057 mm)

## Valve timing* (Turbo and all engines from 20 HB)
Inlet valve:
    Opens ........................................................................................... 13° BTDC
    Closes .......................................................................................... 47° ABDC
Exhaust valve:
    Opens ........................................................................................... 55° BBDC
    Closes .......................................................................................... 21° ATDC
*\* At 0.012 in (0.30 mm) valve clearance*

## Valve clearance
Clearance (cold) – 20HE09, 11, 13, 14, 15, 36, 37, 99, and HF50, 51,52 ... 0.0118 to 0.0126 in (0.30 to 0.32 mm)
Adjust if less than:
    20HE09, 11, 13, 14, 15 and 20HF50, 51, 52 .............................. 0.010 in (0.25 mm)
    20HE36, 37, 99 ............................................................................ 0.008 in (0.20 mm)

## Lubrication system (all models)
System pressure (minimum):
    Idling ........................................................................................... 10 lbf/in² (0.7 bar)
    Running ........................................................................................ 55 lbf/in² (3.8 bar)

**Oil filter (Turbo models)** ............................................................. Champion B103

*Cooling system*
Test pressure................................................................................... 11 to 14 lbf/in² (0.8 to 1.0 bar)

*Fuel and exhaust systems*
**Carburettor models**
Fuel pump (later models):
  Type .......................................................................................... Pierburg electric self-priming, centrifugal
  Delivery pressure ...................................................................... 4.0 lbf/in² (0.3 bar)

**Carburettor models (1984 to 1988)**
Identification:
  Manual gearbox ......................................................................... FZX 1438 to 1465
  Automatic transmission ........................................................... FZX 1439
Piston damper .................................................................................. LZX 1511
Idling speed:
  Manual gearbox ......................................................................... 600 to 700 rpm
  Automatic gearbox .................................................................... 775 to 875 rpm
Fast idle speed:
  Manual gearbox ......................................................................... 1050 to 1150 rpm
  Automatic transmission ........................................................... 1150 to 1250 rpm
Fuel octane rating ............................................................................ 97 RON leaded

**Carburettor models (1988)**
*As 1984 to 1988 models except for the following*
Idling speed:
  Automatic transmission ........................................................... 700 to 800 rpm
Fuel octane rating ............................................................................ 95 RON unleaded or 97 RON leaded

**Carburettor models (1989-on)**
*As 1984 to 1988 models except for the following*
Type.................................................................................................. SU HIF 44E with electronic idle speed control and cold drive enrichment
Identification:
  Manual gearbox ......................................................................... FZX 1499
  Automatic transmission ........................................................... FZX 1500
Piston needle ................................................................................... BGU
Base idle speed:
  Manual and automatic ............................................................. 550 to 650 rpm
Throttle position speed.................................................................... 700 to 750 rpm
ECU controlled speed ..................................................................... 725 rpm
Fuel octane rating ............................................................................ 95 RON unleaded or 97 RON leaded

**Electronic fuel injection (MG EFi, 1984 to 1986)**
Fuel pressure regulator:
  Make.......................................................................................... Lucas 4RV 73239 or 73279
  Fuel pressure:
    With ECU 84399 and pressure regulator 73239 ............. 43.5 lbf/in² (3.0 bar) above manifold depression
    With ECU 84498 or 84413 and pressure regulator 73279... 36.2 lbf/in² (2.5 bar) above manifold depression
Throttle potentiometer ................................................................... JZX 2134
Fuel ECU ......................................................................................... Lucas 11CU 84399, 84498 or 84413
CO mixture:
  With ECU 84399 and pressure regulator 73239................... 1.0 and 1.5%
  All other versions .................................................................... 2.0 to 2.5%
Fuel octane rating ............................................................................ 97 RON leaded

**Electronic fuel injection (Vanden Plas EFi, 1985 to 1986)**
*As MG EFi except for the following*
Fuel pressure regulator:
  Make.......................................................................................... Lucas 4RV 73279
  Fuel pressure........................................................................... 36.2 lbf/in² (2.5 bar) above manifold depression
Fuel ECU ......................................................................................... Lucas 11CU 84498 or 84413
CO mixture ....................................................................................... 2.0 to 2.5%

**Electronic fuel injection (all models, 1987-on)**
*As earlier models except for the following*
Fuel pressure regulator:
  Make.......................................................................................... Lucas 8RV 73310
  Fuel pressure........................................................................... 36.2 lbf/in² (2.5 bar) above manifold depression
Fuel ECU:
  Manual gearbox ......................................................................... Lucas 11CU-84598 AUU 1313
  Automatic transmission ........................................................... Lucas 11CU-84725 AUU 1337
ECU controlled idle speed:
  Manual gearbox ......................................................................... 700 to 790 rpm
  Automatic transmission ........................................................... 705 to 795 rpm
  Setting with air valve closed .................................................. 600 to 650 rpm
CO mixture ....................................................................................... 2.0 to 2.5%

## Electronic fuel injection (1989-on, unleaded fuel specification without MEMS MPi)
*As for 1987-on models except for the following*
Fuel octane rating ........................................................................................ 95 RON unleaded or 97 RON leaded

## Electronic fuel injection (1989-on, unleaded fuel specification with MEMS MPi)
Fuel pressure regulator:
    Make ................................................................................................ AUU 1630
    Fuel pressure ................................................................................. 36.2 lbf/in² (2.5 bar) above manifold depression
Throttle body ............................................................................................. MHB 10018
MEMS ECU .................................................................................................. AUU 1566
Fuel injectors ............................................................................................. ADU 8204 or AUU 1656
Fuel temperature thermistor (manual gearbox only) ................................ ADU 9396 or AUU 1251
ECU controlled idle speed .......................................................................... 725 to 775 rpm
CO Mixture .................................................................................................. 2.0 to 3.5%
Fuel octane rating ...................................................................................... 95 RON unleaded, 97 RON leaded

## Electronic fuel injection (1991-on, unleaded fuel specification with catalytic converter)
Fuel pressure regulator:
    Make ................................................................................................ MKW 10005
    Operating pressure (nominal) ........................................................ 34.8 to 39.2 lbf/in² (2.4 to 2.7 bar) above manifold depression
Fuel temperature sensor ............................................................................ ADU 8225
Fuel ECU .................................................................................................... MEQ 10072
Potentiometer voltage (throttle closed) .................................................... 300 to 340 mV
Base idle speed .......................................................................................... 725 to 775 rpm
ECU controlled idle speed (hot) ................................................................. 800 to 900 rpm
CO mixture .................................................................................................. 0.5% maximum
Fuel octane rating (minimum) .................................................................... 95 RON unleaded only

## Turbocharged models (1985 to 1988)
Fuel pump:
    Type ................................................................................................. Electric self-priming centrifugal
    Make ................................................................................................ Lucas 4FP
    Delivery pressure ........................................................................... 4.0 lbf/in² (0.3 bar)
Carburettor:
    Type ................................................................................................. SU HIF 44E variable choke with electronic mixture control
    Specification number ...................................................................... FZX 1440 or FZX 1496
    Piston spring colour ....................................................................... Green
    Jet size ............................................................................................ 0.100 in
    Needle identification ...................................................................... BGB or BGQ
    Adjustment data:
        Fast idle rod minimum clearance ........................................... 0.005 in (0.13 mm)
        Throttle lever lost motion gap ................................................ 0.060 to 0.080 in (1.52 to 2.03 mm)
        Float height ............................................................................. 0.040 to 0.060 in (1.00 to 1.52 mm)
        Idling speed ............................................................................ 800 to 900 rpm
        Fast idle speed ....................................................................... 1200 to 1300 rpm
        CO content ............................................................................. 1.5 to 2.5%
Turbocharger:
    Turbocharger waste gate opening pressure .................................. 10.0 lbf/in² (0.7 bar)
    Turbocharger dump valve opening pressure ................................. 12.0 lbf/in² (0.8 bar)
    Turbocharger shaft radial clearance ............................................. 0.003 to 0.006 in (0.08 to 0.15 mm)
    Turbocharger shaft axial endfloat ................................................. 0.005 to 0.004 in (0.013 to 0.081 mm)
Intercooler thermostatic valve:
    Maximum opening temperature ..................................................... 129°F (49°C)
    Minimum closing temperature ....................................................... 105°F (41°C)
Fuel octane rating ...................................................................................... 97 RON leaded

## Turbocharged models (1989)
*As earlier models except for the following*
Carburettor:
    Type ................................................................................................. SU HIF 44E with electronic idle speed control and cold drive enrichment
    Specification number ...................................................................... FZX 1501
    Piston damper ................................................................................. LZX 1511
    Piston needle .................................................................................. BGQ
Base idle speed .......................................................................................... 600 to 700 rpm
Throttle position speed ............................................................................... 750 to 800 rpm
ECU controlled speed ................................................................................. 785 rpm
CO mixture .................................................................................................. 1.5 to 3.0%

## Turbocharged models (1990-on)
Type ............................................................................................................ SU HIF 44E with electronic idle speed control and cold drive enrichment
Specification number .................................................................................. FZX 1501

## Turbocharged models (1990-on) (continued)

| | |
|---|---|
| Piston damper | LZX 2085 |
| Piston needle | BGQ |
| Piston spring colour | Green |
| Jet size | 0.09 in |
| Base idle speed | 600 to 700 rpm |
| Throttle position speed | 775 rpm |
| CO mixture | 1.5 to 3.0% (hot) |
| Fuel octane rating (minimum) | 95 RON, leaded or unleaded |

## Torque wrench settings

| | lbf ft | Nm |
|---|---|---|
| Turbocharger to exhaust manifold | 27 | 37 |
| Exhaust elbow to turbocharger | 16 | 22 |

## *Ignition system*
## Carburettor models (1984 to 1988)

| | |
|---|---|
| Ignition coil: | |
|    Type | GCL 139 or GCL 143 |
|    Primary resistance at 20°C (68°F) | 0.71 to 0.81 ohm |
|    Current consumption (idling) | 0.5 amp |
| Programmed ignition: | |
|    Distributor cap | Lucas 544-03944 |
|    Distributor rotor | Lucas 544-04286 |
|    Crankshaft sensor: | |
|       Manual gearbox | Lucas 84198 |
|       Automatic transmission | Lucas 84199 |
| Ignition timing (at 1500 rpm)* | |
|    Vacuum connected | 37 to 42° BTDC |
|    Vacuum disconnected | 20 to 26° BTDC |
| Spark plugs: | |
|    Type | Champion RN9YCC or RN9YC |
|    Gap | 0.8 mm (0.032 in) |

*\* Non-adjustable, for checking purposes only*

## Carburettor models (1988, unleaded fuel specification)

*As 1984 to 1988 models except for the following*

| | |
|---|---|
| Programmed ignition ECU | Lucas AB17-84792 |
| Ignition timing (at 1500 rpm)*: | |
|    Vacuum connected | 25 to 35° BTDC |
|    Vacuum disconnected | 17° BTDC |

*\* Non-adjustable, for checking purposes only*

## Carburettor models (1989-on)

*As 1984 to 1988 models except for the following*

| | |
|---|---|
| Programmed ignition: | |
|    Ignition and fuel ECU | AUU 1527 |
| Ignition timing (at less than 1500 rpm, throttle switch closed and vacuum connected) | 15 to 25° BTDC |
| Spark plugs: | |
|    Type | Champion RN7YCC or RN7YC |
|    Gap | 0.8 mm (0.032 in ) |

## Electronic fuel injection models (MG EFi, 1984 to 1986)

| | |
|---|---|
| Ignition coil: | |
|    Type | GCL 139 or GCL 143 |
|    Primary resistance at 20°C (68°F) | 0.71 to 0.81 ohms |
|    Current consumption (idling) | 0.5 amp |
| Programmed ignition: | |
|    Distributor cap | Lucas 544-03944 |
|    Distributor rotor | Lucas 544-04286 |
|    Direction of rotation | Anti-clockwise |
|    Knock sensor | Lucas or Lamerholm VP50/1- M12 |
|    Crankshaft sensor | Lucas 84198 |
|    Ignition ECU | Lucas AB17-84381 |
| Ignition timing (at 1500 rpm)*: | |
|    Vacuum connected | 40° BTDC |
|    Vacuum disconnected | 15° BTDC |
| Spark plugs: | |
|    Type | Champion RN7YCC or RN7YC |
|    Gap | 0.8 mm (0.032 in) |

*\* Non-adjustable, for checking purposes only*

## Electronic fuel injection models (Vanden Plas EFi, 1985 to 1986)
*As MG EFi except for the following*
Crankshaft sensor:
Manual gearbox ............................................................................... Lucas 84198
Automatic transmission.................................................................... Lucas 84199

## Electronic fuel injection models (all models, 1987 to 1988)
*As earlier models except for the following*
Ignition coil...................................................................................... GCL 143
Ignition timing (at 1500 rpm)*:
Vacuum connected ......................................................................... 40° BTDC
Vacuum disconnected..................................................................... 14° BTDC
*\* Non-adjustable, for checking purposes only*

## Electronic fuel injection models (1989-on, unleaded fuel specification without MEMS MPi)
*As 1989-on models except for the following*
Ignition ECU ..................................................................................... Lucas AB17-80109
Ignition timing (at 1500 rpm):*
Vacuum connected ......................................................................... 40° BTDC
Vacuum disconnected..................................................................... 17° BTDC
*\* Non-adjustable, for checking purposes only*

## Electronic fuel injection models (1989-on, unleaded fuel specification with MEMS MPi)
Ignition coil...................................................................................... GCL 143
Programmed ignition:
Distributor cap ............................................................................... AUU 1186
Distributor rotor ............................................................................. AUU 1690
Direction of rotation ...................................................................... Anti-clockwise
Knock sensor .................................................................................. ADU 8229
Crankshaft sensor:
Manual gearbox............................................................................ ADU 7343
Automatic transmission ............................................................... ADU 7342
Ignition timing (at less than 1500 rpm, throttle switch closed, vacuum
connected)* .................................................................................... 18° + 8° — 6° BTDC
Spark plugs:
Type ................................................................................................ Champion RN7YCC or RN7YC
Gap ................................................................................................. 0.8 mm (0.032 in)
*\* Non-adjustable, for checking purposes only*

## Electronic fuel injection models (1991-on, unleaded fuel specification with catalytic converter)
Ignition coil:
Type ................................................................................................ NEC 10013 or 10014
Ignition timing (at 1500 rpm):
Vacuum connected ......................................................................... 26 to 36° BTDC
Vacuum disconnected..................................................................... 16 to 18° BTDC
Programmed  ignition:
Distributor cap ............................................................................... NJD 10005
Distributor rotor ............................................................................. AUU 1690
Knock sensor .................................................................................. ADU 8229
Crankshaft sensor:
Manual gearbox............................................................................ ADU 7342
Automatic transmission ............................................................... ADU 7343
Ignition ECU ................................................................................... NNN 10016
Spark plugs:
Type ................................................................................................ Champion RN7YCC or RN7YC
Gap ................................................................................................. 0.8 mm (0.032 in)

## Turbocharged models (1985 to 1988)
Ignition coil:
Type ................................................................................................ GCL 139 or GCL 143
Primary resistance at 20°C (68°F)................................................. 0.71 to 0.81 ohm
Current consumption (idling) ......................................................... 0.5 amp
Programmed ignition:
Distributor cap ............................................................................... Lucas 544-03944
Distributor rotor ............................................................................. Lucas 544-04286
Direction of rotation ...................................................................... Anti-clockwise
Crankshaft sensor .......................................................................... Lucas 84198
Ignition ECU ................................................................................... Lucas AB17-84233
Ignition timing (at 2000 rpm)*:
Vacuum connected ......................................................................... 47° BTDC
Vacuum disconnected..................................................................... 24° BTDC
Spark plugs
Type ................................................................................................ Champion RN7YCC or RN7YC
Gap ................................................................................................. 0.8 mm (0.032 in)
*\*Non-adjustable, for checking purposes only*

## Turbocharged models (1989)
*As earlier models except for the following*
Programmed ignition:

Ignition and fuel ECU ............................................................ AUU 1528
Ignition timing (at less than 1500 rpm, vacuum connected)* ................... 15 to 25° BTDC
*Non-adjustable, for checking purposes only*

## Turbocharged models (1990-on)
Ignition coil:

Type ........................................................................ AUU 1326 or ADU 8779
Primary resistance at 20°C (68°F) ........................................... 0.71 to 0.81 ohm
Current consumption (idling) ................................................ 0.5 amp
Ignition timing (at 1500 rpm maximum and throttle switch closed):

Vacuum connected ........................................................... 20° BTDC (non-adjustable)
Programmed ignition:

Distributor cap ............................................................. NJD 10005
Distributor rotor ........................................................... AUU 1690
Direction of rotation ....................................................... Anti-clockwise
Knock sensor ................................................................ ADU 8229
Crankshaft sensor ........................................................... ADU 7342
Ignition and fuel ECU ....................................................... MEQ 10069
Spark plugs:

Type ........................................................................ Champion RN7YCC or RN7YC
Gap ......................................................................... 0.8 mm (0.032 in)

## *Clutch*
Clutch pedal free play ........................................................ 0.5 to 1.2 in (12.0 to 28.0 mm)

## *Manual gearbox*
## General
Gearbox code numbers:

All models except MG ........................................................ L6 AR
MG EFi models up to 1986 .................................................... G6 AR
MG EFi models, 1986 on ...................................................... K6 AR
MG Turbo models ............................................................. K7 AR

Gearbox ratios, up to 1988:

| | **Gearbox code numbers** | | | |
|---|---|---|---|---|
| | *L6 AR* | *G6 AR* | *K6 AR* | *K7 AR* |
| 1st | 2.92:1 | 3.25:1 | 2.92:1 | 2.92:1 |
| 2nd | 1.75:1 | 1.89:1 | 1.75:1 | 1.75:1 |
| 3rd | 1.14:1 | 1.33:1 | 1.24:1 | 1.22:1 |
| 4th | 0.85:1 | 1.04:1 | 0.96:1 | 0.94:1 |
| 5th | 0.65:1 | 0.85:1 | 0.77:1 | 0.76:1 |
| Reverse | 3.00:1 | 3.00:1 | 3.00:1 | 3.00:1 |
| Final drive ratio | 3.937:1 | 3.875:1 | 3.937:1 | 3.647:1 |

Gearbox ratios, 1989-on (as above except for the following):

| | **Gearbox code number** |
|---|---|
| | *K6 AR* |
| 1st | 2.92:1 |
| 2nd | 1.75:1 |
| 3rd | 1.22:1 |
| 4th | 0.94:1 |
| 5th | 0.76:1 |
| Reverse | 3.00:1 |

Countershaft gear endfloat ..................................................... 0.002 to 0.003 in (0.03 to 0.08 mm)

## *Automatic transmission*
## General
Type ........................................................................... ZF (4HP14) four-speed, hydraulically-controlled, with top gear mechanical 'lock-up'
Gear ratios

1st .......................................................................... 2.41:1
2nd .......................................................................... 1.37:1
3rd .......................................................................... 1.00:1
4th .......................................................................... 0.74:1
Reverse ...................................................................... 2.83:1
Final drive ratio ............................................................ 3.83:1
Fluid type/specification ....................................................... Dexron TQII type ATF (Duckhams Uni-Matic or D-Matic)
Fluid capacity:

From dry ..................................................................... 10.5 pints (6.0 litres)
At service fluid change ...................................................... 3.5 pints (2.0 litres)

| Torque wrench settings | lbf ft | Nm |
|---|---|---|
| Driveplate to torque converter | 24 | 32 |
| Selector cable clamp bolt | 6 | 8 |
| Starter inhibitor switch | 31 | 42 |
| Fluid cooler centre screw | 37 | 50 |
| Sump pan screws | 7 | 10 |
| Drain plugs | 11 | 15 |
| Valve block retaining screws | 7 | 10 |
| Fluid strainer retaining screws | 7 | 10 |
| Brake band adjusting screw | 7 | 10 |
| Brake band adjusting screw locknut | 59 | 80 |
| Dipstick/filler tube union nut | 18 | 25 |
| Torque converter housing to engine: | | |
| M10 bolts | 38 | 52 |
| M12 bolts | 67 | 91 |

*Braking system*
## Front brakes
| | |
|---|---|
| Minimum disc thickness | 0.826 in (21.0 mm) |
| Maximum disc thickness variation | 0.0008 in (0.020 mm) |
| Maximum drive flange run-out | 0.0025 in (0.06 mm) |

## Rear brakes
| | |
|---|---|
| Wheel cylinder diameter: | |
| Saloon models | 0.70 in (17.8 mm) |
| Estate models | 0.75 in (19.05 mm) |

*Electrical system*
## Alternator (1988-on)
| | |
|---|---|
| Type | Lucas A127/65 |
| Maximum output | 65 Amps |
| Brush length: | |
| New | 0.67 in (17.0 mm) |
| Minimum | 0.20 in (5.0 mm) |

## Starter motor
| | |
|---|---|
| Tyoe and application: | |
| Manual gearbox models | Lucas M79 or 9M90 |
| Automatic transmission | Lucas M78R or 9M90 |
| Minimum brush length | 0.15 in (3.8 mm) |
| Commutator minimum diameter | 1.134 in (28.8 mm) |

*Suspension and steering*
## Front suspension
| | |
|---|---|
| Coil spring free length – standard: | |
| Saloon models (1988-on): | |
| Turbo and MGi | 15.47 in (393.0 mm) |
| Vanden Plas | 16.69 in (424.0 mm) |
| All other models | 15.16 in (385.0 mm) |
| Estate models (1988 to 1989): | |
| Vanden Plas | 15.83 in (402.0 mm) |
| All other models | 15.47 in (393.0 mm) |
| Estate models (1989-on): | |
| Manual gearbox | 15.83 in (402.0 mm) |
| Automatic transmission | 15.47 in (393.0 mm) |
| Trim height (measured from the centre of the front hub to the edge of the wheel arch): | |
| Saloon models (1988) | 14.8 to 15.8 in (376.0 to 402 mm) |
| Estate models (1988) | 15.2 to 16.2 in (386.0 to 412.0 mm) |
| Saloon models (1989-on) | 14.5 to 15.5 in (370 to 396 mm) |
| Estate models (1989-on) | 15.0 to 16.0 in (380 to 406 mm) |
| Maximum front hub bearing side play | 0.060 in (1.5 mm) measured at wheel rim |

## Rear suspension
| | |
|---|---|
| Coil spring free length – standard: | |
| Saloon models: | |
| All models up to VIN 131623 | 13.5 in (343.0 mm) |
| From VIN 131624: | |
| Turbo and MGi | 13.82 in (351.0 mm) |
| All other models | 14.53 in (396.0 mm) |
| Estate models with standard suspension | 14.53 in (369.0 mm) |
| Estate models with self-levelling suspension and Vanden Plas models | 15.43 in (392.0 mm) |

### Rear suspension (continued)

Trim height (measured from the centre of the rear hub to the edge of the wheel arch):
<br>   Saloon models
<br>      All models up to VIN 131623 ............................................. 14.0 to 15.0 in (354 to 380 mm)
<br>      All models from VIN 131624 ............................................. 14.5 to 15.5 in (370 to 396 mm)
<br>   Estate models ............................................................................. 14.97 to 15.97 in (380 to 406 mm)

### Steering

Camber angle
<br>   Up to 1988 ................................................................................. 0° 8' positive to 0° 34' negative
<br>   1988-on
<br>      Strut type A ....................................................................... 0° 6' positive to 0° 36' negative
<br>      Strut type B ....................................................................... 0° 9' positive to 0° 51' positive
<br>Castor angle (1988-on) ............................................................... 0° 7' positive 1° 07' positive
<br>Steering axis (1989-on):
<br>   Strut type A ............................................................................... 12° to 13°
<br>   Strut type B ............................................................................... 12° 03' to 13° 03'
<br>Toe setting (1988-on):
<br>   Turbo and MGi .......................................................................... 0° 8' to 0° 24' toe-out
<br>   All other models ....................................................................... parallel ± 0° 8'

### Roadwheels and tyres (1986 models)

Wheel size:
<br>   Standard wheel .......................................................................... 120 x 365 mm TD type
<br>   Alloy wheel ................................................................................. 135 x 365 mm TD type
<br>Tyre size:
<br>   Standard ...................................................................................... 180/65 R365 84S or 84T steel braced radial ply
<br>   With alloy wheels except Turbo models ............................. 180/65 R365 84T or 84H steel braced radial ply
<br>   Turbo models ............................................................................. 190/65 R365 88H TD steel braced radial ply

### Roadwheels and tyres (1987 and 1988 models)

Wheel size:
<br>   All models except Turbo ......................................................... 5½J x 14
<br>   Turbo ........................................................................................... 135 x 365 mm TD type
<br>Tyre size:
<br>   All models except EFi and Turbo ......................................... 185/65 SR
<br>   EFi models .................................................................................. 185/65 HR
<br>   Turbo ........................................................................................... 190/65 VR 365

### Roadwheels and tyres (1989-on models)

Wheel size:
<br>   Standard wheel .......................................................................... 5½J x 14
<br>   Models with alloy wheels:
<br>      2.0 and GSi ....................................................................... 5½J x 14
<br>   2.0 GTi (optional) and MG ..................................................... 6 x 15
<br>Tyre size:
<br>   Standard wheel .......................................................................... 185/65 TR 14
<br>   Models with alloy wheels
<br>      2.0 and GSi ....................................................................... 185/65 TR 14
<br>      2.0 GTi (optional) and MG ........................................... 195/55 VR 15

### Tyre pressures (cold)

1986 to 1988 models:
<br>   Saloon:
<br>      Front ................................................................................... 28 psi (2.0 bar)
<br>      Rear .................................................................................... 28 psi (2.0 bar)
<br>   Estate:
<br>      Front ................................................................................... 28 psi (2.0 bar)
<br>      Rear .................................................................................... 31 psi (2.2 bar)
<br>1989-on models
<br>   Saloons with 185/65 x 14 tyres
<br>      Front ................................................................................... 26 psi (1.8 bar)
<br>      Rear .................................................................................... 28 psi (2.0 bar)
<br>   Saloons with 195/55 x 15 tyres
<br>      Front ................................................................................... 28 psi (2.0 bar)
<br>      Rear .................................................................................... 28 psi (2.0 bar)
<br>   Estates with 185/65 x 14 tyres:
<br>      Front ................................................................................... 28 psi (2.0 bar)
<br>      Rear .................................................................................... 31 psi (2.2 bar)
<br>   Estates with 195/55 x 15 tyres:
<br>      Front ................................................................................... 28 psi (2.0 bar)
<br>      Rear .................................................................................... 31 psi (2.2 bar)

| **Torque wrench settings** | **lbf ft** | **Nm** |
|---|---|---|
| Front anti-roll bar bush-to-lower arm nut: | | |
| 10 mm nut ........................................................................ | 13 | 18 |
| 12 mm nut ........................................................................ | 30 | 40 |
| Driveshaft retaining nut: | | |
| Castellated nut................................................................ | 140 | 190 |
| Staked nut ....................................................................... | 150 | 203 |
| Steering wheel bolt............................................................. | 24 | 33 |

## *General dimensions, weights and capacities*

### Dimensions – **1989-on**

| | |
|---|---|
| Overall width (including mirrors)............................................... | 76.0 in (1930 mm) |
| Overall height (at kerb weight): | |
| Saloon............................................................................... | 56.0 in (1422 mm) |
| Estate (including integral roof rack)..................................... | 60.0 in (1524 mm) |
| Ground clearance (at kerb weight): | |
| Saloon: | |
| Excluding Turbo ............................................................ | 6.0 in (153.0 mm) |
| Turbo ............................................................................ | 6.5 in (162.0 mm) |
| Estate ................................................................................ | 6.5 in (162.0 mm) |
| Wheelbase ........................................................................... | 101.25 in (2570 mm) |
| Track: | |
| With 185/65 x 14 tyres: | |
| Front .............................................................................. | 57.5 in (1460 mm) |
| Rear................................................................................ | 56.75 in (1439 mm) |
| With 195/55 x 15 tyres: | |
| Front .............................................................................. | 58.25 in (1480 mm) |
| Rear................................................................................ | 57.5 in (1459 mm) |
| Turning circle | |
| All except Turbo................................................................ | 34 ft 6 in (10.5 m) |
| Turbo ................................................................................ | 35 ft 1 in (10.7 m) |

### Kerb weights (approximate) – **1989-on**

| | |
|---|---|
| Saloon: | |
| Manual gearbox: | |
| L............................................................................... | 2320 lb (1055 kg) |
| SL............................................................................. | 2365 lb (1075 kg) |
| GTi............................................................................ | 2375 lb (1080 kg) |
| GSi............................................................................ | 2355 lb (1070 kg) |
| Automatic transmission: | |
| L............................................................................... | 2365 lb (1075 kg) |
| SL............................................................................. | 2410 lb (1095 kg) |
| GTi............................................................................ | 2420 lb (1110 kg) |
| GSi............................................................................ | 2400 lb (1090 kg) |
| MGi ................................................................................... | 2376 lb (1080 kg) |
| MG Turbo ........................................................................... | 2420 lb (1100 kg) |
| Estate: | |
| Manual gearbox: | |
| L............................................................................... | 2455 lb (1115 kg) |
| SL............................................................................. | 2500 lb (1135 kg) |
| GTi............................................................................ | 2500 lb (1135 kg) |
| GSi............................................................................ | 2530 lb (1150 kg) |
| Automatic transmission: | |
| L............................................................................... | 2500 lb (1135 kg) |
| SL............................................................................. | 2540 lb (1155 kg) |
| GTi............................................................................ | 2540 lb (1155 kg) |
| GSi............................................................................ | 2575 lb (1170 kg) |
| Maximum gross vehicle weight: | |
| Saloon............................................................................... | 3435 lb (1560 kg) |
| Estate ................................................................................ | 3740 lb (1700 kg) |
| Maximum towing weight to restart on 12% (1 in 8) gradient: | |
| Saloon and Estate ............................................................. | 2690 lb (1220 kg) |
| MG ................................................................................... | 2755 lb (1250 kg) |

### Capacities

| | |
|---|---|
| Engine oil (refill with filter change): | |
| Turbo models (up to 1988)................................................. | 6.25 pt (3.6 litres) |
| All models (1988-on)......................................................... | 8.5 pt (4.9 litres) |
| Manual gearbox (1988-on)..................................................... | 4.0 pt (2.2 litres) |
| Automatic transmission (1988-on): | |
| Refill ................................................................................ | 3.5 pt (2.0 litre) |
| Total capacity .................................................................. | 10.5 pt (6.0 litres) |
| Final drive ........................................................................ | 1.25 pt (0.75 litre) |

## 3   Routine maintenance

1   The maintenance schedule at the beginning of this manual is still relevant, but note the following additions.
2   On Turbo models change the engine oil and renew the oil filter every 6000 miles (10 000 km) or 6 months, whichever occurs first. At the same service interval top up the carburettor piston damper.
3   On models fitted with a fuel evaporative emission control (EVAP) system, inspect the system components and connections for condition and security every 12 000 miles (20 000 km) or 12 months, whichever occurs first.

## 4   Engine

### Engine and gearbox assembly (MG Turbo models) – removal and refitting

1   Drain engine oil as described in Chapter 1, cooling system as described in Chapter 2 and gearbox oil as described in Chapter 6.
2   Disconnect the battery negative and positive terminals. Always disconnect the negative terminal first and reconnect last.
3   Refer to Chapter 11 and remove the bonnet.
4   Refer to Section 6 of this Supplement and remove the air cleaner.
5   Refer to Chapter 9 and remove the starter motor.
6   Disconnect the hoses from the turbo intercooler pipes assembly and disengage the clutch cable from the retaining clip. Remove the intercooler pipes assembly from its mounting on the engine.
7   Disconnect the coil HT lead at the distributor cap, and the LT leads at the coil terminals.
8   Remove the battery, followed by the battery tray complete with coil.
9   Remove the cooling fan assembly from the radiator.
10   Disconnect the clutch cable from the operating lever and engine mounting bracket, as described in Chapter 5.
11   Disconnect the wiring harness multi-plugs from the ignition electronic control unit and from the main wiring harness.
12   Disconnect the earth cable at the gearbox.
13   Detach the carburettor cooling fan motor tube and disconnect the motor wiring at the multi-plug.
14   Undo the retaining bolts, and remove the carburettor cooling fan motor assembly.
15   Disconnect the radiator bottom hose at the water pump outlet.
16   Disconnect the radiator top hose and expansion tank hose at the water outlet elbow.
17   Disconnect the heater hoses at the heater pipes.
18   Slacken the screw securing the accelerator cable to the connector on the carburettor linkage. Withdraw the cable from its support bracket, and move it aside.
19   Disconnect the wiring at the coolant thermistor, carburettor cooling fan and oil pressure switch.
20   Disconnect the carburettor vacuum pipe and the engine breather pipe from the brake servo one-way valve.
21   Undo the brake servo banjo union bolt, withdraw the union, and recover the copper washers.
22   Separate the power-assisted steering fluid hoses from the pipes at the connections behind the sump. Allow the fluid to drain into a suitable container, then seal the pipes and hoses to prevent dirt ingress.
23   Disconnect the fuel and vent hoses at the carburettor, and plug the fuel hose to prevent fuel spillage.
24   Disconnect the boost signal hose, then remove the plenum chamber from the carburettor (refer to Section 6 of this Supplement if necessary).
25   Undo the carburettor stay plate retaining bolts, then remove the stay plate and recover the gasket.
26   Disconnect the wiring multi-plug from the carburettor stepper motor, then remove the carburettor heat shield.
27   Disconnect the turbo wastegate air pressure hose at the T-piece connector.
28   Disconnect the multi-plug from the knock sensor on the cylinder block, and the two reversing lamp wires at their connectors.
29   Undo the speedometer cable retaining plate bolt and lift out the speedometer cable and pinion assembly.

30   Disconnect the crankshaft sensor wiring multi-plug at the rear of the engine.
31   Jack up the front of the car and support it on stands. Remove both front roadwheels.
32   Remove the access panels from inside both front wheel arches.
33   Undo the two nuts and bolts each side which secure the front suspension struts to the swivel hubs.
34   Using a stout screwdriver or flat bar, ease the driveshaft inner constant velocity joints out of the final drive assembly. Once the joints have been released, tip the swivel hub outwards at the top, as far as possible without straining the brake hoses unduly, then withdraw the inner joints fully from their locations.
35   Undo the two bolts securing the alternator heat shield to the cylinder block and remove the shield.
36   Release the retaining wire and disconnect the alternator multi-plug. Disconnect the oil pressure switch wire at the pump.
37   Undo the retaining nuts and separate the exhaust downpipe flange from the manifold. Recover the flange gasket.
38   Remove the cover band then tap out the roll pin securing the gearchange rod to the gearchange shaft. Undo the small bolt and remove the dished washer, steady bar and flat washer from the gearbox case. Slide the gearchange rod rearwards and off the shaft.
39   Undo and remove the two bolts each side securing the rear engine mounting to the gearbox.
40   Undo the two bolts securing the engine front snubber bracket to the gearbox, and the two bolts and eccentric flat washers securing the snubber cup to the chassis member. Remove the snubber cap and bracket assembly.
41   Attach a suitable hoist to the engine using chains or rope slings. Preferably attach the chains to home made brackets fastened to the cylinder head studs. Raise the hoist to just take the weight of the engine.
42   Undo the bolts securing the left-hand engine mounting bracket to the gearbox, and the through-bolt and nut securing the mounting to the bracket on the chassis member. Lower the engine slightly and remove the mounting. Recover the two flat washers located each side of the mounting rubber.
43   Undo the bolts securing the mounting bracket to the chassis member and remove the bracket.
44   Undo the screw securing the cooling system expansion tank to its support bracket and the bolts securing the bracket to the chassis member. Remove the bracket.
45   Undo the bolts securing the right-hand engine mounting to the chassis member. The bolts are accessible from above and from under the wheel arch.
46   Make a final check that everything attaching the engine and gearbox to the car has been disconnected and moved well clear.
47   Raise the engine and gearbox carefully, moving them slightly to clear any protrusions. When the unit has been raised sufficiently, draw the hoist forward to bring the engine and gearbox over the front body panel then lower the assembly to the floor.
48   Refitting is a straightforward reverse of the removal sequence, bearing in mind the following points:

(a)   Refit all the engine mounting bolts loosely and then tighten them in the sequence – right-hand mounting, left-hand mounting, rear mounting, snubber bracket. When all the mountings have been tightened, centralise the snubber cup around the snubber rubber and tighten the cup bolts

(b)   Reconnect the clutch cable with reference to Chapter 5

(c)   Adjust the accelerator cable as described in Chapter 3

(d)   Refill the cooling system as described in Chapter 2 and the gearbox with oil as described in Chapter 6. Refill the engine with oil as described in Chapter 1

(e)   Fill and bleed the power-assisted steering as described in Chapter 10

### Cylinder head (MG Turbo models) – removal and refitting

49   Disconnect the battery negative terminal.
50   Drain the cooling system as described in Chapter 2.
51   Remove the air cleaner as described in Section 6 of this Supplement.
52   Disconnect the hoses from the turbo intercooler pipes assembly, and disengage the clutch cable from the retaining clip. Remove the intercooler pipes assembly from its mounting on the engine.
53   Slacken the power-assisted steering pump adjustment and pivot

bolts, move the pump towards the engine and remove the drivebelt. Release the fluid hoses from the sump bracket to the cylinder head. Lift off the pump with hoses still attached and move it aside. Ensure that there is no fluid spillage and that the hoses are not strained.

54    Detach the carburettor cooling fan motor tube and disconnect the motor wiring at the multi-plug.

55    Slacken the screw securing the accelerator cable to the connector on the linkage. Withdraw the cable from its support bracket and move it aside.

56    Disconnect the radiator top hose and expansion tank hose at the water outlet elbow.

57    Disconnect the wiring at the coolant thermistor and carburettor cooling fan switch.

58    Disconnect the carburettor vacuum pipe and the engine breather pipe from the brake servo one-way valve.

59    Undo the brake servo banjo union bolt, withdraw the union and recover the copper washers.

60    Disconnect the fuel and vent hoses at the carburettor and plug the fuel hose to prevent spillage.

61    Disconnect the boost signal hose, then remove the plenum chamber from the carburettor (refer to Section 6 of this Supplement if necessary).

62    Undo the carburettor stay plate retaining bolts, then remove the stay plate and recover the gasket.

63    Disconnect the carburettor stepper motor multi-plug, then remove the carburettor heat shield.

64    Refer to Section 6 of this Supplement if necessary and remove the carburettor and inlet manifold.

65    Undo the nuts securing the exhaust manifold to the cylinder head.

66    Jack up the front of the car and support it on stands. Remove the right-hand side front roadwheel and the access panel from inside the wheel arch.

67    Disconnect the wiring at the oil pressure switch and remove the switch and switch extension.

68    Detach the turbocharger oil feed pipe from the oil pump housing.

69    Disconnect and remove the turbocharger oil drain hose.

70    Release the exhaust manifold and remove the manifold gasket.

71    Turn the crankshaft so that the dimple on the camshaft sprocket rear face is aligned with the hole in the timing belt cover backplate. The notch in the crankshaft pulley should now be aligned with the corresponding notch in the deflector bracket below the water pump. In this position the crankshaft is at 90° BTDC, with No 1 piston on compression. Avoid turning the crankshaft for the remainder of the procedure.

72    Remove the timing belt cover, slacken the two tensioner retaining nuts and slip the timing belt off the camshaft sprocket.

73    Undo the nuts securing the carburettor stay brackets, intercooler pipes bracket and vacuum hose to the cylinder head studs.

74    Undo the bolt securing the dipstick tube to the cylinder head.

75    Detach the breather hose from the oil filler cap and remove the oil filler tube.

76    Remove the distributor cap and spark plug leads, followed by the rotor arm and shield.

77    Progressively slacken the cylinder head retaining bolts in the reverse sequence to that shown in Fig. 12.11 then remove the bolts.

78    Lift the cylinder head off the engine and recover the gasket. If the head is stuck, tap it upwards with a hide or plastic mallet to free it.

79    Before refitting the cylinder head, refer to paragraphs 184 to 190 in this Section regarding cylinder head gasket fitting.

80    Ensure that the crankshaft and camshaft are still positioned at 90° BTDC as described in paragraph 71.

81    Lower the cylinder head assembly onto the gasket, then refer to paragraphs 193 to 195 for cylinder head bolt tightening procedures.

82    Refit all the wiring, pipes, hoses and components to the cylinder head using the reverse sequence of removal. Adjust the timing belt tension as described in Chapter 1, Section 11.

83    Adjust the power-assisted steering pump drivebelt as described in Chapter 10, and the accelerator cable as described in Chapter 3.

84    Refit the air cleaner as described in Section 6 of this Supplement and refill the cooling system as described in Chapter 2.

## Cylinder head (MG Turbo models) – overhaul

85    The procedure is basically the same as that described for normally aspirated engines in Chapter 1, Section 16, but note that the valve

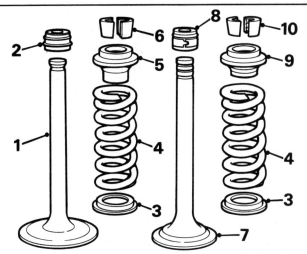

**Fig 12.1 Valve components – MG Turbo models (Sec 4)**

| 1 Inlet valve | 6 Collets |
| 2 Oil seal | 7 Exhaust valve |
| 3 Spring seat | 8 Oil seal |
| 4 Valve spring | 9 Spring top cup |
| 5 Spring top cup | 10 Collets |

collets, spring top cup, oil seal and valve stem are of a different configuration on Turbo models. Refer to Fig 12.1 for details. Store the components in their fitted order after removal.

86    Check the valve head protrusion above the valve seat, and compare the figure with the dimension given in the Specifications of this Supplement. If the protrusion is less than specified, even when checking with a new valve, it will be necessary to renew the valve seat insert. This, and renewal of the valve guides, can only be carried out by a dealer or suitably equipped engineering works.

## Valve stem oil seals – all later models

87    To reduce oil loss past the valve stems, a modified valve seal is fitted to all engines from VIN 554319. It differs from the earlier seal in having a combined valve seal lower seat. New inlet valves (with modified chrome flashing) are fitted to suit the new valve seal and it is essential that the new seal (part number UAM 7497) and inlet valve (part number UAM 7713) are used together (not mixed with the earlier seal/valve). The manufacturer recommends the use of special tool number 18G 1577 for removal and refitting of the seal. Note that on Turbo models, the new seal is only fitted to the inlet valves.

## Oil pump and filter (MG Turbo models) – general

88    The arrangement of components on the oil pump housing on Turbo models is slightly different to that of standard versions.

89    The oil filter is located at the base of the housing and can be reached easily from below.

90    The oil pressure relief valve is situated on the upper face of the housing and is retained by a split pin.

91    The overhaul procedures are not affected and remain as described in Chapter 1.

## Piston and connecting rod assemblies (MG Turbo models) – modifications

92    On Turbo models, the pistons are retained on the connecting rods by fully floating gudgeon pins which are retained by circlips.

93    After removing the piston and connecting rod assemblies from the engine using the procedures described in Chapter 1, extract the retaining circlips and lightly tap out the gudgeon pin using a brass drift.

94    Should wear be apparent on the small end bush in the connecting rod, this can be renewed separately. As a press is required for this operation and the new bush must be reamed to size after installation, the work should be entrusted to a dealer or engineering works.

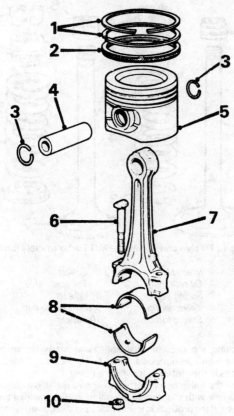

**Fig 12.2 Piston and connecting rod components – MG Turbo models (Sec 4)**

| | |
|---|---|
| 1 Compression rings | 6 Connecting rod bolt |
| 2 Oil control ring | 7 Connecting rod |
| 3 Circlips | 8 Big-end bearing shells |
| 4 Gudgeon pin | 9 Connecting rod cap |
| 5 Piston | 10 Connecting rod nut |

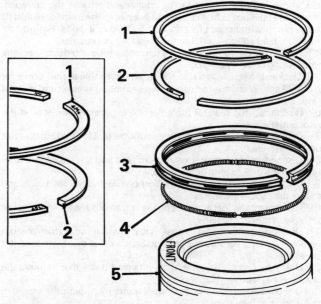

**Fig 12.3 Piston ring fitting details – MG Turbo models (Sec 4)**

| | |
|---|---|
| 1 Top compression ring | 4 Expander spring |
| 2 Second compression ring | 5 Piston |
| 3 Oil control ring | |

95   Also on Turbo models, the oil control piston ring is of the one-piece scraper type incorporating an expander spring (Fig. 12.3). When refitting the piston rings, position the oil control ring gap at 90° to the expander spring joint, and position all the ring gaps at 90° apart and away from the thrust side of the engine.

## Engine (20 HB onwards) – general description

96   During the latter part of 1985 various minor modifications were carried out to the 20 H engine and the unit was redesignated 20 HB.
97   The changes include larger diameter rear oil seals for the crankshaft and camshaft, shorter cylinder block, repositioned oil filter, oil pressure relief valve and alternator, modified timing belt tooth profile with changes to the sprockets and tensioner, and a modified water pump.
98   The 20 HB engine has had minor changes over the years, corresponding to HC, HD and HE. Apart from the changes described in this and the following Sections of this Supplement, all other operations remain unaffected by these modifications.

## Timing belt (20 HB onwards) – removal, refitting and adjustment

99   Disconnect the battery negative terminal.
100   Remove the alternator and water pump drivebelt and, where fitted, the power-assisted steering pump drivebelt.
101   Remove the timing belt cover.
102   Using a spanner or socket on the crankshaft pulley bolt, turn the crankshaft until the dimple in the rear edge of the camshaft sprocket is in line with the hole in the timing belt cover backplate. The notch in the crankshaft pulley should also be aligned with the pointer on the oil pump housing. In this position the crankshaft is at 90° BTDC with No 1 piston on its compression stroke. Avoid turning the crankshaft during subsequent operations.

103   Release the cooling system expansion tank cap slowly, using a cloth if the engine is hot.
104   Place a suitable container beneath the water pump, then disconnect the bottom hose at the pump outlet.
105   Slacken the timing belt tensioner and remove the timing belt from the sprockets.
106   If the timing belt is to be re-used, mark it with an arrow, using chalk, to indicate its running direction, and store it on edge while it is off the engine.
107   Renew the timing belt if there is any sign of cracking, fraying or uneven wear on the belt or teeth, or if there is any sign of oil contamination.
108   Check that the tensioner is fitted to the outer of the two tapped holes in the water pump boss and reposition it if necessary.
109   Check that the crankshaft and camshaft sprockets are still positioned as described in paragraph 102, then ensure that the sprocket teeth and the tensioner are clean and dry.
110   Fit the timing belt and centralize it on the sprockets. If a new belt is being fitted, equalize the tension on both sides by flexing both lengths inwards together, at a point midway between the two sprockets.
111   Mark the timing belt cover backplate with two parallel lines, one in line with the belt and the second 0.3 in (8.0 mm) to the outside of the first line. Make the marks on the backplate midway between the camshaft sprocket and water pump pulley (Fig. 12.5).
112   Move the tensioner in against the belt and tighten the retaining bolt. The tensioner can be moved by engaging the end of a socket bar in the square hole in the tensioner body. If the tensioner 'overcentres' without achieving tension on the belt, reposition it in the inner hole in the water pump boss.
113   Turn the crankshaft through two complete revolutions in the normal direction of rotation, then realign the crankshaft pulley notch with the timing pointer. Check that the camshaft sprocket dimple and backplate hole are also still aligned. If not, slacken the tensioner, move the belt one tooth either way as necessary on the camshaft sprocket, then retighten the tensioner. Repeat the above procedure until the belt is positioned correctly.
114   To set the tension accurately, attach a spring balance to the timing belt, using a suitable shaped length of stiff wire or thin rod positioned midway between the camshaft sprocket and water pump pulley.
115   Pull the spring balance until the belt is deflected as far as the outer of the two lines previously marked on the backplate. Observe the reading on the spring balance and compare this with the timing belt tension figures given in the Specifications (Chapter 1).

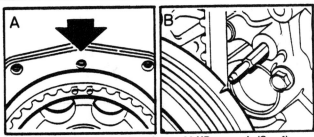

**Fig 12.4 Timing mark locations – 20 HB onwards (Sec 4)**

A  *Camshaft sprocket dimple aligned with hole in timing belt cover backplate (arrowed)*
B  *Crankshaft pulley notch aligned with pointer*

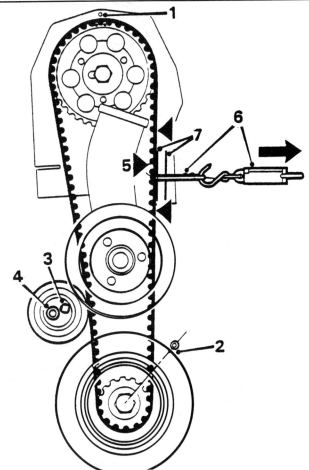

**Fig 12.5 Timing belt tension adjustment – 20 HB onwards (Sec 4)**

1  *Camshaft sprocket timing marks*
2  *Crankshaft pulley timing marks*
3  *Square adjustment hole in tensioner*
4  *Tensioner retaining bolt*
5  *Tension checking point*
6  *Spring balance and shaped wire*
7  *Parallel lines drawn on cover backplate*

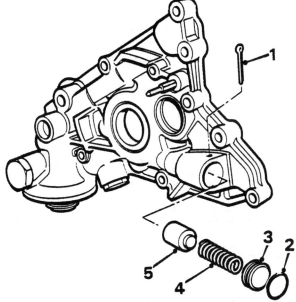

**Fig 12.6 Oil pump housing and pressure relief valve – 20 HB onwards (Sec 4)**

1  *Split pin*      4  *Spring*
2  *O-ring*       5  *Plunger*
3  *Plug*

116    If adjustment is necessary, alter the position of the tensioner as required,
117    Refit the timing belt cover, alternator and where fitted, power-assisted steering drivebelts, and the battery negative terminal. Adjust the drivebelt tensions as described in the relevant Chapters of this Manual.

### Valve clearances (20 HB onwards) – checking and adjustment
118    The procedure is identical to that for the 20 H engine described in Chapter 1, Section 12, but note the modified arrangement of the timing belt tensioner, timing marks and revised belt tensioning procedure described previously in this Section.

### Camshaft and tappets (20 HB onward) – removal and refitting
119    The procedure is identical to that for the 20 H engine described in Chapter 1, Section 13, but note the modified arrangement of the timing belt tensioner, timing marks and revised belt tensioning procedure described previously in this Section.

### Oil pump and housing (20 HB onwards) – removal and refitting
120    Disconnect the battery negative terminal.
121    Put the transmission in gear and firmly engage the handbrake. Using a socket and long handle, slacken the crankshaft pulley retaining bolt. If the engine is not in the car, engage a strip of angle iron between the flywheel teeth and one of the adaptor plate dowels to prevent

rotation of the crankshaft. Return the transmission to neutral after slackening the bolt.
122    Remove the timing belt as described previously in this Section.
123    Unscrew the pulley retaining bolt and remove the one-piece pulley and sprocket assembly. Carefully lever it off, using two screwdrivers if it is tight.
124    Release the adjusting arm from its bracket and move the alternator clear of the oil pump housing. Remove the adjustment arm bracket from the water pump and oil pump housing.
125    Disconnect the oil pressure switch wire and, where fitted, the oil cooler pipes from the pump housing.
126    Undo the retaining bolts and withdraw the pump housing from the crankshaft and crankcase.
127    If required, the oil filter can be removed by unscrewing it from the pump housing. Oil pump overhaul is identical to that for the 20 H engine, and reference should be made to Chapter 1, Section 19.
128    To refit the pump housing, first ensure that all traces of old gasket are removed from the housing and engine mating faces, then place a new gasket in position.
129    Apply some grease to the lips of the crankshaft front oil seal and carefully refit the pump housing.
130    Fit the retaining bolts and tighten them to the specified torque.
131    Reconnect the oil pressure switch wire and, where fitted, the oil cooler pipes.
132    Refit the adjustment arm bracket and the alternator adjustment arm.
133    Refit the crankshaft pulley and sprocket assembly, but do not fully tighten the bolt at this stage.

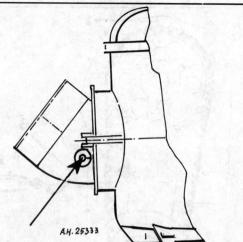

**Fig 12.7 Air duct elbow hole drilling position (arrowed) (Sec 4)**

134   Refit and tension the timing belt as described previously in this Section.
135   Using the same procedure as for removal, tighten the pulley retaining bolt to the specified torque.
136   Reconnect the battery negative terminal.

### Engine mountings (20 HB onwards) – modifications
137   On 20 HB engines, the front mounting (front snubber) has been modified and the removal and refitting procedure is as follows.
138   Slacken the front mounting through-bolt and undo the two bolts securing the mounting to the front chassis member.
139   Undo the two bolts securing the mounting bracket to the gearbox, and remove the mounting and the through-bolt.
140   Refitting is the reverse sequence to removal, but tighten the through-bolt last with the mounting in position.

### Engine oil emulsification – rectification
141   Where engine oil emulsification ('salad cream' effect) is a problem, especially in cold weather, the following modification which introduces an air bleed between the air cleaner and oil filler tube will reduce or eliminate the problem. Austin Rover dealers stock the necessary parts.

**Carburettor models (except Turbo)**
142   Remove the air cleaner assembly.
143   Drill a 0.3 in (8.0 mm) hole in the air duct elbow as shown in Fig. 12.7.
144   Clean away all swarf then fit the plastic bush into the drilled hole.
145   Insert the short length of tube into the bush and refit the air cleaner.
146   Remove the oil filler tube and clean it thoroughly.
147   Drill a 0.2 in (5.0 mm) hole in the oil filler tube at the position shown in Fig. 12.8.
148   Cut a 1.0 in (25.0 mm) length of brake pipe and braze it into the drilled hole in the oil filler tube. On later models the oil filler tube already incorporates the air bleed connection and all that is necessary is to remove the blanking cap.
149   Fit the lagging to the oil filler tube using cable ties to secure it in position, then refit the oil filler tube.
150   Fit the long length of pipe between the connections just made, ensuring that the upper end protrudes into the air cleaner elbow by 0.2 in (5.0 mm) and that it slopes down to the oil filler tube connection over its full length.
151   Remove and discard the oil separator and fit a new, lagged type separator in its place.
152   Clean out all the engine ventilation pipes and fit a new oil filler cap.
153   If the emulsification problem is of long standing, or severe, renew the engine oil and filter.

**EFi models**
154   Remove and discard the original air cleaner and fit a modified type – consult your dealer.
155   On vehicles with engine prefix 20HA, remove and clean the oil filler tube and use a piece of tubing as described in paragraph 145.

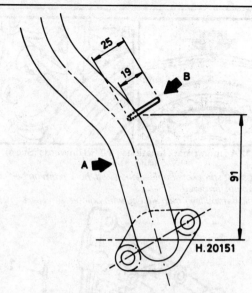

**Fig 12.8 Oil filler tube drilling diagram (Sec 4)**

A   Oil filler tube          B   Tubing

*All dimensions in mm*

156   Fit lagging using cable ties, and refit the oil filler tube.
157   On engines with prefix 20HB or 20HC, remove the oil filler tube. If it already has an air bleed connection, clean the tube and lag it. If it does not, fit a modified tube – consult your dealer.
158   Connect a bleed hose between the oil filler tube and air cleaner stub, making sure that the hose slopes downwards throughout its length.
159   Fit a new lagged type oil separator and two new gaskets.
160   If the emulsification problem is of long standing, or severe, renew the engine oil and filter.

**Turbo models**
161   Remove the air cleaner from its bracket.
162   Remove the outlet adaptor and seal from the air cleaner body and fit a modified type and seal – consult your dealer. Refit the air cleaner.
163   On engines with prefix 20HA, carry out the modification described in paragraphs 147 to 149.
164   On engines with prefix 20HB or 20HC, remove the oil filler tube. If it has an air bleed connection, clean the tube and fit lagging. If the tube does not have an air bleed connection, change it for a modified type – consult your dealer.
165   Fit a bleed hose between the air cleaner and the oil filler tube, making sure that the hose slopes downwards throughout its length.
166   Fit a modified oil separator with new gaskets.
167   Change the engine breather pipe non-return restrictor for a new type; also change the regulator valve and the engine oil filler cap for modified types.
168   If the emulsification problem is of long standing, or severe, renew the engine oil and filter.

### Valve timing – Turbo models
169   Some engines have been produced with two timing marks on the crankshaft pulley wheel.
170   When setting or checking valve timing the mark on the outer edge of the rear of the pulley wheel (nearest to the cylinder block) must be aligned with the mark on the water pump bracket.
171   In this position, the mark on the front outer edge of the pulley wheel will be at, approximately, the 12 o'clock position.

### Oil filter leak (automatic transmission) – rectification
172   On automatic transmission models it is possible for the engine stone guard to foul against the engine oil filter and cause an oil leak.
173   To prevent this happening the following modification to the stone guard can be carried out.
174   Raise the front of the vehicle onto axle stands and turn the steering to full right-hand lock.
175   Remove the stone guard.

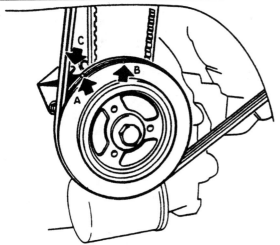

**Fig 12.9 Valve timing marks on crankshaft pulley wheel – Turbo models (Sec 4)**

A Mark on outer edge of rear of pulley
B Mark on front outer edge
C Mark on water pump bracket

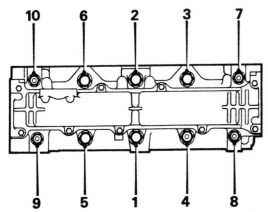

**Fig 12.11 Cylinder head bolt tightening sequence (Sec 4)**

176  Drill a 0.25 in (6.0 mm) hole in the stone guard in line with the existing hole but in the centre of the thicker, outer edge (see Fig. 12.10).
177  Refit the stone guard using the new fixing hole.
178  Renew the oil filter if necessary. Check and top up the engine oil.
179  Remove the vehicle from the axle stands.

*Cylinder head gasket (all models) – oil leak*
180  An oil leak from the oil gallery area of the cylinder head can be caused by one or all of the following:

(a)  *Distorted head gasket*
(b)  *Non-standard head bolts fitted*
(c)  *Cylinder head or cylinder block mating faces damaged*

181  Where this condition exists, the cylinder head should be removed, the appropriate parts inspected and a new cylinder head gasket fitted (according to type) as follows.
182  On all pre 1990 models, obtain the correct cylinder head gasket – UAM 2287 for all models except the Turbo, and UAM 9534 for Turbo models.
183  A later modification to align the oilways in the cylinder head with those in the cylinder block was introduced in 1990 at VIN 425366 (previously the oilways were offset). This modification necessitated the fitting of a specific head gasket (according to engine type) from engine numbers 20HC76 115197, 20HD05 109265, and 20HD08 100793. Part numbers are as follows.

Non-Turbo models ............................................................LVB 10003
Turbo models ...................................................................LVB 10011

**Note:** *Because the oil feed hole positions in the later gaskets differ from those of the earlier type, it is essential that they are not used on earlier models.*

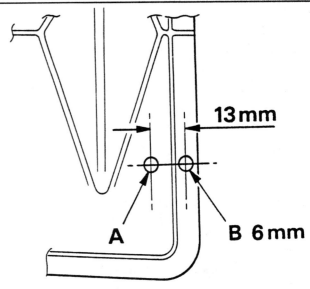

**Fig 12.10 Hole drilling diagram in stone guard on automatic transmission models (Sec 4)**

A  Existing hole
B  New hole

*All dimensions in mm*

184  Check the mating surfaces of both the cylinder head and the cylinder block for flatness. This can be done by placing a straight edge on the cylinder head/block faces in the longitudinal, transverse and diagonal planes and measuring between the straight edge and head/block face. Tolerances were not available at time of writing, but would be in the order of 0.008 in (0.2 mm) for the cylinder head and half that for the cylinder block. If these figures are obviously exceeded, the advice of your Austin Rover dealer should be sought.
185  Also check the head and block faces for damage, especially in the area of the oil gallery. Any damage must be rectified and again the advice of your dealer should be sought.
186  Carefully clean the cylinder head and block mating surfaces, checking that no debris is left in the oil ways or bolt holes.
187  Some early engines were produced with cylinder head bolts which had a flanged head and no separate washers. The flanged type bolts must be discarded, and new standard bolts with separate washers obtained.
188  Clean the threads of the bolts and the internal threads in the cylinder block thoroughly, removing all grime, rust and old oil.
189  Screw the bolts into the cylinder block by hand to check that the threads are undamaged and the bolts screw in freely. The threads can be cleaned using an M11 x 1.5 mm tap. Also ensure that both faces of the bolt washers are clean, using emery cloth if necessary to do this.
190  Check that the 'Viton' oil sealing ring in the cylinder head gasket is intact.
191  Check the cylinder head and block mating faces are still clean.
192  Locate the cylinder head gasket on the cylinder block then carefully lower the cylinder head in place.
193  Lightly oil the cylinder head bolt threads and underside of the bolt washers.
194  Fit and tighten the cylinder head bolts as described in later paragraphs.
195  The remaining cylinder head refitting procedure is as described in Chapter 1, Section 14 or 15.

**5  Cooling system**

*Cooling system filling (20 HB engine onwards) – general*
1  The radiator bottom hose on later models incorporates a bleed plug, and the revised filling procedure is as follows.

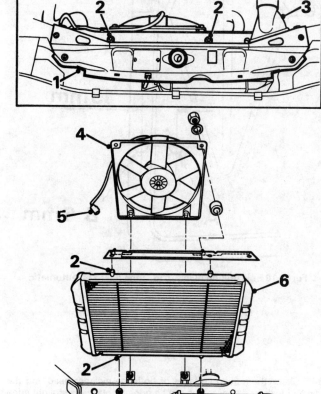

**Fig 12.12 Radiator and cooling fan components – MG Turbo models (Sec 5)**

1  Front body panel
2  Mounting lugs
3  Air cleaner cold air intake
4  Fan and cowl assembly
5  Wiring multi-plug
6  Radiator

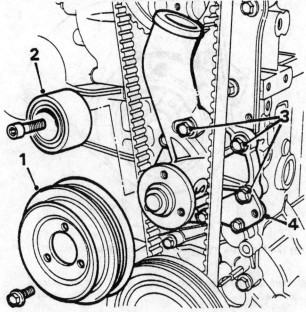

**Fig 12.13 Water pump attachment – 2.0 HB engine onwards (Sec 5)**

1  Water pump pulley
2  Timing belt tensioner
3  Pump retaining bolts
4  Alternator adjustment arm bracket

2    If the system has been flushed, refit any hoses or components that were removed for this purpose.

3    Unscrew the plug from the radiator bottom hose bleed point.

4    Slowly fill the system through the expansion tank with the appropriate mixture of water and antifreeze (see Chapter 2), until the coolant flows from the bleed point.

5    Refit the bleed plug and continue filling until the expansion tank is full.

6    Start the engine and run it at a fast idle for approximately one minute. During this time compress the top hose several times to release any air pockets in the system.

7    Stop the engine, top up the expansion tank to the indicated level, then refit the filler cap.

### Radiator (MG Turbo models) – general

8    The procedures described in Chapter 2, Section 7 are applicable, but note that, on Turbo models, the radiator cooling fan is positioned on the engine side of the radiator. The fan and cowl engage with lugs at the bottom and are secured to a mounting plate at the top. The complete assembly can be lifted out after removal of the front body panel.

### Water pump (20 HB engine onwards) – removal and refitting

9    Disconnect the battery negative terminal.

10   Drain the cooling system as described in Chapter 2.

11   If power-assisted steering is fitted, remove the drivebelt as described in Chapter 10, then undo the steering pump mounting bracket bolts and move the pump aside.

12   Remove the water pump and alternator drivebelt as described later in this Section.

13   Refer to Section 4 and remove the timing belt.

14   Undo the bolts and remove the alternator adjustment arm bracket.

15   Undo the three bolts and remove the water pump pulley.

16   Disconnect the radiator bottom hose at the water pump.

17   Undo the timing belt tensioner retaining bolt and remove the tensioner.

18   Undo the remaining water pump retaining bolts and remove the pump from the engine.

19   Before refitting, clean away all traces of old sealant from the mating faces.

20   Apply a bead of RTV silicone sealant to the pump mating face, and position the pump on the engine. Fit and tighten the retaining bolts to the specified torque.

21   Refit the alternator adjustment arm bracket.

22   Refit the water pump pulley.

23   Refer to Section 4 and refit the timing belt.

24   Refit the alternator drivebelt as described later in this Section.

25   If power-assisted steering is fitted, place the pump in position and secure the mounting bracket with the retaining bolts. Refit the drivebelt and adjust its tension as described in Chapter 10.

26   Reconnect the battery negative terminal, then fill the cooling system as described earlier in this section.

### Radiator cooling fan assembly (MG Turbo models) – removal and refitting

27   Disconnect the battery negative terminal.

28   Disconnect the fan motor wiring multi-plug and release the wiring from the clips.

29   Undo the two upper retaining nuts and washers, then lift out the fan and cowl assembly.

30   Extract the retaining circlip and remove the fan from the motor shaft.

31   Undo the three fan motor and cover retaining bolts and remove the motor.

32   Refitting is the reverse sequence to removal.

### Water pump and alternator drivebelt – removal, refitting and adjustment

33   On Turbo and later models the alternator has been repositioned

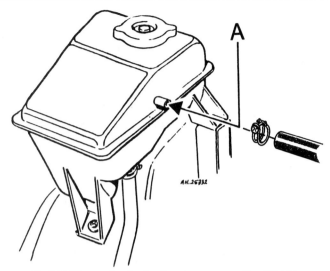

**Fig 12.14 Expansion tank return hose connection (Sec 5)**

*A Hole must be $\frac{3}{16}$ in (5.0 mm)*

and is now on the left-hand (front facing) side of the engine. The procedures described in Chapter 2, Section 12 are still applicable apart from this difference.

### Cooling system – pressure testing

34 When using pressure test equipment to check the cooling system for leaks it is important that the recommended test pressures are not exceeded (see Specifications).

### Engine (all models) – overheating

35 Overheating of the engine may be caused by the hole in the coolant expansion tank return hose connection being too small.
36 This can be checked by disconnecting the return hose and attempting to insert a $\frac{3}{16}$ in (5.0 mm) drill into the hole.
37 If the drill enters the hole then it is the correct size.
38 If the drill will not enter the hole then use the drill to enlarge the hole. Drill slowly and apply a spot of grease to the drill to collect swarf.
39 Re-connect the return hose on completion.

### 6 Fuel and exhaust systems

### SU HIF 44 carburettor – adjustment

1 Certain later models are fitted with a carburettor which incorporates a progressive throttle cam (Fig. 12.15).
2 When carrying out the adjustments described in Chapter 3, Section 10, the lost motion clearance should be set by bending the tag (8).

### Electric fuel pump (all models) – general

3 All later Montego models covered by this manual are fitted with an electric fuel pump identical to the type previously used on MG EFi models only. Removal and refitting procedures are contained in Chapter 3, Section 21, but on carburettor engines ignore all references to depressurisation of the fuel system. When refitted, ensure that the pump is positioned so that its outlet hose and connector are vertically aligned with the hose connector at the top.

### Operation on unleaded fuel

4 It is essential that only fuel of the correct type and octane rating is

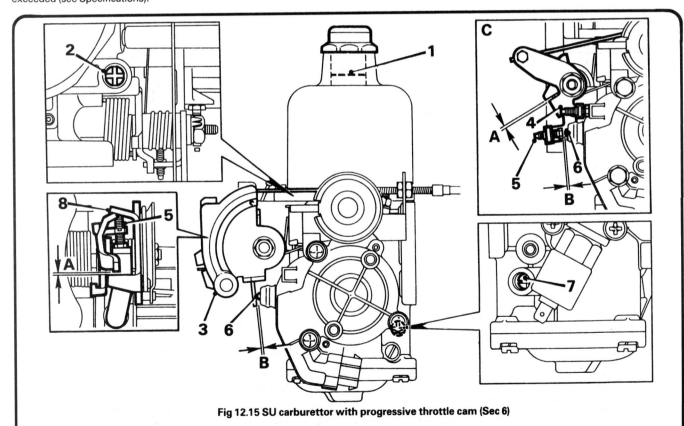

**Fig 12.15 SU carburettor with progressive throttle cam (Sec 6)**

| | | |
|---|---|---|
| 1 Piston damper oil level | 4 Throttle lever adjusting screw | 7 Mixture adjusting screw | B Clearance – fast idle rod |
| 2 Idle speed adjusting screw | 5 Fast idle adjusting screw | 8 Lost motion adjusting tag | C Linkage with throttle lever |
| 3 Progressive throttle cam | 6 Fast idle pushrod | A Clearance – lost motion gap | |

214

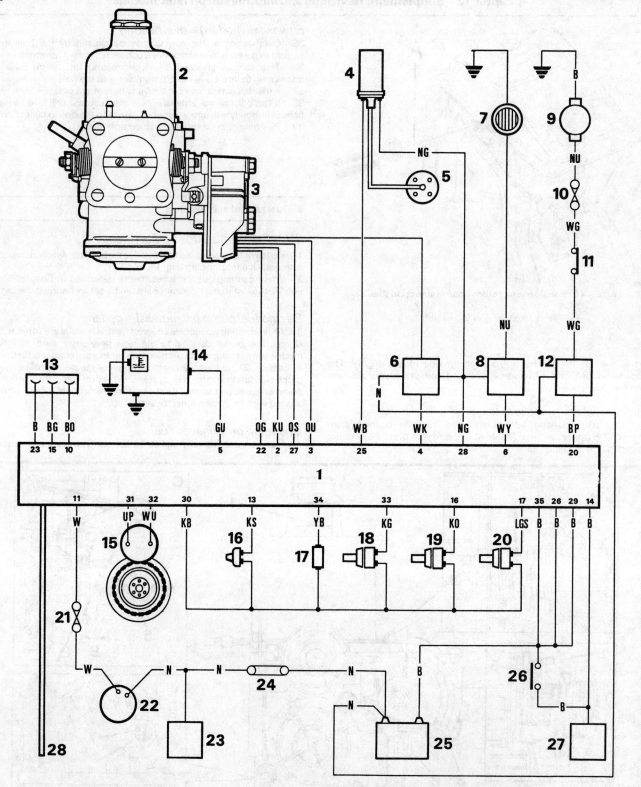

**Fig 12.16 Electronically Regulated Ignition and Carburation system schematic wiring diagram – 1989-on models (Sec 6)**

1 ECU
2 Carburettor
3 Stepper motor
4 Ignition coil
5 Distributor cap
6 Main relay
7 Inlet manifold heater
8 Manifold heater relay
9 Electric fuel pump
10 Fuse No 17 in fusebox
11 Inertia switch
12 Fuel pump relay
13 Engine tuning and diagnostic socket
14 Temperature gauge
15 Crankshaft sensor
16 Throttle pedal switch
17 Ambient air temperature sensor
18 Coolant thermistor
19 Inlet air temperature sensor
20 Knock sensor
21 Fuse No 1 in fusebox
22 Ignition switch
23 Ignition relay
24 Fusible link
25 Battery
26 Automatic transmission inhibitor switch
27 Starter relay
28 Manifold vacuum tube

used. Some models are able to use either leaded or unleaded fuel whilst on others, only one fuel type is suitable. Serious damage can occur to the engine (and where applicable, the catalytic converter) if the incorrect fuel type is used. Refer to the Specifications for the recommended type and octane rating.

5    Some 1988 and all 1989-on carburettor models (excluding Turbo) can run on unleaded fuel without adjustment to the ignition system. Similarly, some 1989-on fuel injection models can use unleaded fuel without adjustment. *Catalytic converter models must only be run on unleaded fuel.* Early models with engines originally designed to run on leaded fuel may have already been converted (or are suitable for conversion) to run on unleaded fuel. Consult your Rover dealer if in doubt about which fuel type to use.

### *Cold air intake tube (carburettor models, 1989-on) – modification*

6    The cold air intake tube from the front cross panel to the air cleaner assembly has been modified to prevent it from collapsing inward, causing poor performance and high fuel consumption.
7    The modified tubes are fitted to all models from VIN 478996, and these modified tubes can be fitted to earlier models if this problem has occurred.

### *Air cleaner assembly (carburettor models, non-Turbo 1989-on) – removal and refitting*

8    Disconnect the cold air intake tube between the front cross panel and the air cleaner assembly.

9    Recover the intake mesh if it should fall out.
10    Disconnect the air temperature control tube from the thermac unit and the breather tube from the air cleaner.
11    Disconnect the two remaining air tubes from the air cleaner.
12    Remove the bolts securing the air cleaner and withdraw the assembly.
13    The filter can be removed from the cleaner assembly after unclipping the top cover.
14    Refitting is a reversal of removal.

### *Electronically Regulated Ignition and Carburation system (ERIC) – general description*

All non-Turbo carburettor models from 1989 are fitted with the Electrically Regulated Ignition and Carburation system (ERIC). This is a complete engine management system which meets a very high standard of emission control.

Using a single Engine Control Unit (ECU), which effectively combines the fuel ECU and Ignition ECU modules used in the programmed systems of earlier models. ERIC controls and synchronises all fuel and ignition requirements of the engine.

The ECU receives its information from the same sources as the earlier systems, and acts on the carburettor and ignition system in the same manner.

For the purposes of this Supplement, the fuelling aspects of the system are dealt with in this Section and the ignition aspects in Section 7.

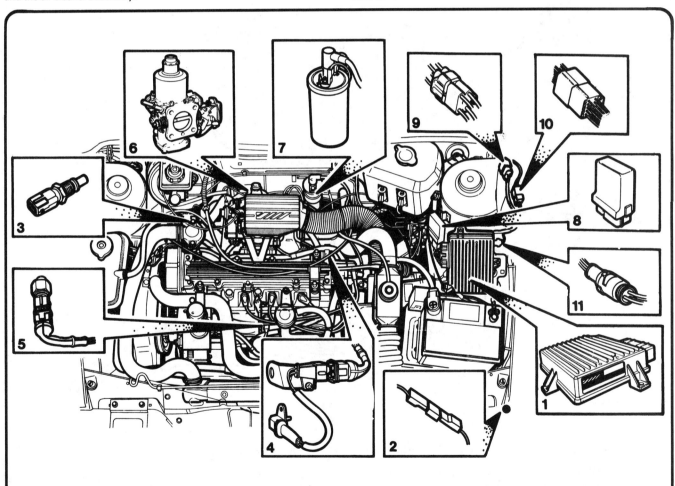

**Fig 12.17 ERIC component location diagram (Sec 6)**

| | | |
|---|---|---|
| *1 ECU* | *4 Crankshaft sensor* | *8 Engine multi-function unit (contains* |
| *2 Ambient air temperature* | *5 Knock sensor* | *main, fuel pump, manifold heater* |
| *sensor – behind horns* | *6 Carburettor* | *and starter relays)* |
| *3 Coolant temperature sensor* | *7 Ignition coil* | *9 Four-way harness connector* |
| | | *10 Thirteen-way harness connector* |
| | | *11 Engine diagnostic and tuning socket* |

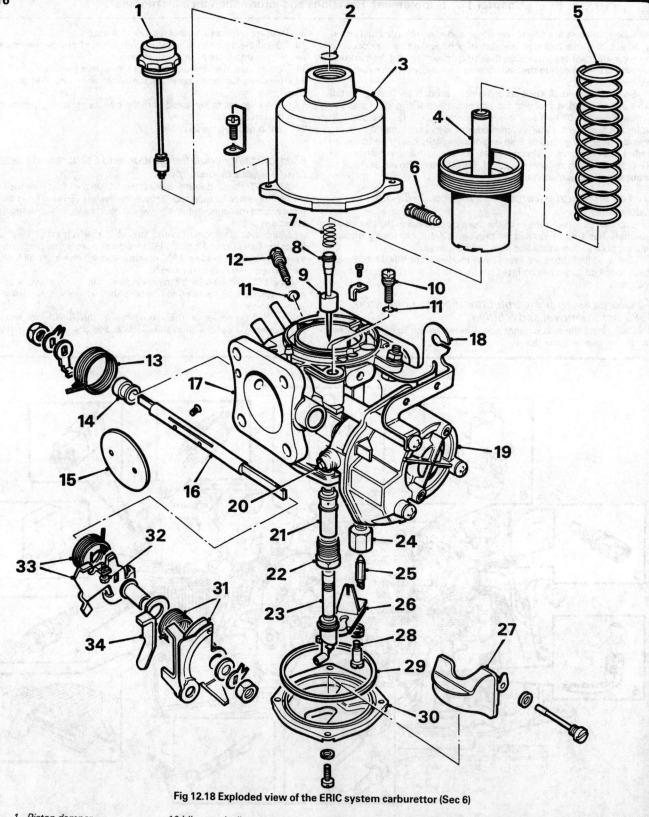

**Fig 12.18 Exploded view of the ERIC system carburettor (Sec 6)**

1  Piston damper
2  Spring ring – piston
3  Suction chamber
4  Piston
5  Piston spring
6  Needle retaining spring
7  Needle bias spring
8  Jet needle
9  Needle guide
10  Idle speed adjustment screw
11  Seal
12  Mixture adjusting screw
13  Throttle return spring
14  Throttle spindle seal
15  Throttle disc
16  Throttle spindle
17  Carburettor body
18  Mounting bracket
19  Mixture control stepper motor
20  Fast idle push rod
21  Jet bearing
22  Jet bearing nut
23  Jet assembly
24  Float needle seat
25  Float needle
26  Bi-metal jet lever
27  Float
28  Jet retaining screw
29  Float chamber cover seal
30  Float chamber cover
31  Throttle cam and return
    spring
32  Throttle position screw
33  Lost motion link and return
    spring
34  Fast idle lever

**Fig 12.19 Throttle position speed adjustment on ERIC system carburettor (Sec 6)**

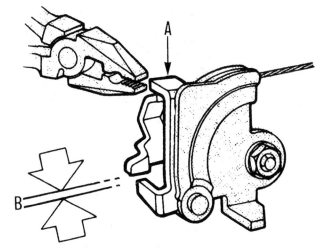

**Fig 12.20 Lost motion clearance adjustment on ERIC system carburettor (Sec 6)**

*Bend the lever (A) to obtain a clearance (B) of 0.138 to 0.158 in (3.5 to 4.0 mm)*

## ERIC system – component location, removal and refitting

15   The layout of the ERIC system components and their location is shown in Fig. 12.17.
16   The removal and refitting of the components is as described in Chapters 3 and 4.

## ERIC system – tuning, adjustment and fault diagnosis

17   The accurate tuning, adjustment and fault diagnosis of the ERIC system requires the use of special test equipment which is likely only to be available to Austin Rover dealers.
18   For this reason there is no alternative but to take the vehicle to an Austin Rover dealer for tuning, adjustment and fault diagnosis, and subsequent rectification.
19   The ERIC system is equipped with a diagnostic and tuning socket connector into which the special test equipment can be connected for speedy testing and fault diagnosis. The ECU is equipped with a memory interface in which are stored intermittent faults which can be read off on the test equipment, so simplifying the test procedure and pin-pointing faults.

## ERIC system carburettor – overhaul, tuning and adjustment

20   The carburettor fitted to models with the ERIC system is basically the same as the carburettor described in Chapter 3, and can be overhauled using the same procedure, bearing in mind the following differences.
21   There is no fuel shut-off valve.
22   The throttle linkage is slightly different, and the lost motion clearance is adjusted by bending the tag on the throttle lever as described later.
23   Without the special tuning and diagnostic equipment mentioned earlier the carburettor cannot be adjusted accurately. Approximate settings can be obtained following the procedure given in Chapter 3, Section 10, but note also the following.
24   Base idle is set to the specified speed using the idle speed adjustment screw, and the CO content is set using the mixture screw (both these screws are as shown in Fig. 3.4 in Chapter 3).
25   Throttle position speed is adjusted on the fast idle adjustment screw.
26   Lost motion clearance is adjusted by bending the tag on the fast idle lever and must be set to the specified clearance shown in Fig. 12.20.

## ECU failure (EFi models, pre-1989) – short circuit

27   Short circuiting of the battery to the airflow meter body by a spanner or such like can cause an earth track on the printed circuit board of the ECU, resulting in failure.
28   The importance, therefore, of always disconnecting the battery negative terminal first, and reconnecting it last cannot be over-emphasised.

## Fuel system (fuel injection models) – description

29   Later models are available with an electronic fuel injection system identical to that used on MG EFi models. Reference should be made to Chapter 3, Part B (and this section) for all fuel and exhaust system operations on fuel injection models.

## Fuel hoses (Turbo and fuel injection models) – general

30   The following precautions must be observed when dealing with the flexible fuel hoses on cars equipped with electronic fuel injection systems or turbochargers.
31   The hoses are designed to withstand higher fuel temperatures and incorporate an inner lining of 'Viton' material. Standard hoses supplied for use on carburettor engines are not suitable for use on the fuel injection or Turbo systems.
32   The hoses should never be cut to length, as there is a risk of damaging the inner lining. Pre-cut hoses in lengths to suit all applications are available from dealer parts stockists.
33   When a hose is being fitted to a steel pipe or pipe stub, ensure that the hoses, pipe, or stub are perfectly clean, dry and free from oil, grease or other lubricant. The hoses must be fitted dry.
34   Do not push the hose on so far as to cause the end of the hose to fold back due to contact with the pipe or stub end.
35   Hoses must engage pipes or stubs by at least 1.0 in (25 mm). Where clips are used for retention, they must grip the flat section of the hose and never be positioned over any of the raised swages on the pipe. Ensure that 0.04 to 0.1 in (1.0 to 3.0 mm) of hose protrudes beyond the end of the clip.
36   Where the hose is being fitted to a pipe or stub with an abutment face, the hose should just touch the face.

## Fuel pressure regulator (1987-on models) – removal and refitting

37   Disconnect the battery negative lead.
38   Depressurise the fuel system as described in Chapter 3.
39   Place rags under the fuel pressure regulator and disconnect the spill return pipe.
40   Disconnect the vacuum hose from the regulator, then remove the regulator mounting nuts.
41   Slacken the screw which secures the support bracket to the

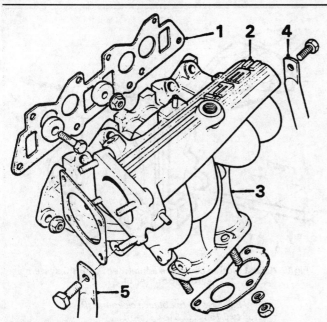

**Fig 12.21 Inlet and exhaust manifolds on 1987-on fuel injection models (Sec 6)**

| | |
|---|---|
| 1 Gasket | 4 Support strut (RH) |
| 2 Inlet manifold | 5 Support strut (LH) |
| 3 Exhaust manifold | |

manifold and move the bracket aside. Pull the regulator from the fuel rail.

42   Refitting is a reversal of removal but make sure that the regulator sealing ring is in good condition.

## Inlet and exhaust manifolds (1987-on fuel injection models) – removal and refitting

*For models with MEMS MPi see later in this Section*

43   Disconnect the battery negative lead.

44   Depressurise the fuel system as described in Chapter 3.

45   Disconnect the vacuum hose from the inlet chamber; also the multi-plug from the air valve stepper motor.

46   Disconnect the wiring plugs from the fuel injectors, fuel temperature switch and coolant temperature sensor.

47   Place rags under the fuel regulator valve to absorb spilled fuel, and disconnect the spill return hose.

48   Extract the two screws which hold the fuel injector rail to the manifold.

49   Extract the screws which hold the fuel pressure regulator to the manifold. Place the fuel rail and injector assembly to one side. Plug the hose and regulator to prevent loss of fuel.

50   Extract the screws which hold the accelerator cable bracket to the manifold, then release the accelerator cable trunnion from the quadrant, and disconnect the throttle housing from the inlet manifold flange. Peel off the gasket.

51   Unscrew the servo vacuum banjo bolt and retain the washers.

52   Slacken the manifold fixing nuts and bolts which are accessible from above; also slacken the exhaust manifold nut located below the thermostat housing.

53   Slacken the bolts which secure the support struts to the manifold, then move the stays and coolant pipe brackets aside.

54   Apply the handbrake fully and raise the front of the vehicle. Alternatively, place the vehicle over an inspection pit.

55   If the exhaust manifold is to be renewed, disconnect the exhaust downpipes.

56   Unscrew the engine and transmission support strut bolts.

57   Working from underneath the engine, remove the bolts, clamps and washers which secure the inlet manifold. Withdraw the manifold and gasket.

58   Remove the exhaust manifold and gasket.

59   Refitting is a reversal of removal but observe the following points.

60   Use new gaskets.

61   Tighten all nuts and bolts to the specified torque.

62   On completion, depress the inertia fuel cut-off switch button.

## Modular Engine Management System (MEMS MPi) – general description

The Modular Engine Management system (MEMS) is fitted to some fuel injected models from 1989. It controls a MultiPoint fuel injection system (MPi) and a programmed ignition system. The main features of the system are as follows:

One Electronic Control Unit (ECU) controls all the fuelling and ignition requirements of the engine. The ECU incorporates short circuit protection and can also store information on certain intermittent faults for interrogation by computerised test equipment.

The ECU utilises a speed/density method of airflow measurement to calculate fuel delivery. This method, using the engine as a pre-calibrated vacuum pump with its characteristics stored in the ECU, measures the air inlet temperature and inlet manifold pressure, allowing the correct amount of fuel per air density/speed to be injected.

Should certain elements of the system fail, the ECU can implement a back-up facility, allowing the system to operate at a reduced level of performance until the fault can be rectified.

A diagnostic socket allows tuning or fault diagnosis to be carried out on special test equipment without disconnecting the ECU harness.

A throttle pedal switch triggers the ECU between the main fuelling map when the switch is open (accelerator pedal depressed) and idle speed control when the switch is closed (accelerator pedal released). The throttle pedal switch also controls an over run fuel cut-off.

The ECU, in determining optimum ignition timing, receives information from the crankshaft sensor (engine speed and crankshaft position), manifold absolute pressure transducer (engine load), knock sensor (detonation), coolant temperature thermistor (engine temperature) and from the accelerator pedal (throttle pedal switch), to determine main fuelling or idle speed control.

It can be seen that the system is similar to the earlier programmed fuel injection system described in Chapter 3, the main differences being the method of determining airflow (there is no airflow meter) and the combining of the fuel and ignition ECUs into one unit.

Because of the complexity of the system and the need for computerised test equipment, there is no alternative but to take the vehicle to an Austin Rover dealer for tuning and fault diagnosis, although certain operations which could be performed by the DIY mechanic follow in later paragraphs.

## Modular Engine Management System – component renewal

**Warning:** *Before any part of the fuel line between the pump and the pressure regulator is disconnected, the fuel system must be depressurised as described here. Read also the warning note in Section 16 of Chapter 3.*

**Fuel system – depressurising**

63   Remove the fusebox cover and operate the inertia switch by pulling the plunger up so that the red portion of its stem is visible.

64   If running the engine would be a problem, also disconnect one of the LT cables from the coil.

65   Operate the starter and allow the engine to run until it stops, or crank the engine on the starter for at least 20 seconds.

66   Disconnect the battery.

67   Using a suitable container to catch residual fuel, disconnect the fuel line.

68   On completion of the work required, re-make all disturbed connections and re-set the inertia switch.

**Air cleaner**

69   Disconnect the hot and cold air intake hoses from the air cleaner.

70   Disconnect the crankcase breather tube.

71   Remove the screws securing the air cleaner assembly and lift it sufficiently to disconnect the air inlet temperature sensor, then withdraw the complete air cleaner assembly.

72   To renew the air filter unclip and lift off the top cover.

73   Refitting is a reversal of removal.

**Fuel line filter**

74   Refer to Chapter 3, Section 20.

**Inertia switch**

75   Remove the fusebox cover.

76   Remove the panel from the underside of the facia on the driver's side.

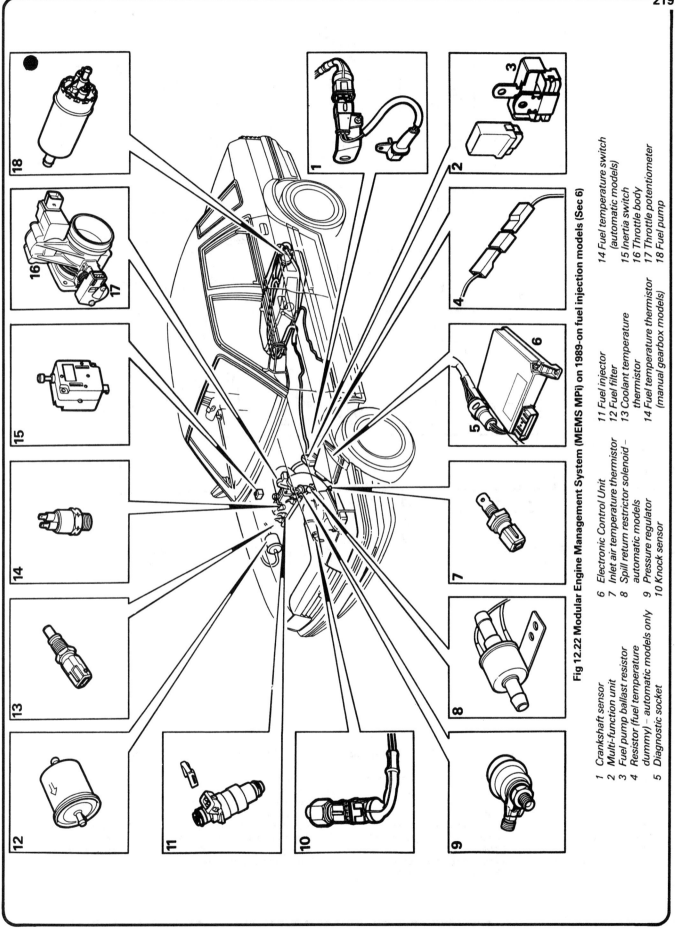

Fig 12.22 Modular Engine Management System (MEMS MPi) on 1989-on fuel injection models (Sec 6)

1  Crankshaft sensor
2  Multi-function unit
3  Fuel pump ballast resistor
4  Resistor (fuel temperature dummy) – automatic models only
5  Diagnostic socket
6  Electronic Control Unit
7  Inlet air temperature thermistor
8  Spill return restrictor solenoid – automatic models
9  Pressure regulator
10 Knock sensor
11 Fuel injector
12 Fuel filter
13 Coolant temperature thermistor
14 Fuel temperature thermistor (manual gearbox models)
14 Fuel temperature switch (automatic models)
15 Inertia switch
16 Throttle body
17 Throttle potentiometer
18 Fuel pump

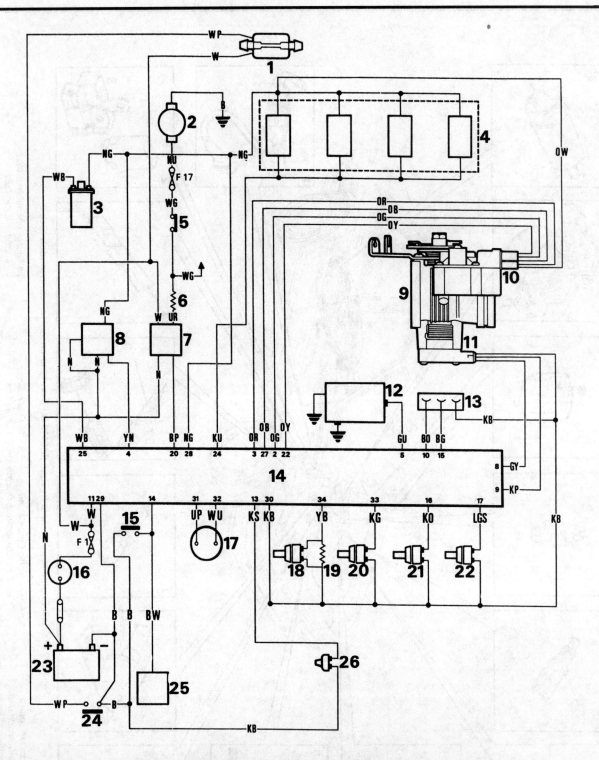

**Fig 12.23 Modular Engine Management System – schematic wiring diagram (Sec 6)**

1  Spill return restrictor
   solenoid
2  Fuel pump
3  Ignition coil
4  Injectors
5  Inertia switch
6  Ballast resistor – 1 ohm
7  Fuel pump relay – inside
   multi-function unit

8  Main relay – inside multi-function
   unit
9  Throttle body
10 Stepper motor
11 Throttle potentiometer
12 Temperature gauge and LED
13 Diagnostic socket
14 Electronic Control Unit
15 Inhibitor switch – automatic models

16 Ignition switch
17 Crankshaft sensor
18 Fuel temperature thermistor –
   manual models
19 Resistor – 6.8 kohm – fuel
   temperature dummy – automatic
   models
20 Coolant temperature thermistor
21 Intake air temperature thermistor

22 Knock sensor
23 Battery
24 Fuel temperature switch –
   automatic models
25 Starter relay – inside
   multi-function unit
26 Throttle pedal switch

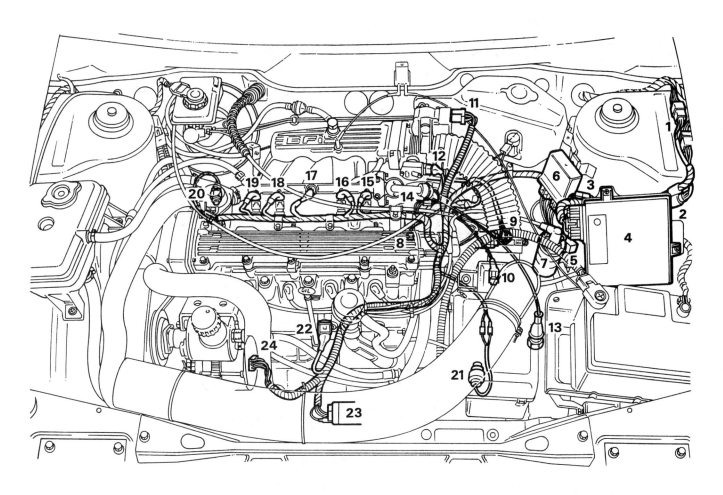

**Fig 12.24 Component location – MEMS MPi system, 1989-on (Sec 6)**

| | | | |
|---|---|---|---|
| 1 Main harness connection | 8 Crankshaft sensor | 14 Spill return restrictor solenoid | 19 Number four injector |
| 2 Diagnostic socket | 9 Starter solenoid (manual models) | 15 Number one injector | 20 Coolant temperature thermistor |
| 3 Resistor (6.8 kohm, fuel temperature dummy – automatic models) | 10 Intake air temperature sensor | 16 Number two injector | 21 Reverse lamp switch (manual models) |
| | 11 Stepper motor | 17 Fuel temperature thermistor | |
| 4 MEMS ECU | 12 Throttle potentiometer | (manual models) or fuel | 22 Knock sensor |
| 5 Ballast resistor (1 kohm) | 13 Reverse lamp and | temperature switch (automatic | 23 Starter motor |
| 6 Multi-function unit | starter/inhibitor switch | models) | 24 Alternator |
| 7 Ignition coil | (automatic models) | 18 Number 3 injector | |

77    Remove the two screws securing the inertia switch and its bracket to the fusebox.

78    Withdraw the switch and bracket, disconnect the plug and remove the assembly from the fusebox.

79    Check the operation of the switch by depressing the plunger then striking the rear of the switch with the palm of the hand. The plunger should spring out and lock in the raised position. Renew the switch if this is not the case.

80    Refit in reverse order.

**Accelerator cable**

81    Unclip the square nut on the accelerator cable end bracket then release the inner cable from the throttle lever.

82    Release the accelerator cable from the clip.

83    Remove the underside panel from the facia on the driver's side.

84    Remove the clip securing the inner cable to the accelerator pedal.

85    Release the inner cable from the pedal and the outer cable from the bulkhead and pull the cable into the engine compartment to remove it.

86    Refit in reverse, adjusting the cable as follows.

87    Screw the square adjusting nut fully onto the threaded portion of the outer cable.

88    Connect the inner cable to the throttle lever.

89    Pull the outer cable away from the throttle lever until all slack and lost motion are taken up.

90    Keeping the cable in this position, screw the square nut along the threaded portion of the outer cable until it just touches the support bracket.

91    Without moving the square nut, clip it into its slot in the bracket.

**Throttle pedal switch**

92    Remove the underside panel from beneath the facia on the driver's side.

93    Release the accelerator cable from the accelerator pedal as described above.

94    Remove the accelerator return spring.

95    Unscrew the pedal securing nut and remove the pedal.

96    Remove the clip securing the switch to the pedal, disconnect the cables and remove the switch, noting the washer.

97    Begin refitting by fitting the washer to the switch, fit the switch to the bracket and secure in position with the clip.

98    Connect the two cables.

99    Refit the accelerator pedal and fit and tighten the nut.

100    Connect the accelerator cable to the pedal and fit the retaining clip.

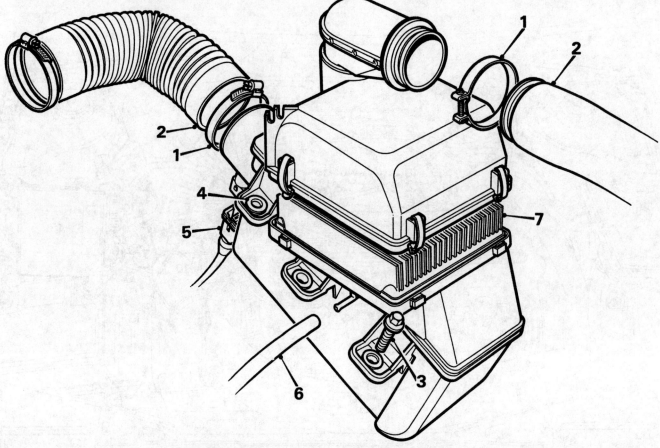

**Fig 12.25 Air cleaner assembly on MEMS MPi models (Sec 6)**

| | | |
|---|---|---|
| 1  Hose clips | 3  Securing screws | 5  Inlet air temperature sensor multi-plug | 7  Air cleaner element |
| 2  Air inlet hoses | 4  Rubber grommets | 6  Crankcase breather hose | |

101    Fit the pedal return spring and refit the underside panel.

**Electronic Control Unit (ECU)**

102    Disconnect the battery.

103    Disconnect the multi-plug and vacuum pipe from the ECU.

104    Remove the unit securing screws and withdraw the ECU.

105    Refit in reverse order.

**Fuel temperature switch (automatic models only)**

106    Disconnect the two electrical connections and unscrew the switch from the fuel rail.

107    Refit in reverse order.

**Injectors**

108    The procedure is as described in Chapter 3, Section 33, but the fuel temperature switch on automatic models must also be removed.

**Spill return restrictor solenoid (automatic models only)**

109    Pull out the tag on the rear of the multi-plug and release the plug from the bracket.

110    Disconnect the harness multi-plug from the solenoid.

111    Disconnect the inlet and outlet hoses from the solenoid by compressing the clips and pulling the hoses out – plug the hoses to prevent fuel leakage.

112    Remove the solenoid securing screws and withdraw the solenoid and bracket.

113    Refit in reverse order.

**Fuel temperature thermistor (manual models only)**

114    Refer to Chapter 3, Section 30.

**Intake air temperature sensor**

115    Remove the air cleaner assembly as described earlier.

116    Unscrew the air temperature sensor from the air cleaner.

117    Refit in reverse order.

**Fig 12.26 Inertia switch on MEMS MPi models (Sec 6)**

| | |
|---|---|
| 1  Securing screws | 3  Inertia switch |
| 2  Multi-plug | |

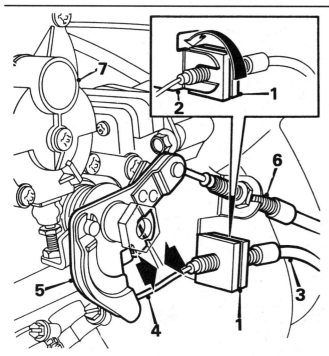

**Fig 12.27 Accelerator cable adjustment on MEMS MPi models (Sec 6)**

| | | | |
|---|---|---|---|
| 1 | Square adjusting nut | 5 | Throttle lever |
| 2 | Support bracket | 6 | Kickdown cable |
| 3 | Outer accelerator cable | 7 | Throttle housing |
| 4 | Inner accelerator cable | | |

**Fig 12.29 Throttle housing and potentiometer on MEMS MPi models (Sec 6)**

| | | | |
|---|---|---|---|
| 1 | Potentiometer multi-plug | 4 | Stepper motor multi-plug |
| 2 | Potentiometer | 5 | Air inlet hoses |
| 3 | Potentiometer securing screws | 6 | Throttle housing securing nuts |

**Throttle potentiometer**

118    Disconnect the multi-plug from the potentiometer.

119    Remove the securing screws and withdraw the potentiometer from the throttle housing.

120    Refit in reverse order, ensuring that the potentiometer lever is to the left of the throttle lever.

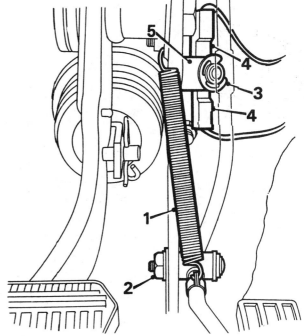

**Fig 12.28 Throttle pedal switch on MEMS MPi models (Sec 6)**

| | | | |
|---|---|---|---|
| 1 | Pedal return spring | 4 | Electrical connectors |
| 2 | Securing nut | 5 | Throttle pedal switch |
| 3 | Clip | | |

**Throttle housing**

121    Disconnect the battery.

122    Disconnect the multi-plugs from the stepper motor and throttle potentiometer.

123    Disconnect the air inlet hose from the throttle housing.

124    Disconnect the accelerator cable from the throttle lever as described earlier.

125    On automatic models, disconnect the kickdown cable from the throttle housing.

126    Using a suitable container or some rags, disconnect the coolant hoses from the throttle housing.

127    Disconnect the crankcase breather hoses.

128    Unscrew the four nuts securing the throttle housing to the inlet manifold and remove the throttle housing.

129    Refit in reverse order, ensuring the mating surfaces of the throttle housing and the inlet manifold are clean, and adjust the accelerator cable as described earlier.

130    Top up the cooling system.

**Fuel pipes and hoses**

131    Before disconnecting any fuel pipes, depressurise the system as described earlier and read the precautions on fuel spillage in Section 16 of Chapter 3, and in the *Safety First* section at the beginning of this manual.

**Fuel rail**

132    Depressurise the fuel system as described earlier.

133    Disconnect the battery.

134    Place absorbent rag beneath the fuel rail.

135    Disconnect the multi-plugs from the fuel temperature thermistor (manual models) and the leads on the fuel temperature switch (automatic models).

136    Disconnect the fuel inlet hose union at the fuel rail and plug the hose end to minimise fuel spillage.

137    Disconnect the fuel return hose from the pressure regulator and again plug the hose end.

138    Disconnect the vacuum hose between the pressure regulator and the inlet manifold.

139    Disconnect the fuel injector multi-plugs and remove the clips securing the injectors to the fuel rail.

140    Remove the three screws securing the fuel rail to the manifold.

141    Holding the fuel injectors to prevent them being dislodged, ease the fuel rail from the manifold.

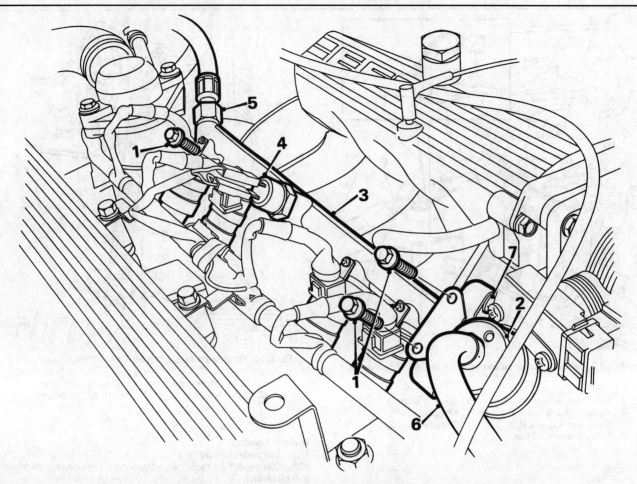

**Fig 12.30 Fuel rail and pressure regulator on MEMS MPi models (Sec 6)**

| | | | |
|---|---|---|---|
| 1 | Fuel rail securing screws | 3 | Fuel rail |
| 2 | Pressure regulator | 4 | Fuel temperature |
| | | | switch/thermistor |

| | |
|---|---|
| 5 | Fuel inlet union |
| 6 | Fuel return hose |

| | |
|---|---|
| 7 | Pressure regulator securing |
| | screws |

142   Remove the O-ring seals from the injectors.
143   Remove the screws and separate the regulator from the fuel rail.
144   Refitting is a reversal of removal, using new O-ring seals on the injectors.

**Fuel pressure regulator**
145   Depressurise the fuel system as described earlier.
146   Disconnect the battery.
147   Position absorbent rag beneath the pressure regulator.
148   Disconnect the fuel return hose from the regulator, then plug the hose end.
149   Remove the screws securing the pressure regulator to the fuel rail.
150   Withdraw the regulator from the fuel rail and disconnect the vacuum hose.
151   Remove the O-ring seal from the pressure regulator.
152   Refit in reverse order, using a new O-ring seal.

**Fuel pump**
153   Refer to Chapter 3, Section 21.

**Fuel tank**
154   Refer to Chapter 3, Section 22.

**Fuel pump ballast resistor**
155   Contained within the multi-function unit, removal and refitting of the fuel pump ballast resistor is self-evident.

**Resistor (fuel temperature dummy) automatic models only**
156   Contained within the multi-function unit, removal and refitting of the resistor is self-evident.

**Crankshaft sensor**
157   Refer to Chapter 4, Section 4.

**Knock sensor**
158   Refer to Chapter 4, Section 5.
**Coolant temperature thermistor**
159   Refer to Chapter 2, Section 8.

*Modular Engine Management System – pressure test*
**Note:** *A pressure gauge capable of reading up to approximately 50 psi (3.5 bar) will be required, together with suitable adaptors and tee-pieces, to connect the pressure gauge into the fuel line between the fuel filter and fuel rail. An adaptor is available from Austin Rover dealers.*
160   Depressurise the system as described earlier, then re-set the inertia switch.
161   Connect the pressure gauge as described above.
162   Disconnect the vacuum hose from the pressure regulator.
163   Start the engine and allow it to idle.
164   The pressure should be 34 to 38 psi (2.4 to 2.6 bar).
165   Keep the engine running and connect the vacuum hose to the pressure regulator, when the pressure should drop to about 26 psi (1.8 bar).
166   If the pressure does not drop, the likely cause is a blocked, kinked or leaking vacuum hose.
167   Switch off the engine. The pressure should not fall by more than 10 psi (0.7 bar) during the first minute.

*Inlet and exhaust manifolds (MEMS MPi models) – removal and refitting*
168   Remove the fuel rail as described earlier.

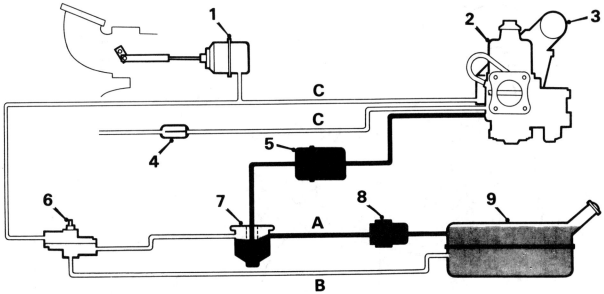

**Fig 12.31 Fuel circuit layout – MG Turbo models (Sec 6)**

| | | | |
|---|---|---|---|
| 1 Wastegate actuator | 4 Carburettor vent valve | 7 Vapour separator | A Fuel supply line |
| 2 Carburettor | 5 Fuel filter | 8 Fuel pump | B Fuel return line |
| 3 Plenum chamber | 6 Fuel pressure regulator | 9 Fuel tank | C Turbocharger boost pressure |

169  Disconnect the vacuum servo banjo union.

170  Disconnect the air inlet hose to the throttle housing.

171  Remove the bolts securing the throttle housing to the manifold and release the housing from the manifold.

172  Remove the bolts from the manifold support stays.

173  Raise the front of the vehicle onto axle stands then remove the front right-hand road wheel and the inner wing access panel.

174  Disconnect the exhaust downpipes from the manifold, but note that if only the inlet manifold or manifold gasket is to be removed, the exhaust downpipes do not have to be disconnected.

175  Remove the two lower inlet manifold nuts, recovering their washers.

176  Note the position of the coolant pipe and accelerator cable brackets, then undo and remove the four screws/nuts securing the inlet manifold, again recovering their washers.

177  Release the inlet manifold from the gasket and withdraw the manifold.

178  Refitting is a reversal of removal, ensuring the pressure regulator and ECU vacuum hoses are reconnected, and that the inertia switch is re-set.

### Fuel system (MG Turbo models) – description

179  The fuel system on Turbo models consists of a centrally mounted fuel tank, electric fuel pump, air cleaner, SU variable choke side draught carburettor with electronic mixture control system, exhaust driven turbocharger and intercooler.

180  The fuel tank is fitted with a two-way valve in the filler neck to prevent a vacuum being formed in the tank as fuel is used, and to release tank pressure due to thermal expansion of the fuel. A swirl pot is also incorporated in the base of the tank. Fuel from the fuel return line enters the swirl pot via a tapered nozzle within a larger diameter tube in the side of the swirl pot. This creates a depression within the swirl pot, causing the surrounding cooler fuel to be drawn in. The fuel pick-up pipe is located in this area, and cool fuel is thus drawn through the pick-up tube and supplied to the fuel pump.

181  Fuel line pressure provided by the high pressure full flow pump is maintained by a pressure regulator valve at 4.0 lbf/in² (0.3 bar) above the air pressure in the plenum chamber. Excess fuel is returned to the tank via the fuel return line. Fuel at line pressure passes through a vapour separator incorporating a fine mesh screen, then to an in-line filter, and

then to the carburettor float chamber. Excess fuel and vapour is returned to the tank via the pressure regulator valve.

182  Intake air is drawn into the side of the air cleaner body, passed through the filter element, and is then ducted to the turbocharger compressor. After passing through the compressor, the air is ducted to the intercooler intake which contains a temperature sensitive flap valve. If the ducted air is below 45°C, the flap valve remains closed, the air bypasses the intercooler and is directed to the plenum chamber and into the carburettor. If the air temperature is above 45°C, the flap valve opens and the air circulates around the intercooler matrix before entering the plenum chamber. The intercooler performs the same function as the radiator in the cooling system, except that it is air that is cooled, not water. When the turbocharger is in operation and there is boost pressure in the inlet manifold, crankcase ventilation gases also enter the intake air stream via an inlet in the base of the air cleaner body. When there is no boost pressure and a depression exists in the inlet manifold, a one-way valve opens and the gases are drawn directly into the inlet manifold.

183  The turbocharger is bolted to the exhaust manifold and consists of three housings: exhaust, centre and compressor. Located within these housings are the turbine and compressor wheels attached to a common shaft, the fully floating bearings and lubrication ports, and the wastegate valve. The wastegate actuator is attached externally to the compressor housing and operates the wastegate valve according to pressure in the plenum chamber. The turbocharger is connected to the engine lubrication system, and is supplied with a high volume flow of oil for bearing lubrication and heat dissipation.

184  The SU variable choke carburettor is identical in its design and operation to the unit used on normally aspirated engines, with the exception of the fuel shut-off circuit which is not used on Turbo models. A full description of the carburettor and its operation are given in Chapter 2, Section 9.

185  The system functions as follows. With the ignition switched on, the fuel pump is isolated and receives its initial electrical supply on actuation of the starter solenoid. With the engine running, the pump electrical supply is provided by the oil pressure switch. If the oil pressure should drop, if the engine stalls, or in the event of an accident, the pump will be electrically isolated.

186  As engine speed increases, exhaust gases flowing over the blades of the turbine cause the turbine and compressor to spin, such that the intake air passing through the compressor is pressurised up to 10 lbf/in² (0.7 bar) at between 2500 and 4500 engine rpm. At maximum

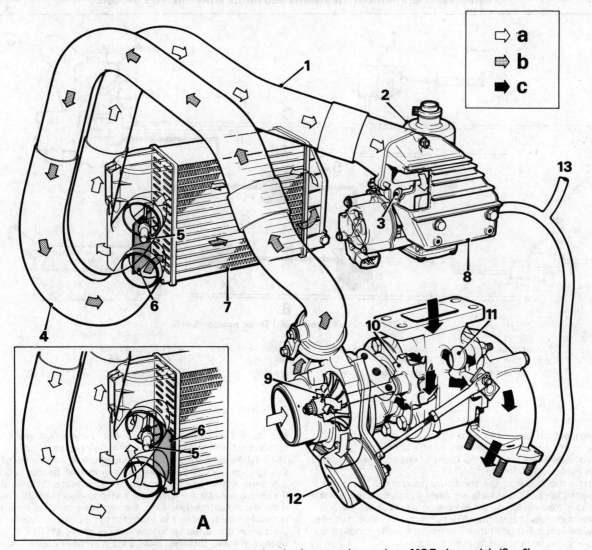

Fig 12.32 Turbocharger and intercooler pipe layout and operation – MG Turbo models (Sec 6)

| | | |
|---|---|---|
| 1 Intercooler-to-carburettor pipe | 6 Flap valve | 10 Turbine | A Intercooler flap valve shut |
| 2 Carburettor | 7 Intercooler matrix | 11 Wastegate valve | a Air below 45°C |
| 3 Dump valve | 8 Plenum chamber | 12 Wastegate actuator | b Air above 45°C |
| 4 Turbocharger-to-intercooler pipe | 9 Compressor | 13 To fuel pressure regulator | c Exhaust gas |
| 5 Flap valve temperature sensor | | | |

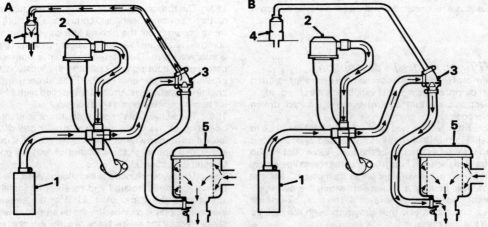

Fig 12.33 Crankcase ventilation system layout and operation – MG Turbo models (Sec 6)

| | | | |
|---|---|---|---|
| A Operation under manifold depression | 1 Crankcase oil separator | 3 Regulator valve | 5 Air cleaner |
| B Operation under manifold pressure | 2 Oil filler cap | 4 Inlet manifold one-way valve | |

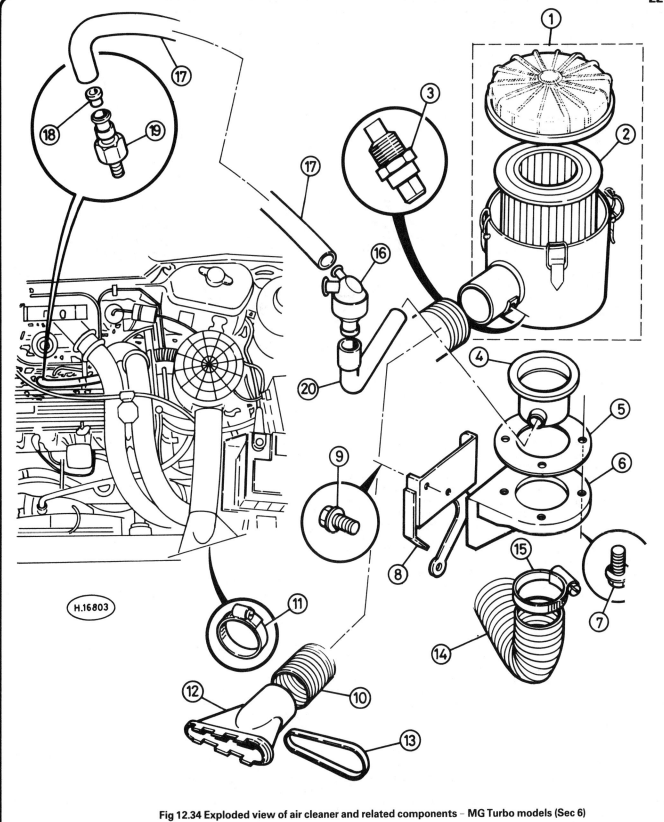

H.16803

**Fig 12.34 Exploded view of air cleaner and related components – MG Turbo models (Sec 6)**

1  Air cleaner assembly
2  Filter element
3  Air temperature sensor
4  Outlet pipe
5  Rubber seal

6  Air cleaner bracket
7  Bolt
8  Mounting bracket
9  Bolt
10  Cold air intake tube

11  Hose clip
12  Intake pipe
13  Sealing ring
14  Turbocharger intake tube
15  Hose clip

16  Regulator valve
17  Breather hose
18  Restrictor
19  One-way valve
20  Breather hose

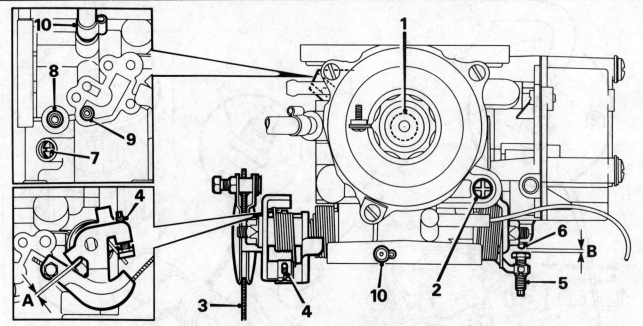

**Fig 12.35 Carburettor adjustment points – MG Turbo models (Sec 6)**

| | | | |
|---|---|---|---|
| 1 Piston damper oil level | 4 Throttle lever adjustment screw | 7 Mixture adjustment screw | 10 Bypass connection |
| 2 Idle speed adjustment screw | 5 Fast idle adjustment screw | 8 Fuel inlet tube | A Lost motion gap |
| 3 Accelerator cable | 6 Fast idle pushrod | 9 Float chamber breather tube | B Fast idle clearance |

boost pressure exhaust gas flow over the turbine can cause the turbine and compressor wheels to rotate at speeds up to 130 000 rpm. Intake air under pressure from the compressor enters the plenum chamber via the intercooler, where its pressure is sensed by a dump valve in the plenum chamber housing. When the pressure reaches approximately 10 lbf/in$^2$ (0.7 bar), the wastegate actuator is operated via a sensing line from the actuator to the dump valve. The wastegate valve is opened reducing exhaust gas flow to the turbine and thereby controlling compressor delivery pressure. The pressurised intake air enters the carburettor from the plenum chamber, and then enters the cylinder via the inlet manifold in the conventional manner.

187    The operation of the carburettor and its electronic mixture control system is described in Chapter 2, but additionally on Turbo models, the fuel ECU further activates the stepper motor during engine warm-up, causing it to enrich the fuel mixture when accelerating and then return to its original setting over a three second period. To prevent fuel evaporation due to high underbonnet temperatures when the engine is switched off, a carburettor cooling fan directs air through a duct around the base of the carburettor. The fan operates when the temperature of the coolant exceeds 90°C and will operate for 10 to 15 minutes.

188    The remainder of this Section describes fuel system procedures applicable to Turbo models where differences occur from normally aspirated versions. For all other fuel system operations, reference should be made to Chapter 3.

### Air cleaner and element (MG Turbo models) – removal and refitting

189    To renew the air cleaner element, spring back the clips and lift off the air cleaner cover.

190    Withdraw the element and wipe clean the inside of the air cleaner body and cover.

191    Fit a new element, locate the cover on the air cleaner body and secure with the retaining clips.

192    To remove the air cleaner assembly, release the cold air intake tube from the body front panel and withdraw the tube from the air cleaner body. Disconnect the sensor wiring.

193    Undo the bolts securing the air cleaner bracket to the mounting bracket on the engine.

194    Lift up the air cleaner, slacken the clamp and detach the outlet tube at the base of the air cleaner.

195    Disconnect the breather hose and remove the air cleaner from the engine.

196    Refitting is the reverse sequence to removal.

### Fuel tank (MG Turbo models) – removal and refitting

197    The procedures are the same as described in Chapter 3, Section 5, but additionally there is a fuel system return hose on Turbo models.

### SU carburettor (MG Turbo models) – adjustments

198    The procedure is the same as given in Chapter 3, Section 10, but note the following additional points:

(a)    Remove the plenum chamber as described later in this Section
(b)    Slacken the retaining clamp to allow the piston damper to be unscrewed, and tighten the clamp after refitting the damper
(c)    The location of the various adjusting screws is shown in Fig. 12.35

### Carburettor heat shield (MG Turbo models) – removal and refitting

199    Disconnect the battery negative terminal.

200    Slacken the retaining clip and disconnect the cooling fan air hose from the heat shield.

201    Undo the bolts securing the support stays to the carburettor support bracket.

202    Disconnect the pressure hose, then remove the plenum chamber and the carburettor support bracket.

203    Disconnect the carburettor fuel and breather hoses and plug the fuel hose after removal.

204    Disconnect the carburettor stepper motor wiring plug.

205    Release the heat shield wiring grommet and remove the heat shield.

206    Refitting is the reverse sequence to removal, but it is essential that the air pressure sensing hose (5) (Fig. 12.40) does not foul the exhaust shield. If necessary, turn the wastegate actuator through 180° in its support bracket.

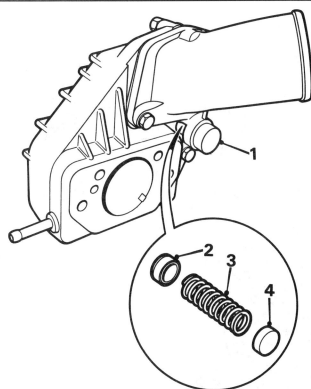

**Fig 12.36 Plenum chamber and dump valve components –
MG Turbo models (Sec 6)**

1  Plenum chamber intake pipe
   and dump valve
2  Dump valve

3  Spring
4  Spacer (where fitted)

219   Disconnect the fuel inlet hose, float chamber vent hose, and vacuum hose at the carburettor. Plug the fuel hose after removal.
220   Undo the carburettor retaining nuts and lift off the carburettor, insulation block, abutment bracket and gaskets from the inlet manifold.
221   Refitting is the reverse sequence to removal, but ensure that all mating faces are perfectly clean, and use new gaskets on all mating faces. Connect the accelerator cable with a small amount of free play, and carry out the carburettor adjustments described earlier in this Section after refitting.

### SU carburettor (MG Turbo models) – dismantling, overhaul and reassembly
222   The procedures given in Chapter 3, Section 12 are valid for Turbo models, but note that the fuel shut-off valve and vacuum switch are not fitted. Other minor detail changes are shown in Fig. 12.37.
223   For the fuel system to operate satisfactorily, the carburettor must be leak-free when subjected to full turbocharger boost pressure. To ensure that this is the case, the unit must be pressure tested after any major dismantling. As this test entails the use of special equipment, have the work carried out by a dealer before refitting the carburettor.

### Vapour separator (MG Turbo models) – removal and refitting
224   Disconnect the battery negative terminal.
225   Place absorbent rags beneath the vapour separator located on the engine compartment bulkhead.
226   Slacken the clamp bolts and withdraw the vapour separator and fuel filter from their locations.
227   Disconnect the fuel outlet hose from the top of the separator, and the inlet and regulator hoses from the side, then remove the vapour separator from the car. Plug the fuel hoses after removal.
228   Refitting is the reverse sequence to removal, but ensure that the regulator hose is connected to the nozzle marked 'R'.

### Plenum chamber (MG Turbo models) – removal and refitting
207   Disconnect the battery negative terminal.
208   Disconnect the pressure hose at the plenum chamber, and undo the bolts securing the heat shield to the plenum chamber, and the plenum chamber to the carburettor.
209   Slacken the clip and disconnect the intake tube at the plenum chamber intake pipe, then remove the plenum chamber assembly.
210   To remove the dump valve, undo the intake pipe retaining bolts, lift off the pipe and gasket, then remove the spacer (where fitted), spring and dump valve from the plenum chamber.
211   Thoroughly clean all mating faces and obtain new gaskets prior to reassembly and refitting.
212   Locate the dump valve, spring and spacer (where fitted) in the plenum chamber, place a new gasket on the mating face, and secure the intake pipe with the retaining bolts.
213   Retain the plenum chamber-to-carburettor gasket in place using the through-bolts, engage the intake pipe with the intake tube, and secure the tube with the clip positioned 0.25 in (6.0 mm) from the end.
214   Secure the plenum chamber, and refit the heat shield and the pressure hose. Reconnect the battery.

### SU carburettor (MG Turbo models) – removal and refitting
215   Disconnect the battery negative terminal.
216   Remove the heat shield, carburettor bracket and plenum chamber as described previously in this Section.
217   Disconnect the stepper motor wiring plug.
218   Slacken the retaining screw and disconnect the accelerator cable from the connector on the carburettor linkage.

### Fuel pressure regulator (MG Turbo models) – removal and refitting
229   Disconnect the battery negative terminal.
230   Place absorbent rags beneath the regulator located alongside the air cleaner assembly.
231   Disconnect the air pressure hose from the top of the regulator.
232   Disconnect the fuel pressure hose from the side of the regulator, then undo the bolts and withdraw the regulator from its mounting bracket.
233   Disconnect the fuel return hose and remove the unit from the car. Plug the fuel hoses after removal. Do not alter the position of the adjusting screw on the top of the regulator, which is preset during manufacture.
234   Refitting is the reverse sequence to removal.

### Intercooler unit (MG Turbo models) – removal and refitting
235   Disconnect the battery negative terminal and remove the radiator grille.
236   Undo the three bolts each side securing the front body panel in position. Release the air cleaner cold air intake tube, then withdraw the panel and place it to one side, leaving the bonnet release cable still attached.
237   Slacken the clips securing the air inlet and outlet tubes to the intercooler pipes assembly.
238   Undo the right-hand side mounting bracket retaining screws and withdraw the intercooler assembly. Detach the tubes and remove the intercooler from the car.
239   Refitting is the reverse sequence to removal.

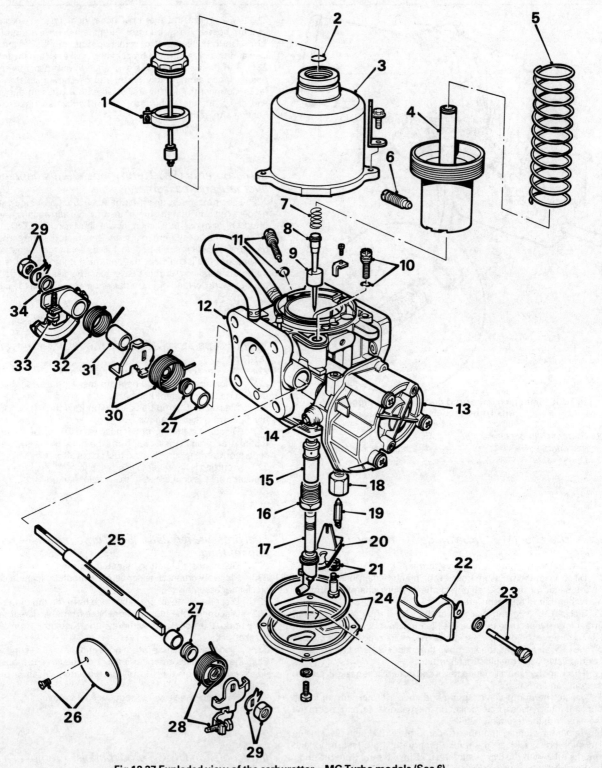

**Fig 12.37 Exploded view of the carburettor – MG Turbo models (Sec 6)**

| | | | |
|---|---|---|---|
| 1 | Piston damper and clip | 10 | Idle speed adjustment screw and seal |
| 2 | Retaining circlip | 11 | Mixture adjustment screw and seal |
| 3 | Suction chamber | 12 | Carburettor body |
| 4 | Piston | 13 | Mixture control stepper motor |
| 5 | Piston spring | 14 | Fast idle pushrod |
| 6 | Needle guide locking screw | 15 | Jet bearing |
| 7 | Needle bias spring | 16 | Jet bearing nut |
| 8 | Jet needle | 17 | Jet assembly |
| 9 | Needle guide | 18 | Float needle seat |

| | |
|---|---|
| 19 | Float needle |
| 20 | Bi-metal jet lever |
| 21 | Jet lever retaining screw |
| 22 | Float |
| 23 | Float pivot spindle and seal |
| 24 | Float chamber cover and seal |
| 25 | Throttle spindle |
| 26 | Throttle plate |
| 27 | Throttle spindle bearing and seal |

| | |
|---|---|
| 28 | Fast idle adjustment screw and lever spring |
| 29 | Spindle retaining nut and lock tab |
| 30 | Lost motion link and spring |
| 31 | Bush |
| 32 | Progresive lever adjustment |
| 33 | Throttle lever adjustment |
| 34 | Washer |

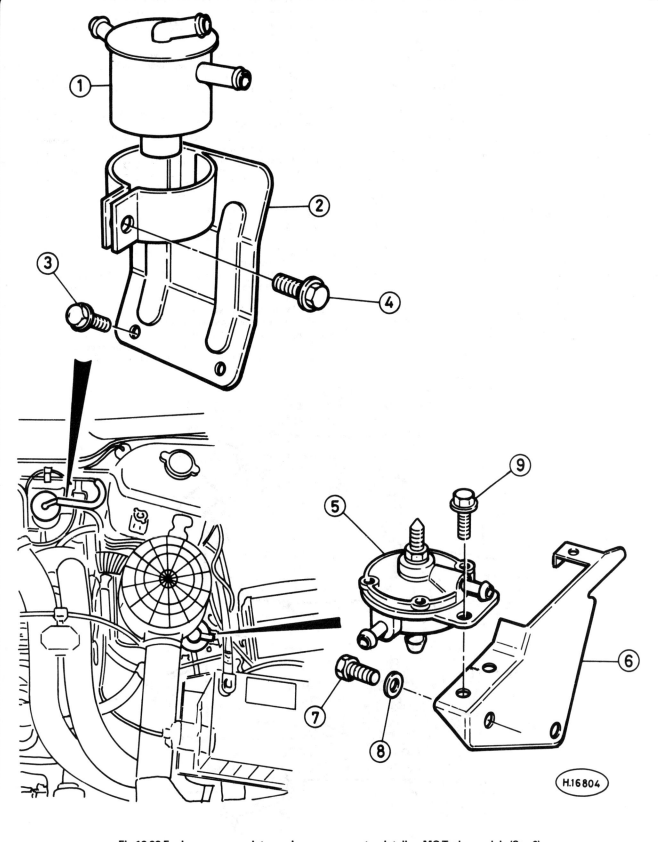

**Fig 12.38 Fuel pressure regulator and vapour separator details – MG Turbo models (Sec 6)**

1  Vapour separator    4  Clamp bolt    6  Mounting bracket    8  Washer
2  Mounting bracket    5  Fuel pressure regulator    7  Bolt    9  Bolt
3  Bolt

H.16804

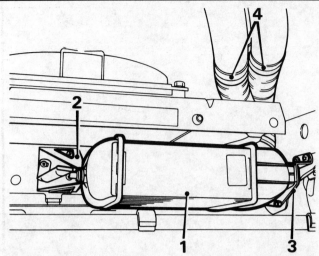

**Fig 12.39 Intercooler mounting details – MG Turbo models (Sec 6)**

1  Intercooler unit
2  Right-hand mounting bracket
3  Left-hand mounting bracket
4  Inlet and outlet hoses

### Carburettor cooling fan (MG Turbo models) – removal and refitting

240   Disconnect the battery negative terminal.
241   From under the right-hand wheel arch, release the intake tube retaining strap.
242   Slacken the clip and disconnect the outlet tube from the fan motor elbow.
243   Detach the fan motor wiring connector from the clip and disconnect the multi-plug.
244   Undo the retaining screws and remove the motor and bracket assembly from the car.
245   Remove the intake tube and the bracket assembly from the motor.
246   Refitting is the reverse sequence to removal.

### Carburettor cooling fan delay unit (MG Turbo models) – removal and refitting

247   The operation of the carburettor cooling fan is controlled by a delay unit in conjunction with a thermostatic switch located in the thermostat housing.
248   To remove the delay unit, disconnect the battery negative terminal and remove the fusebox cover from below the facia on the right-hand side.
249   Withdraw the two relays at the bottom of the panel.
250   Undo the screws securing the delay unit to the bottom of the fusebox, lower the unit and disconnect the wiring multi-plug.
251   Remove the delay unit from its mounting bracket.
252   Refitting is the reverse sequence to removal.

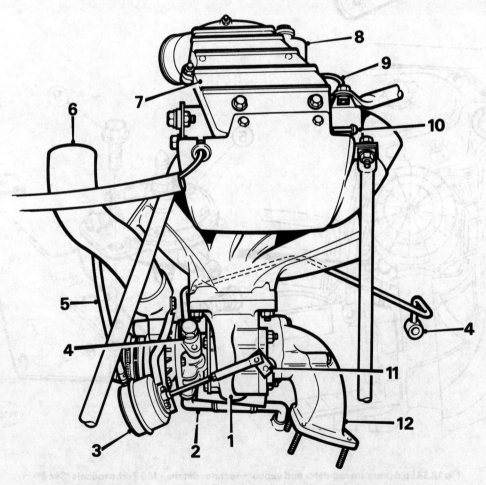

**Fig 12.40 Turbocharger and related components – MG Turbo models (Sec 6)**

| | | | |
|---|---|---|---|
| 1  Turbocharger | 4  Oil feed pipe | 7  Plenum chamber | 10 Air pressure sensing tube |
| 2  Oil drain elbow | 5  Pressure hose | 8  Carburettor | 11 Wastegate lever |
| 3  Wastegate actuator | 6  Compressor outlet hose | 9  Bypass hose | 12 Exhaust elbow |

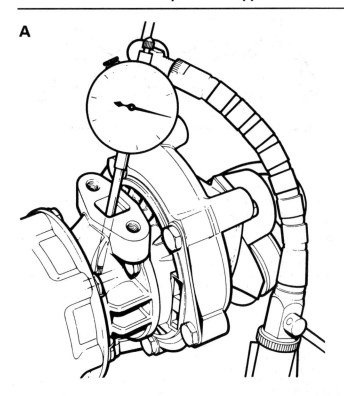

**A**

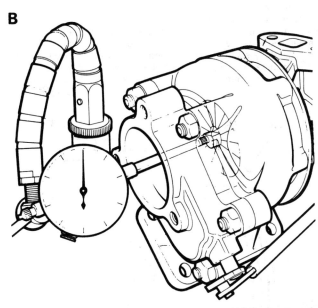

**B**

**Fig 12.41 Checking turbocharger clearances – MG Turbo models (Sec 6)**

*A  Radial clearance*      *B  Axial endfloat*

### Carburettor cooling fan thermostatic switch (MG Turbo models) – testing, removal and refitting

253    The thermostatic switch is located in the thermostat housing, and operates the cooling fan in conjunction with the delay unit whenever the coolant temperature exceeds 160°F (70°C).
254    If the operation of the switch or cooling fan is suspect, the same test procedure as that given for the radiator cooling fan can be used, and reference should be made to Chapter 2, Section 11.
255    If the switch is to be renewed, follow the procedure given in Chapter 2, Section 13 for the coolant temperature thermistor.

### Turbocharger exhaust elbow (MG Turbo models) – removal and refitting

256    Jack up the front of the car and support it on stands. Remove the right-hand front roadwheel and the access panel under the wheel arch.
257    Undo the four bolts securing the anti-roll bar clamps to the subframe, remove the clamps and lower the anti-roll bar.
258    Undo the nuts securing the exhaust front pipe to the exhaust elbow, lower the pipe and recover the gasket.
259    Knock back the locking tabs and undo the exhaust elbow retaining bolts. Remove the elbow from under the car and recover the gasket.
260    Refitting is the reverse sequence to removal, but use new gaskets and locking tabs and tighten all bolts to the specified torque.

### Turbocharger oil feed pipe (MG Turbo models) – removal and refitting

261    Jack up the front of the car and support it on stands. Remove the right-hand front roadwheel and the access panel under the wheel arch.
262    Disconnect the wire at the oil pressure switch and unscrew the switch and switch extension (where fitted).
263    Disconnect the oil pipe bracket from the manifold.
264    Unscrew the oil feed pipe banjo union bolt or union nut at the oil pump and, where fitted, recover the sealing washers.
265    Remove the intercooler pipes and the turbocharger compressor outlet tube. Place rags over the outlet to prevent dirt ingress.
266    Remove the carburettor left-hand support stay.
267    Undo the oil feed pipe banjo union bolt or union nut at the turbocharger and, where fitted, recover the sealing washers.
268    Remove the oil feed pipe from the car.
269    Refitting is the reverse sequence to removal, but where banjo unions are fitted, renew the sealing washers and support the union as the bolt is tightened.

### Turbocharger oil drain elbow (MG Turbo models) – removal and refitting

270    Jack up the front of the car and support it on stands. Remove the right-hand front roadwheel and the access panel from under the wheel arch.
271    Undo the two nuts and bolts securing the front suspension strut to the swivel hub and separate the strut from the hub.
272    Place a container beneath the driveshaft inner constant velocity joint to catch any oil released from the differential during subsequent operations.
273    Using a flat bar or large screwdriver, lever between the inner constant velocity joint and the differential housing to release the joint from the differential sun gear.
274    Tip the swivel hub outwards and withdraw the constant velocity joint from the differential. Move the driveshaft to one side to provide access to the oil drain elbow.
275    Undo the heat shield retaining bolt and remove the heat shield.
276    Slacken the clip and disconnect the hose from the oil drain elbow.
277    Undo the bolts and remove the elbow and gasket.
278    Refitting is the reverse sequence of removal, but use a new gasket on the elbow mating face and tighten all bolts to the specified torque. Top up the gearbox oil level with reference to Chapter 6.

### Turbocharger (MG Turbo models) – removal and refitting

279    The turbocharger is removed together with the inlet and exhaust manifolds as a complete assembly, and can then be separated with the assembly on the bench. The procedures are described in the next sub-section.
280    With the unit removed, wipe it clean with rags but do not use any solvents.
281    Check the axial and radial clearances of the shaft and bearings using a dial test indicator as shown in Fig. 12.41. If the clearances exceed the specified limits, the turbocharger must be renewed, as overhaul is not possible.
282    Check the general condition of the waste gate and check for free movement. To check the operation, a regulated air pressure of 10 lbf/in² (0.7 bar) must be applied to the waste gate actuator, which should then cause the pushrod to extend 0.5 in (13.0 mm). The turbocharger will require renewal if this is not the case.
283    Before fitting a new turbocharger, plug the oil drain elbow and fill the oil inlet pipe port with clean engine oil. Rotate the compressor by hand to prime the shaft bearings, then drain the oil.

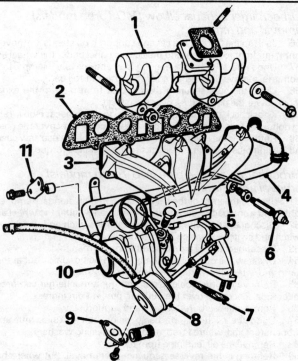

Fig 12.42 Inlet and exhaust manifold attachments – MG Turbo models (Sec 6)

1   Inlet manifold
2   Gasket
3   Exhaust manifold
4   Heater pipe
5   Turbocharger oil feed pipe
    and unions

6   Oil pressure switch and extension
7   Flange gasket
8   Wastegate actuator
9   Oil drain elbow
10  Turbocharger
11  Bracket

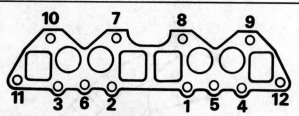

Fig 12.43 Manifold bolt tightening sequence – MG Turbo models (Sec 6)

## Inlet and exhaust manifolds (MG Turbo models) – removal and refitting

284   Disconnect the battery negative terminal.
285   Drain the cooling system as described in Chapter 2.
286   Remove the air cleaner and carburettor as described previously in this Section.
287   Slacken the clips and disconnect the air intake and outlet tubes at the turbocharger compressor.
288   Remove the outlet pipe from the turbocharger compressor.
289   Undo the nuts and bolts and disconnect the four manifold support struts.
290   Slacken the clips and disconnect the heater hoses from the pipes at the bulkhead connection and thermostat housing. Undo the nuts and remove the heater pipe from the manifold.
291   Disconnect the fuel vapour separator outlet hose, undo the retaining bolts and move the vapour separator and bracket aside.
292   Undo the brake servo banjo union bolt, withdraw the union and recover the copper washers. Disconnect all remaining vacuum and breather hoses at the inlet manifold.
293   Remove the turbocharger oil drain elbow and oil feed pipe as described earlier in this Section.
294   Undo the nuts securing the exhaust front pipe to the turbocharger, and separate the flange joint. Recover the gasket.
295   Carefully seal all openings and ports in the turbocharger with clean rags or tape, taking care to ensure absolute cleanliness.
296   Undo the nuts, bolts and clamp washers securing the two

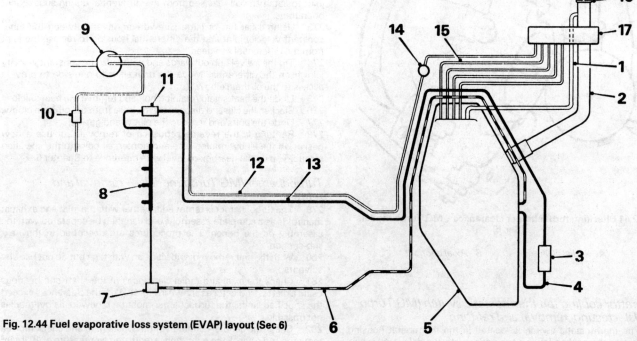

Fig. 12.44 Fuel evaporative loss system (EVAP) layout (Sec 6)

1   Fuel tank breather
2   Fuel filler pipe
3   Fuel pump
4   Fuel feed pipe
5   Fuel tank

6   Fuel return pipe
7   Pressure regulator
8   Injectors
9   Charcoal canister
10  Purge valve

11  Fuel filter
12  Fuel tank vent pipe
13  Fuel feed pipe
14  2-way valve

15  Fuel tank to vapour separation
    tank pipes
16  Fuel filler cap
17  Separation tank

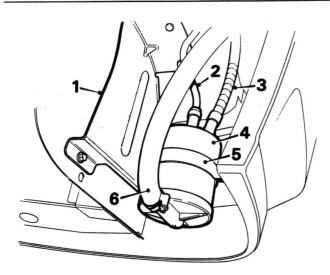

**Fig. 12.45 Charcoal canister and connections (Sec 6)**

| | | | |
|---|---|---|---|
| 1 | Mounting bracket | 4 | Charcoal canister |
| 2 | Vapour pipe | 5 | Clamp |
| 3 | Purge pipe | 6 | Drain pipe |

manifolds, then carefully withdraw the inlet manifold and the exhaust manifold with turbocharger, and remove them from the car. Recover the gasket.

297  With the manifolds on the bench, release the lock wire and locktabs, undo the turbocharger retaining nuts, then remove the unit from the manifold. Recover the gasket and cover the opening in the turbocharger.

298  Before refitting, clean all mating faces thoroughly and obtain new gaskets and seals to replace all those disturbed during removal.

299  Refitting is the reverse sequence to removal, bearing in mind the following points:

(a)  Apply a high temperature anti-seize compound to the threads of all nuts, bolts and studs before fitting

(b)  Tighten all nuts and bolts to the specified torque where applicable, and lock the turbocharger to manifold studs using locking wire. Observe the tightening sequence shown in Fig. 12.43 when tightening the inlet and exhaust manifold nuts and bolts

(c)  Refill the cooling system with reference to Section 5 of this Supplement and Chapter 2

### Catalytic converter – general

300  Some models are fitted with a catalytic converter in the exhaust system, the function of which is to minimise the pollutants passed into the atmosphere through the exhaust system. A 'closed-loop' system is fitted in which an oxygen sensor, screwed into the catalytic converter casing, measures the oxygen content passing through the exhaust and sends this information to the ECU. The sensor feeds a varying amount of voltage to the ECU in accordance with the amount of oxygen in the system. The ECU then makes the necessary adjustments to the fuel and ignition systems to provide the most efficient mixture level.

301  The catalytic converter is connected to the exhaust downpipe at its front end and to the exhaust intermediate section at the rear. It has an integral ceramic and precious metals filter to reduce the harmful exhaust emissions at very high temperatures. It is a fragile device which can be easily damaged. With this in mind, it is imperative that the following precautionary observations be noted and adhered to whenever working on, or in the vicinity of, the catalytic converter in order to ensure its efficiency and maximum service life.

### Catalytic converter – precautions

302  The catalytic converter is a reliable and simple device which needs no maintenance in itself, but there are some facts of which an owner should be aware if the converter is to function properly for its full service life.

(a)  DO NOT use leaded petrol in a car equipped with a catalytic converter – the lead will coat the precious metals, reducing their converting efficiency and will eventually destroy the converter.

(b)  Always keep the ignition and fuel systems well-maintained in accordance with the manufacturer's schedule – particularly, ensure that the air cleaner filter element, the fuel filter (where fitted) and the spark plugs are renewed at the correct interval – if the intake air/fuel mixture is allowed to become too rich due to neglect, the unburned surplus will enter and burn in the catalytic converter, overheating the element and eventually destroying the converter.

(c)  If the engine develops a misfire, do not drive the car at all (or at least as little as possible) until the fault is cured – the misfire will allow unburned fuel to enter the converter, which will result in its overheating, as noted above.

(d)  DO NOT push- or tow-start the car – this will soak the catalytic converter in unburned fuel, causing it to overheat when the engine does start – see (b) above.

(e)  DO NOT switch off the ignition at high engine speeds – if the ignition is switched off at anything above idle speed, unburned fuel will enter the (very hot) catalytic converter, with the possible risk of its igniting on the element and damaging the converter.

(f)  DO NOT use fuel or engine oil additives – these may contain substances harmful to the catalytic converter.

(g)  DO NOT continue to use the car if the engine burns oil to the extent of leaving a visible trail of blue smoke – the unburned carbon deposits will clog the converter passages and reduce its efficiency; in severe cases the element will overheat.

(h)  Remember that the catalytic converter operates at very high temperatures – hence the heat shields on the car's underbody – and the casing will become hot enough to ignite combustible materials which brush against it. DO NOT, therefore, park the car in dry undergrowth, over long grass or piles of dead leaves.

(i)  Remember that the catalytic converter is FRAGILE – do not strike it with tools during servicing work, take great care when working on the exhaust system, ensure that the converter is well clear of any jacks or other lifting gear used to raise the car and do not drive the car over rough ground, road humps, etc., in such a way as to 'ground' the exhaust system.

(j)  In some cases, particularly when the car is new and/or is used for stop/start driving, a sulphurous smell (like that of rotten eggs) may be noticed from the exhaust. This is common to many catalytic converter-equipped cars and seems to be due to the small amount of sulphur found in some petrols reacting with hydrogen in the exhaust to produce hydrogen sulphide ($H_2S$) gas; while this gas is toxic, it is not produced in sufficient amounts to be a problem. Once the car has covered a few thousand miles the problem should disappear – in the meanwhile, a change of driving style or of the brand of petrol used may effect a solution.

(k)  The catalytic converter, used on a well-maintained and well-driven car, should last for between 50 000 and 100 000 miles – from this point on, careful checks should be made at all specified service intervals of the CO level to ensure that the converter is still operating efficiently – if the converter is no longer effective it must be renewed.

(l)  Do not overfill the engine oil level. Excess oil may enter the exhaust system, causing the catalytic converter to be clogged and excess HC emissions will result as it is being burnt off.

(m)  Do not apply an exhaust sealant to any of the system joints forward of the catalytic converter. Any excess sealant compressed into the system when tightening the joints may be transferred into the converter.

### Fuel evaporative loss control system (EVAP) – general

303  The function of the evaporative fuel loss control system is to prevent the fuel vapours that are produced in the fuel tank when the engine is turned off, escaping to the atmosphere. This is achieved by transferring the vaporized fuel to a charcoal canister. When the engine is restarted, the stored fuel vapour is transferred from the canister to the induction system and burnt off in the normal combustion process. A schematic diagram of the system is shown in Fig. 12.45.

304  The charcoal canister (located under the right-hand front wing), contains charcoal granules which absorb and store the fuel system vapours. A purge valve, mounted in the engine compartment (at the rear of the right-hand headlamp), prevents purging between the charcoal canister and the throttle body under certain operating conditions which would otherwise affect the operation of the engine and/or catalytic converter (where applicable).

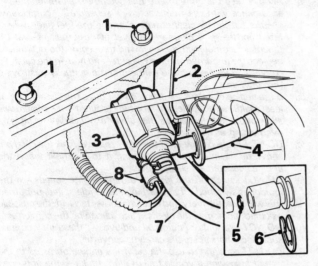

**Fig. 12.46 Purge valve and connections (Sec 6)**

1   Mounting bracket securing        5   O-ring
    bolts                            6   Circlip
2   Mounting bracket                 7   Vacuum pipe
3   Purge control valve              8   Wiring multi-plug
4   Hose connector

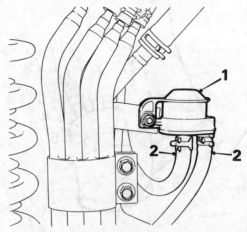

**Fig. 12.47 2-way valve and connections (Sec 6)**

1   2-way valve
2   Hoses (note fitted positions before dismantling)

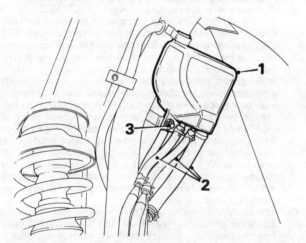

**Fig. 12.48 Vapour separation tank and connections (Sec 6)**

1   Separation tank
2   Hoses
3   Retaining screw

305   The vapours pass from the fuel tank to a separation tank, mounted under the right-hand rear wing which collects fuel vapour, separates the fuel from it and passes the remaining vapour to the charcoal canister via a 2-way valve located under the separation tank. The valve opens under pressure (approx 0.4 lbf/in$^2$) to enable the fuel vapour to pass from the separation tank to the charcoal canister. At the same time, the valve also opens the fuel tank to atmosphere to relieve the depression in the fuel tank. The liquid fuel collected in the separation tank is passed back into the main fuel tank.

## Fuel evaporative loss control system (EVAP) components – removal and refitting

### Charcoal canister

306   Loosen off (but do not remove at this stage) the right-hand front roadwheel, then raise and support the front of the vehicle on axle stands. With the vehicle securely supported, remove the roadwheel.

307   Where applicable, undo the retaining screws and remove the front spoiler.

308   Release the retaining studs and detach the wheel arch liner from the leading edge on the underside of the right-hand front wing.

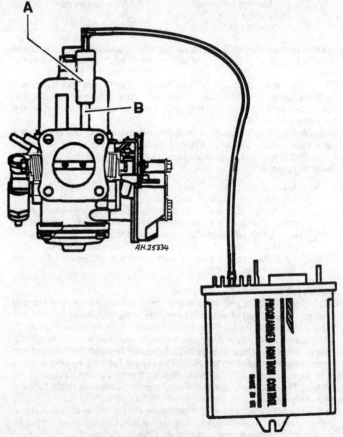

**Fig 12.49 Ignition ECU fuel trap (Sec 7)**

A   Fuel trap              B   Connector

309   Undo the mounting bracket retaining screw and nuts, then lower the canister plate. Loosen off the securing clip and detach the drain pipe from the canister. Compress the vapour and purge pipe hose clips, lower the canister and detach the hoses from their connectors at the top.

310   Refit in the reverse order of removal. Ensure that the hoses are in good condition and are securely connected.

### Purge valve

311   Undo the bolts securing the valve mounting bracket, then move the bracket to enable the multi-plug to be detached.

312 Disconnect the vacuum pipe from the purge valve, then detach the hose connector circlip and remove the connector. The purge valve can now be slid free from its bracket and removed.
313 Refit in the reverse order of removal. Renew the O-ring when fitting it to the hose connector; smear it with clean engine oil to ease assembly. Ensure that all connections are cleanly and securely made.

**2-way valve**
314 Loosen off (but do not remove at this stage) the right-hand rear roadwheel, then raise and support the rear of the vehicle on axle stands. With the vehicle securely supported, remove the roadwheel.
315 Unscrew and remove the 2-way valve retaining screw. Note the fitted positions of the hoses, compress their retaining clips and detach the hoses from the valve. Remove the valve.
316 Refit in the reverse order of removal. Ensure that the hoses are correctly and securely mounted on the valve prior to refitting it into position.

**Vapour separation tank**
317 Loosen off (but do not remove at this stage) the right-hand rear roadwheel, then raise and support the rear of the vehicle on axle stands. With the vehicle securely supported, remove the roadwheel.
318 Unscrew and remove the separation tank retaining screws. Note the fitted positions of the hoses, compress their retaining clips and detach them from the tank. Remove the tank.
319 Refit in the reverse order of removal. Ensure that the hoses are correctly and securely connected to the separation tank prior to refitting it into position.

---

**7 Ignition system**

*Intake air temperature sensor (MG Turbo models) – general*
1 The intake air temperature sensor is located in the air cleaner cold air intake, and consists of an element, the resistance of which changes according to temperature. On Turbo models, the ignition ECU uses the changing resistance of this unit, rather than the coolant temperature thermistor, to determine whether changes to the ignition timing, to cater for alterations in temperature, are necessary.
2 To remove the unit, disconnect the wiring connector and unscrew the sensor from the air cleaner intake.
3 Refit using the reverse sequence to removal.

*Ignition system (fuel injection models) – general*
4 The ignition system components used on later fuel injection models are identical to those used on MG EFi versions. Specific references in the text and Specifications of Chapter 4 to MG EFi models are therefore applicable also to later models, unless indicated otherwise in the Specifications of this Supplement.

*Ignition ECU fuel contamination – rectification*
5 It is possible for fuel to enter the ignition ECU on early Turbo models. This problem can be overcome by fitting a fuel trap as shown in Fig. 12.49. Later models are equipped with the trap during production.

*Rotor arm securing screw seizure – programmed ignition*
6 Seizure of the distributor rotor arm securing screw is due to thread locking compound remaining in the threads of the hole in the camshaft after the screw is removed.
7 Whenever the rotor arm is removed the threads must be cleaned using an M6 tap and die.
8 Apply thread-locking compound to the threads of the securing screw on refitting.

*Electronically Regulated Ignition and Carburation system (ERIC) – general description*
9 As described in Section 6 of this Supplement, the Electronically Regulated Ignition and Carburation system (ERIC) combines the functions of the fuel and ignition systems into one system.
10 Reference should be made to Section 6 for details for the system and to Chapter 4, Part 2 for details of the ignition components of the system which are the same as used on the earlier programmed ignition system.

*Modular Engine Management System (MEMS MPi) – general description*
11 For details of the ignition aspects of the system, refer to Section 6 of this Supplement.

---

**8 Clutch**

*Clutch cable adjustment (all models) – checking*
1 Whenever the clutch cable is disturbed for any reason the following adjustment check must be carried out.
2 Using moderate hand pressure, push the clutch lever on the bellhousing downward (in the opposite direction to its normal travel) through its full stroke.
3 When released the lever should return to its original position.
4 Press the adjuster body into its abutment on the engine bulkhead with one hand, and with the other pull the outer cable away from the bulkhead until resistance is felt, holding the outer cable at a point just behind the spring.
5 Release both parts of the cable.
6 This should ensure the self-adjusting mechanism is correctly set. If the adjuster fails to operate correctly, refer to the fault diagnosis Section of Chapter 5.

---

**9 Manual gearbox**

*Countershaft bearings (MG Turbo models) – modifications*
1 On Turbo models, the countershaft is supported at its gearcase end by a roller bearing and single ball race, instead of the double ball race used on all other models. The repair and overhaul procedures are not affected by this modification.

*Gearbox modification – 1989-on*
2 On 1989-on gearboxes, the 5th/reverse gear selector shaft assembly is slightly different, the selector forks being retained on the shaft by a machined collar and not a circlip as previously.

*Gearbox – oil leakage*
3 If an oil leak from the bellhousing-to-gearcase joint in the area of the rear mounting bracket is a problem, the probable cause is the joint faces being pulled apart by the mounting bolts as they are tightened. It can be rectified as follows.
4 Shims are available from Austin Rover dealers for insertion between the mounting and the gearcase to prevent the joint being pulled apart.
5 Note that no more than three shims must be used, and if more than this number are required, the mounting bracket must be renewed.

*Manual gearbox – identification*
6 Originally all models were fitted with a gearbox built by Honda in Japan.
7 Progressively from approximately September 1987, gearboxes built in the UK by Austin Rover were used.
8 A number of internal components differ and are not interchangeable between the two types of gearbox, and it is therefore necessary to identify the origin of the gearbox when ordering spares.
9 Those gearboxes manufactured in Japan have a Gold label and seven digit serial number beginning with 1, those manufactured in the UK have a white label and seven digit serial number beginning with 2..

---

**10 Automatic transmission**

*General description*
1 Introduced on the Montego range in 1986, the ZF four-speed automatic transmission comprises a hydrodynamic torque converter with an integrated torsion damper and a four-speed planetary gear unit.
2 Two spur gear sets with a countershaft transfer the power to the differential.
3 Only 40% of input power is fed hydraulically to the gearbox through the torque converter, while the remaining 60% is transmitted mechanically.

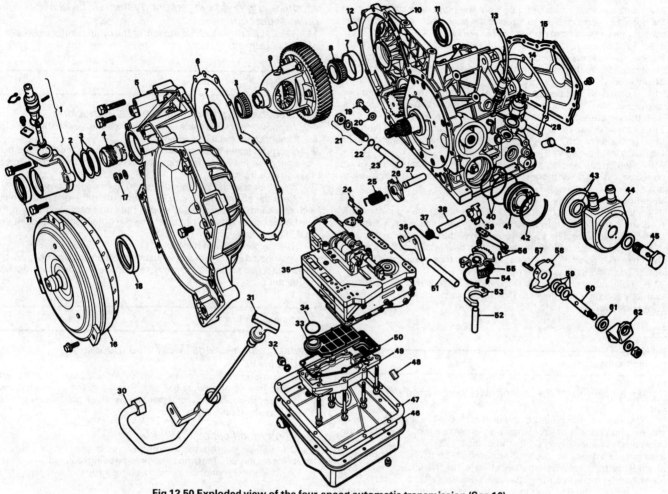

**Fig 12.50 Exploded view of the four-speed automatic transmission (Sec 10)**

| | | | |
|---|---|---|---|
| 1 Speedometer pinion assembly | 15 Gearcase and cover | 30 Filler tube | 46 Sump pan |
| 2 O-ring seal | 16 Torque converter | 31 Dipstick | 47 Gasket |
| 3 Shims | 17 Final drive drain plug | 32 Sump pan drain plug | 48 Magnet |
| 4 Speedometer drivegear | 18 Oil seal | 33 Oil strainer | 49 Fluid strainer housing |
| 5 Torque converter housing | 19 Countershaft bearing locating screw | 34 O-ring seal | 50 Gasket |
| 6 Gasket | 20 Sealing washer | 35 Valve block | 51 Pushrod |
| 7 Bearing outer track | 21 Brake band adjuster | 36 Parking pawl | 52 Park lock cam shaft |
| 8 Taper roller bearing | 22 O-ring seal | 37 Spring | 53 Locking cam |
| 9 Differential and final drive assembly | 23 Adjustment rod | 38 Parking pawl shaft | 54 Spring |
| 10 Gearcase | 24 Retaining plate | 39 Oil diverter | 55 Selector stop plate |
| 11 Oil seal | 25 Spring | 40 O-ring seals | 56 Detent spring and roller |
| 12 Hollow dowel | 26 Kickdown cam | 41 Servo cover | 57 Roll pin |
| 13 Kickdown cable | 27 Kickdown cam shaft | 42 Snap ring | 58 Selector cam |
| 14 Gasket | 28 Starter/inhibitor switch | 43 Fluid cooler seals | 59 Shims |
| | 29 Gearcase breather | 44 Fluid cooler | 60 Selector shaft |
| | | 45 Centre screw | 61 Oil seal |
| | | | 62 Selector lever |

4    In 4th, input power is transmitted 100% mechanically through the converter cover and integral damper. This results in improved economy.

5    A kickdown (downshift) facility is provided for overtaking, actuated by flooring the accelerator pedal.

6    Driver control of the transmission is by a seven position selector lever which allows fully automatic operation, with a hold facility on the first and second gear ratios.

7    Fluid used in the transmission is common to the gearbox and final drive, but separate drain plugs are used.

8    Due to the complexity of the automatic transmission, any repair or overhaul work must be left to a BL dealer or automatic transmission specialist with necessary equipment for fault diagnosis and repair. The contents of the following Sections are therefore confined to supplying general information and any service information and instructions that can be used by the owner.

*Fluid level – checking*

9    At the intervals specified in Routine Maintenance, check the transmission fluid level. This may be done with the engine/transmission hot or cold with the vehicle standing on a level surface.

**Engine/transmission cold**

10    Apply the handbrake and select 'P'. Start the engine, and as soon as it idles evenly, withdraw the dipstick and wipe clean on paper or a non-fluffy cloth.

11    Insert the dipstick fully and immediately withdraw it. The fluid

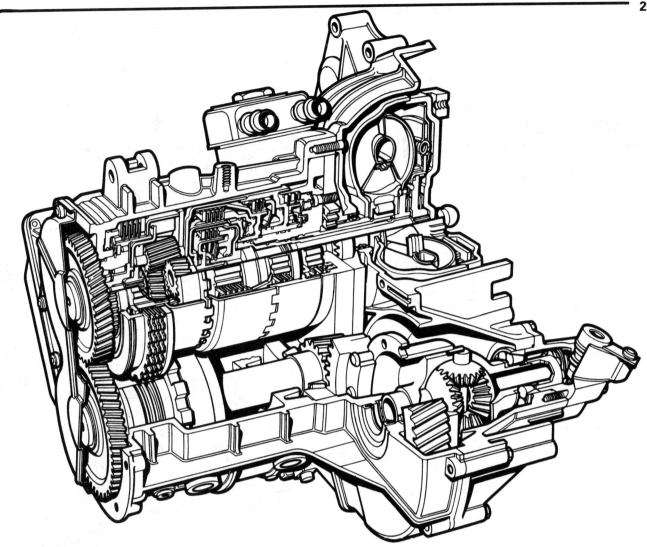

**Fig 12.51 Cutaway view of the four-speed automatic transmission (Sec 10)**

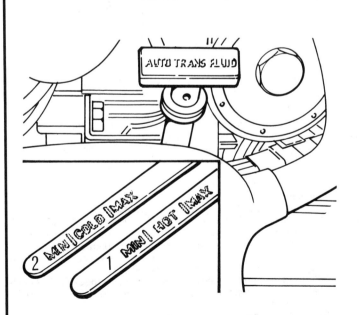

**Fig 12.52 Fluid dipstick (Sec 10)**

*1  Hot side*      *2  Cold side*

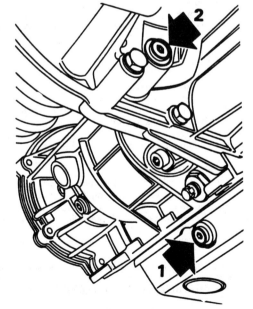

**Fig 12.53 Drain plugs (Sec 10)**

*1  Gearcase drain plug*      *2  Final drive housing drain plug*

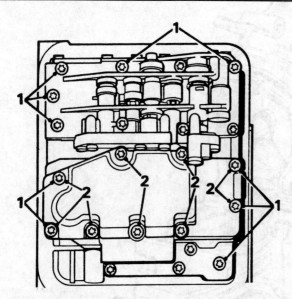

**Fig 12.54 Valve block and fluid filter fixing screws (Sec 10)**

*1   Valve block screws*          *2   Fluid filter screws*

level should be between the MIN and MAX marks on the COLD side of the dipstick.

12   If more fluid is required, top up via the dipstick guide tube to the mid-point between the marks. Note that checking the fluid level cold is only approximate – ideally it should be checked hot.

**Engine/transmission hot**

13   The engine/transmission should be at normal operating temperature, preferably after a run of at least 10 miles (15 kilometres).

14   Level checking and topping-up are as described in paragraphs 10 to 12, but check the level against the HOT side of the dipstick.

*Fluid – renewal*

15   At the intervals specified 'Routine Maintenance', the automatic transmission fluid should be renewed.

16   Place the front end of the vehicle over an inspection pit or raise it on ramps.

17   Remove the undershield.

18   Place a large container under the transmission and unscrew the socket-headed drain plugs from the transmission casing. Allow the fluid to drain.

19   Refit the drain plugs and the undershield.

20   Refill through the dipstick guide tube using the correct grade of fluid until the fluid level is at MAX on the COLD side of the dipstick. The torque converter cannot be drained at routine fluid change, so only 3.5 pints (2.0 litres) will be required.

21   Start the engine and as soon as it idles evenly, move the selector lever slowly through all positions to prime the transmission.

22   Check the fluid level against the COLD side of the dipstick as described in the preceding Section, and top up if necessary.

*Fluid filter – cleaning/renewal*

23   This is not a routine maintenance operation and will normally only be required after a very high mileage, or if metal filings or other contaminants are evident in the fluid at time of regular fluid renewal.

24   Drain the fluid as described in the preceding Section. Refit the drain plug.

25   Release the dipstick guide/fluid filler tube from the transmission sump pan.

26   Extract the sump pan fixing screws, take off the sump pan and discard the gasket.

27   Clean the interior of the sump pan and remove and clean the swarf collecting magnet.

28   Remove the screws which secure the fluid filter, noting carefully their different lengths. Do not touch the valve block screws.

29   Clean the filter mesh with petrol and dry it. If badly contaminated, renew the filter.

30   Clean out the filter housing.

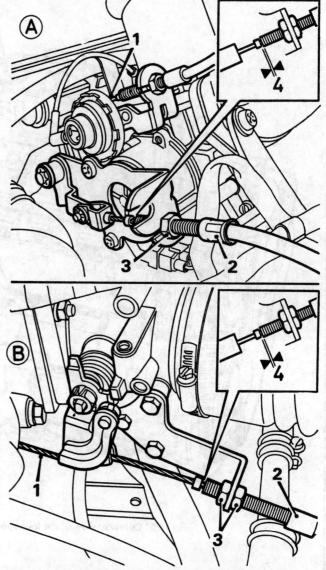

**Fig 12.55 Kickdown cable adjustment (Sec 10)**

*A   Carburettor models*
*B   EFi models*
*1   Throttle cable*

*2   Kickdown cable*
*3   Threaded adjuster locknuts*
*4   Crimped sleeve to adjuster endface clearance (see text)*

31   Fit the filter complete with new O-ring seal and gasket.

32   Tighten the screws to specified torque.

33   Fit the sump pan using a new gasket and then refill the transmission as previously described.

*Kickdown cable – adjustment*

**Pre-1989**

34   Slacken the throttle cable and kickdown cable locknuts at the abutment bracket within the engine compartment, then release the tension in the kickdown cable.

35   Adjust the throttle outer cable at the abutment bracket, until slackness is just eliminated from the inner cable without preloading the throttle lever. Tighten the throttle cable locknut.

36   Hold the throttle in the fully open position, at the same time gently pulling the kickdown cable past the detent until all slackness in the cable is eliminated.

37   Hold the kickdown outer cable to retain this setting, and then tighten the locknut against the abutment bracket.

38   Release the throttle and check that the gap (Fig. 12.55) between the crimped sleeve on the kickdown inner cable and the end of the cable adjuster is between 0.02 and 0.04 in (0.5 and 1.0 mm).

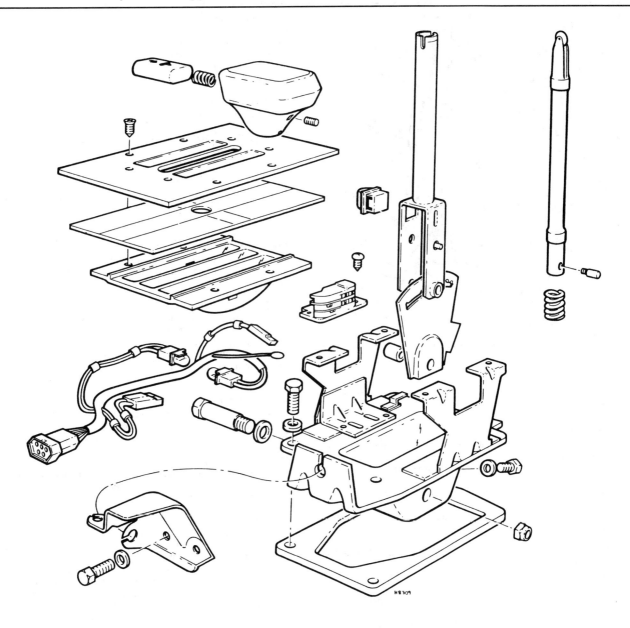

**Fig 12.56 Selector lever assembly – exploded view (Sec 10)**

39    Adjust the bracket locknuts if necessary to obtain the correct gap.

**1989-on**

40    Engage 'P' and run the engine until it reaches normal operating temperature.

41    Check that the engine is idling at the specified speed.

42    Slacken the kickdown cable locknuts on the support bracket and release all tension in the cable.

43    Check that the gap between the crimped sleeve on the inner cable and the end of the cable is as specified for earlier models.

44    If necessary, adjust the position of the crimped sleeve by using the kickdown cable adjustment nuts. Tighten the nuts on completion.

*Kickdown cable – renewal*

45    Record the routing of the cable run so that the new cable can be fitted in the same way.

46    Disconnect the battery and remove the engine compartment undershield.

47    Drain the transmission fluid and remove the sump pan and strainer as previously described.

48    Unscrew and remove the socket-headed valve block screws. The screws are of differing lengths so note their locations. A simple way to do this is to sketch their positions on a sheet of paper, and then push the

screws into the points marked on the paper.

49    Release the kickdown inner cable from the throttle lever, and the outer cable from the abutment bracket.

50    Working at the transmission, disconnect the inner cable from the kickdown cam and the outer cable from the transmission casing.

51    Withdraw the cable.

52    Fit the new cable to the transmission. Connect the inner cable to the cam after having turned the cam back through three quarters of a turn against its spring tension.

53    Connect the inner and outer cables to the throttle lever and abutment bracket respectively.

54    Adjust the cable as described earlier.

55    Refit the valve block, the fluid filter and the sump pan, tightening all screws to specified torque.

56    Refill the transmission with fresh fluid, reconnect the battery and refit the undershield.

*Selector cable – adjustment*

57    Move the selector control lever to 'P'.

58    Slacken the selector cable clamp bolt at the trunnion.

59    Working at the transmission, swivel the selector lever fully anti-clockwise and then tighten the clamp bolt to the specified torque.

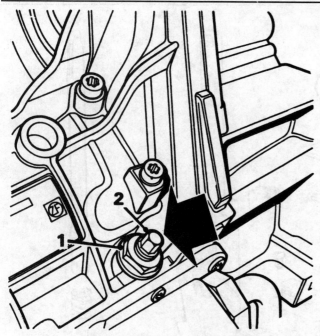

**Fig 12.57 Brake band adjusting locknut (1) and adjuster screw (2) (Sec 10)**

60    Check that the starter motor only operates with the control lever in positions 'N' and 'P'.

### Selector cable – renewal

61    Record the routing of the cable run so that the new cable can be fitted in the same way.
62    Refer to Chapter 11 and remove the centre console.
63    Move the selector lever to 'P'.
64    Extract the spring clip and disconnect the selector cable from the selector lever.
65    Unscrew the locknut and free the outer cable from its support bracket.
66    Attach a drawstring to the cable eye.
67    Slacken the cable mounting nut on the transmission bracket; also slacken the clamp bolt at the trunnion.
68    Release the cable mounting from the transmission bracket and withdraw the cable from the trunnion.
69    Prise out the grommet from the engine compartment rear bulkhead and withdraw the cable into the engine compartment.
70    To fit the new cable, attach the drawstring to the cable eye of the new cable, and use it to pull the cable into the car interior. Peel back the carpet from the left-hand side of the heater ducts and then draw the cable below the ducts to the bracket. Remove the drawstring.
71    Apply grease to both ends of the inner cable and connect it to the control lever and bracket.
72    Refit the centre console and select 'P'.
73    Turn the selector lever on the transmission fully anti-clockwise then pass the cable through the engine compartment and trunnion. Engage it in the bracket, locate the mounting washers and tighten the nut to the specified torque.
74    Tighten the clamp bolt and check that the starter will only operate in 'N' and 'P'.

### Selector lever – removal and refitting

75    Refer to Chapter 11 and remove the centre console. Select 'P'.
76    Extract the spring clip and disconnect the selector cable from the selector lever.
77    Unscrew the locknut and release the outer cable from its bracket. Disconnect the switch leads.
78    Unbolt and remove the selector lever assembly.
79    The assembly may be dismantled for repair after extracting the small Allen screw and removing the selector handle.
80    Remove the selector index panel and light diffuser; withdraw the bulbholders and guide plate.

81    Remove the pivot bolt and the screw which hold the selector lever and quadrant.
82    Remove the locating pin, quadrant and spacer and withdraw the operating rod and spring.
83    Reassembly and refitting are reversals of removal and dismantling, but observe the following points.

*Apply grease to all pivot points*
*Apply thread locking fluid to the locating pin*

### Starter inhibitor/reversing lamp switch – removal and refitting

84    Disconnect the battery and the leads from the switch.
85    Unscrew the switch and remove it together with its sealing washer.
86    Use a new washer when refitting and tighten the switch to specified torque.

### Brake band – adjustment

87    This is not a routine operation, and should only be required as the result of symptoms evident after reference to the fault diagnosis Section.
88    With the vehicle over an inspection pit or raised on ramps, remove the engine compartment undershield.
89    Slacken the brake band adjuster screw locknut and tighten the screw to 7.0 lbf ft (10.0 Nm).
90    Unscrew the screw exactly two turns and tighten the locknut without disturbing the position of the screw.
91    Refit the undershield.

### Automatic transmission – removal and refitting

92    Disconnect the battery negative terminal.
93    Remove the air cleaner assembly, as described in Chapter 3.
94    Disconnect the speedometer cable from its transmission attachment.
95    Drain the cooling system.
96    Unscrew the union nut at the base of the dipstick/fluid filler tube, unbolt the tube support bracket and remove the tube assembly.
97    Refer to Chapter 9 and remove the starter motor.
98    Prise off the left-hand front wheel trim and slacken the wheel nuts. Jack up the front of the car, support it on axle stands and remove the roadwheel.
99    Undo and remove the retaining screws and lift off the access panel from under the wheel arch.
100   From underneath the front of the car, mark the drive flange to inner constant velocity joint flange relationship using paint or a file.
101   Lift off the protective covers and then undo and remove the bolts securing the constant velocity joints to the drive flanges, using an Allen key. Tie the driveshafts out of the way using string or wire.
102   Undo the nut securing the selector cable trunnion to the selector lever. Slip the trunnion out of the lever.
103   Disconnect the kickdown cable from the throttle lever.
104   Disconnect the coolant hoses from the fluid cooler. Cap the open ends of the cooler pipe studs.
105   Turn the crankshaft as necessary using a socket or spanner on the pulley bolt until one of the torque converter retaining bolts becomes accessible through the starter motor aperture. Undo the bolt then turn the crankshaft and remove the remaining two bolts in the same way.
106   Place a jack beneath the engine sump with a block of wood between the jack head and sump.
107   Slacken, but do not remove, the upper front transmission-to-engine retaining bolt then remove all the other remaining bolts securing the transmission to the engine and adaptor plate.
108   Undo the bolts securing the left-hand mounting bracket to the body. Lower the engine slightly, undo the mounting-to-transmission bolts, remove the earth cable and the mounting assembly.
109   Remove the engine front snubber bracket and the rear mounting-to-crossmember assembly.
110   Place a second jack beneath the transmission with interposed block of wood and just take the weight of the unit.
111   Remove the remaining transmission-to-engine bolt, then lower both jacks until the transmission is clear of the body side-member.
112   Withdraw the transmission and torque converter from the engine and remove the assembly from under the car.

113   Refitting the transmission is the reverse sequence to removal, bearing in mind the following points

(a)   Tighten all retaining and mounting bolts to the specified torque where applicable
(b)   Top up or refill the transmission and final drive with the specified lubricants as described earlier in this Chapter
(c)   Align the marks on the drive flanges and inner constant velocity joints before refitting the retaining bolts
(d)   Adjust the front snubber cup position so that it is central around the snubber rubber
(e)   Check and, if necessary, adjust the kickdown cable and selector cable, as described earlier

## Fault diagnosis – four-speed automatic transmission

Although it is desirable to have any faults diagnosed by your dealer with the vehicle operational, the following symptoms may be used as a guide to preliminary diagnosis of some elementary malfunction

| Symptom | Reason(s) |
| --- | --- |
| No drive in 'R' | Selector cable requires adjustment<br>Blocked fluid strainer |
| Vehicle moves with 'N' selected | Selector cable requires adjustment |
| Vehicle starts off in 2nd with 'D' selected | Brake band requires adjustment |
| Transmission noisy after long journeys with transmission slip | Clogged fluid filter |

**Note:** *The majority of other faults are caused by incorrect fluid level or dirt or wear in the valve block.*

## 11   Driveshafts

### Driveshaft retaining nut – modifications

1   Following the introduction of staked driveshaft retaining nuts, the driveshaft removal and refitting procedure for all models is now as follows.

**Driveshafts retained with castellated nut and split pin**
2   The procedures described in Chapter 7, Section 3 are unchanged up to and including paragraph 15. From there, proceed as follows.
3   With the car standing on its wheels, tighten the wheel nuts to the specified torque, then have an assistant firmly depress the footbrake.
4   Tighten the driveshaft retaining nut to the torque wrench setting given in the Specifications at the beginning of this Supplement.
5   If the split pin hole in the constant velocity joint is aligned with one of the castellations in the nut, fit a new split pin and secure by bending over its ends.
6   If the split pin hole does not align with one of the castellations in the nut, continue tightening until it does, but only up to a maximum torque of 158 lbf ft (215 Nm). If the hole is still not aligned, remove the nut and thrustwasher and fit a new thrustwasher of different thickness from the following range:

| | |
| --- | --- |
| *Silver* | *6.1 mm* |
| *Green* | *6.2 mm* |
| *Dark grey* | *6.3 mm* |

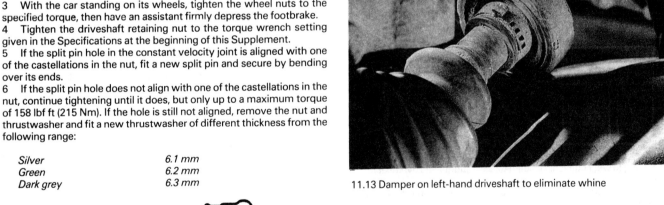

11.13 Damper on left-hand driveshaft to eliminate whine

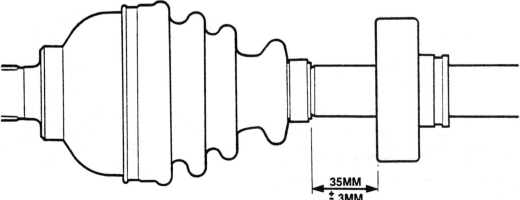

**Fig 12.58 Left-hand driveshaft damper fitting dimension (Sec 11)**

35MM
± 3MM

7    Using the new thrustwasher, repeat the tightening procedure until the split pin can be inserted.

8    Refit the wheel trim, and refill the gearbox with the specified grade of oil with reference to Chapter 6.

**Driveshafts retained by staked retaining nut**

9    The procedure described in Chapter 7, Section 3 is unchanged except for the following.

10    Before undoing the nut, tap up the staking to release it from the driveshaft joint groove. Discard the old nut.

11    When refitting, tighten the new nut to the torque wrench setting given in the Specifications, then tap its flange into the driveshaft joint groove using a punch.

### Transmission whine – rectification

12    Harmonic vibrations in the left-hand driveshaft can cause transmission whine to be transmitted into the vehicle at certain road speeds.

13    On vehicles from VIN 470151 a damper has been fitted to the left-hand driveshaft which eliminates the whine (photo).

14    The damper can be fitted to earlier models using the method described in Chapter 7, Section 5 and set to the dimension given in Fig. 12.58.

---

### 12    Braking system

### Front brake disc checking – general

1    To avoid the possibility of brake judder being caused by a disc which is wearing unevenly, the front brake disc run-out should be checked whenever a disc is removed.

2    Before refitting the disc, check the thickness using a micrometer at four equidistant points around the disc, and at 0.4 in (10.0 mm) in from the outer edge. If the thickness is outside the tolerance given in the Chapter 9 Specifications the disc must be renewed. Under no circumstances may the disc be refaced or machined in any way.

3    With the disc in place, use a dial test indicator to check the run-out with the probe positioned 0.25 in (6.3 mm) in from the outer edge. Slowly rotate the disc and check that the run-out does not exceed the figure given in the Chapter 9 Specifications. If it does, remove the disc, turn it through 180°, refit and check the run-out again. Ensure that complete cleanliness of the disc and drive flange mating faces is maintained, and check that there are no burrs or irregularities on the surface which may affect the readings.

4    If the run-out still exceeds the specified figure, remove the disc and check the drive flange run-out. If this is excessive, renew the drive flange. If the flange run-out is satisfactory, renew the disc.

5    Note also that on all new vehicles, and similarly on new replacement brake discs, a protective coating is applied to combat corrosion during storage.

6    Harsh braking during the running-in period of new discs will cause the coating to wear unevenly leading to noise, judder and overheating of the brake pads.

7    During the running-in period therefore, it is recommended that harsh braking is avoided.

### Rear brake drum – inspection

8    Whenever a brake drum is removed, take the opportunity to inspect it for cracks or grooving caused by the shoes.

9    If these conditions are evident, renew the drum; do not have the drum ground or machined as the original diameter of the drum must not be increased.

### Brake fluid level indicator – malfunction

10    Overfilling of the brake master cylinder reservoir or inversion of the filler cap during servicing can lead to contamination of the low level warning switch contacts, giving rise to illumination of the warning light even when the reservoir is full.

11    If this problem occurs, thoroughly clean the outside of the cap and reservoir using lint-free cloth.

12    Disconnect the switch leads, unscrew the cap (keeping it the right way up) and allowing any excess fluid to drain back into the reservoir.

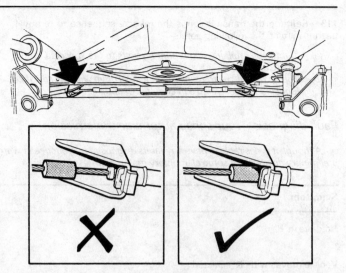

**Fig 12.59 Location and correct fitted position of handbrake cable seal (Sec 12)**

13    Using a piece of cloth to prevent drips, transfer the cap to a clean bench.

14    Prise off the plastic cap and remove the screwed cap.

15    Invert the cap, allowing the float to drop and uncover the switch contacts, which should be cleaned using lint-free cloth.

16    Reassemble the cap and refit to the reservoir in reverse order of removal.

### Brake fluid – approved type

17    Brake hydraulic fluid to DOT 4 specification is now the specified brake system fluid. Austin Rover state that fluid to DOT 3 must no longer be used.

18    DOT 4 fluid has a high boiling point than DOT 3, and is completely compatible with, and can be used to top up systems containing DOT 3 fluid.

19    Note that Duckhams Universal Brake and Clutch fluid meets the specified requirements of DOT 4 fluid.

### Handbrake cable seal – displacement

20    The seal on each handbrake cable where the exposed portion of the inner cable enters the outer cable (see Fig. 12.59) is fitted to prevent dirt and water ingress.

21    During routine servicing operations and when a new cable is fitted, ensure that the seal is correctly fitted. The handbrake must be released (after chocking the wheels) while the check is carried out.

22    It is also possible for the cable to rattle against the sides of the hole in the bracket through which it passes. A rubber sleeve is available for fitting to prevent this problem.

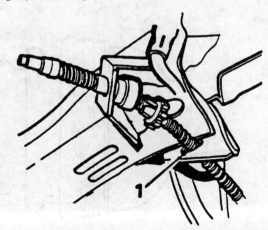

**Fig 12.60 Handbrake cable and body bracket showing the hole (1) in the bracket (Sec 12)**

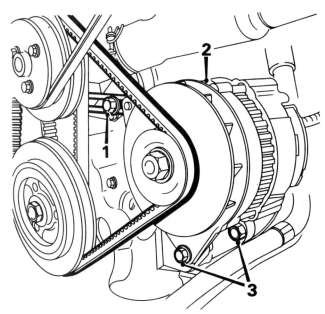

**Fig 12.61 Alternator mountings – later models (Sec 13)**

*1 Adjustment arm bolt*　　*3 Pivot mounting bolts*
*2 Alternator*

## 13 Electrical system

### Alternator – revised positioning

1　On later models the alternator has been repositioned and is located on the left-hand (front facing) side of the engine. Additionally, the heat shield has been deleted. Apart from this, the procedures given in Chapter 9, Section 6 are still applicable. Mounting and adjustment details for Turbo models are shown in Fig. 12.61.
2　Note also that later models have an alternator with a higher output (see Specifications).
3　Servicing and overhaul procedures are unaffected.

### Starter motor (later models)

4　Later models are equipped with different types of starter motors – see Specifications. Removal, overhaul and refitting procedures are given below.

### Starter motor (Lucas type M78R) – removal and refitting

5　Remove the air intake trunking and disconnect the intake air temperature sensor multi-plug.
6　Disconnect the breather hose from the diverter valve (if fitted).
7　Unbolt the air cleaner bracket from the transmission casing and move the air cleaner enough to gain access to the outlet trunking. Release the clip, remove the trunking and the air cleaner.
8　Disconnect the battery and the leads from the starter solenoid.
9　Unscrew the starter motor mounting bolts and withdraw the unit from the vehicle.
10　Refitting is a reversal of removal, tighten the bolts to the specified torque.

### Starter motor (Lucas type M78R) – overhaul

11　With the starter motor removed from the vehicle, clean away external dirt and grease.
12　Disconnect the lead from the solenoid 'STA' terminal, extract the solenoid fixing screws and withdraw the solenoid, at the same time disengaging its plunger from the engaging lever.
13　Extract the screws and remove the nuts which hold the commutator end bracket. Withdraw the bracket from the yoke.
14　Remove the grommet from the yoke and ease the brushbox off the commutator.

15　Remove the brush springs and unclip and remove the earth brushes.
16　Remove the insulation plate and withdraw the brushes complete with bus bar.
17　If the brushes are worn below their specified length, renew them and make sure that they slide freely in their holders.
18　Remove the sun and planet gears and push out the drive and bearing bracket assembly from the drive end bracket.
19　Tap the thrust collar towards the pinion to expose the jump ring, which should be prised from its groove and removed.
20　Remove the collar and drive assembly from the armature shaft.
21　Clean the commutator with a fuel-moistened cloth. If it is badly discoloured polish it with fine glasspaper. Do not undercut the insulators between the segments.
22　This will normally be the limit of economical overhaul. Where the commutator is found to be deeply grooved, or the armature shaft bushes are worn, then it is recommended that a new or factory-reconditioned starter is obtained.
23　Check the drive pinion assembly for wear or damage. Check also that the pinion rotates in one direction only independently of the clutch body.
24　Apply grease to the drive assembly pivot and lever, and fit the drive assembly and the thrust collar to the armature shaft.
25　Fit the jump ring into its groove and tap the thrust collar over the ring.
26　Insert the driveshaft into the drive-end bracket bush.
27　Fit the planet and sun gears.
28　Locate the through-bolts, the intermediate bracket and yoke on the drive-end bracket.
29　Fit the bus bar to the brush box then locate the brushes and the insulation plate. Insert the earth brushes and fit the clips.
30　Offer the assembly over the commutator and fit the brush springs.
31　Slide the armature and brushbox inside the field coils and fit the yoke grommet.
32　Align the commutator end bracket and fit it to the yoke.
33　Engage the solenoid plunger with the engagement lever, fit the solenoid mounting screws then connect the lead to the 'STA' terminal.

### Starter motor (Lucas type M79) – removal and refitting

34　The operations are similar to those described earlier for the type M78R, but refer to Chapter 3 for details of the air cleaner trunking and attachments which must be removed for access.

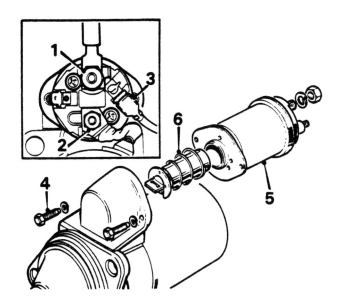

**Fig 12.62 Starter solenoid (Sec 13)**

*1 Battery terminal 'BAT'*　　*4 Retaining screws*
*2 Starter terminal 'STA'*　　*5 Solenoid*
*3 Starter terminal '50'*　　*6 Solenoid plunger*

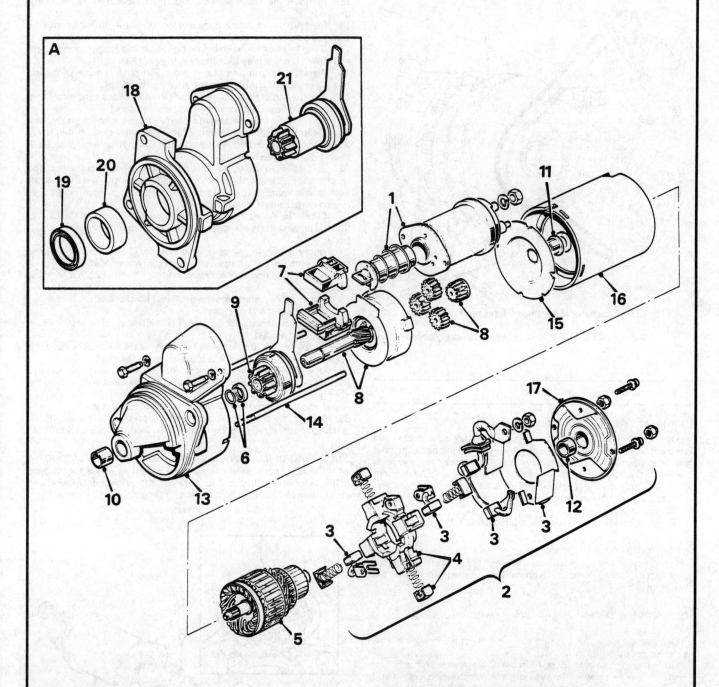

**Fig 12.63 Exploded view of the Lucas type M78R starter motor (Sec 13)**

1 Solenoid and plunger
2 Commutator
3 Brushes
4 Brush springs
5 Armature
6 Jump ring and thrust collar

7 Pivot and grommet
8 Driveshaft, bearing bracket and gears
9 Drive assembly
10 Drive end bracket bush
11 Intermediate bracket bush

12 Commutator end bracket bush
13 Drive end bracket
14 Through-bolts
15 Intermediate bracket
16 Field coil yoke

17 Commutator end bracket
18 Drive end bracket
19 Oil seal
20 Bush
21 Drive assembly

Inset (A) – Drive end bracket (automatic transmission models)

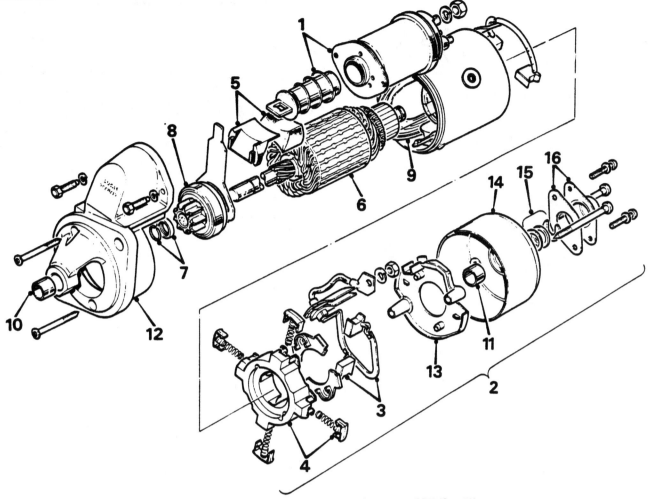

**Fig 12.64 Starter motor components   Lucas type M79 (Sec 13)**

| | | |
|---|---|---|
| 1  Solenoid and plunger | 6  Armature | 11 Commutator end bracket bush | 14 Insulation plate |
| 2  Commutator end bracket | 7  Jump ring and thrust collar | 12 Drive end bracket (manual | 15 Commutator end bracket |
| 3  Brushes | 8  Drive assembly | gearbox) | 16 Circlip and washers |
| 4  Brush springs | 9  Field coils and yoke | 13 Drive end bracket (automatic | 17 Sealing cup and gasket |
| 5  Pivot and packing piece | 10 Drive end bracket bush | transmission) | |

## Starter motor (Lucas type M79) – overhaul

35   Carry out the operations described earlier in paragraphs 11 and 12.
36   Remove the sealing cup and gasket from the commutator end bracket.
37   Extract the circlip and washers from the armature shafts.
38   Note the end bracket to yoke alignment marks, remove the retaining screws and carefully withdraw the end bracket from the yoke.
39   Remove the brush springs and withdraw the earth brushes, then lift the brushbox from the commutator.
40   Remove the insulation plate and withdraw the positive brushes with bus bar.
41   If the brushes are worn below their specified length, renew them and make sure that they slide freely in their holders.
42   Remove the pivot and grommet from the drive-end bracket.
43   Extract the screws and remove the yoke.
44   Withdraw the armature from the drive-end bracket.
45   Tap the thrust collar towards the pinion to expose the jump ring, then prise the ring from its groove and remove it, together with the collar and drive assembly.
46   Carry out the operations described in paragraphs 21 to 26 of this Section.

47   Locate the yoke on the drive-end bracket and fit the retaining screws.
48   Place the bus bar on the brushbox and fit the brushes and the insulation plate. Insert the earth brushes; and fit the clips.
49   Offer the assembly over the commutator and fit the brush springs.
50   Slide the commutator end bracket over the brush springs and engage the grommet. Align the reference marks and fit the screws.
51   Fit the washers and circlip to the armature shaft, then fit the gasket and sealing cup.
52   Engage the solenoid plunger with the engagement lever and secure it with the screws. Make sure that the 'STA' terminal is adjacent to the starter motor. Fit the lead to the 'STP' terminal.

## Relays and control units – location
### 1986 to 1988 models
53   Various relays and control units are housed in the fusebox located in the compartment below the steering column (see Fig. 12.65).
54   The electronic control unit (ECU) for the fuel system, courtesy light delay unit, programmed wash/wipe control unit and oil pressure warning light relay are housed behind a cover in the roof of the glovebox.
55   The electronic control unit (ECU) for the ignition system is located

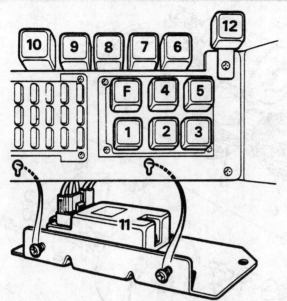

**Fig 12.65 Relays in fusebox – 1986 to 1988 models (Sec 13)**

| | | | |
|---|---|---|---|
| 1 | Ignition | 8 | Interior lamp |
| 2 | Headlamps | 9 | Fuel pump |
| 3 | Auxiliary ignition 2 | 10 | Rear screen wiper |
| 4 | Starter solenoid | 11 | Windscreen wiper delay |
| 5 | Auxiliary ignition 1 | | unit |
| 6 | Heated rear screen | 12 | Dim-dip unit |
| 7 | Manifold heater (carburettor | F | Flasher unit |
| | models)/Cooling fan (Turbo) | | |

on the left-hand valance in the engine compartment.

56    The control unit and relays for the electric windows are located behind the driver's door trim panel.

57    On Turbo models, the radiator cooling fan relay is located on the right-hand suspension tower in the engine compartment.

58    On EFi models, the inertia switch is located behind and to the right of the radio/cassette player aperture, and the fuel system ECU is located beneath the carpet in the front passenger footwell.

59    On carburettor models equipped with an oil pressure switch controlled fuel pump, the fuel pump relay may be located in one of two different places.

*Pre-1986 models: behind panel in roof of glovebox*
*1986-on models: located with other relays adjacent to the fusebox*

60    The central locking switch unit (or relay) is located in the driver's door behind the trim panel.

**1989-on models**
61    As for earlier models except for the following.
62    For fuel and ignition ECUs and relays refer to Section 6.
63    The window lift control unit is located behind the cover in the glovebox.
64    A delay unit mounted on the fusebox in the glovebox allows the electric windows, sunroof and radiator cooling fan to be operated for approximately 45 seconds after the ignition has been switched off.
**Warning:** *In the case of the cooling fan which is switched on by the thermal switch, it means that the fan can cut-in and start running even when the ignition is switched off, so great care must be taken when working on a hot engine which has just been switched off.*
65    A timer unit, also mounted in the fusebox, controls the operation of the heated rear screen, limiting its use to ten minutes.

## Heater fuse (all models, 1987-on) – higher rating

00    Internal modifications to the heater have resulted in the blower motor drawing more current and overloading the existing fuse, which can blow.
67    If this happens repeatedly, with no evidence of malfunction within

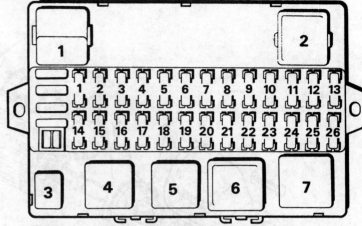

**Fig 12.66 Fuses and relays on fusebox on 1989-on models (Sec 13)**

| Fuse number | Rating (amps) | Function |
|---|---|---|
| 1 | 10 | Engine |
| 2 | 5 | RH side, tail and number plate lights |
| 3 | 5 | Instruments |
| 4 | 10 | Cigar lighter |
| 5 | 10 | RH headlamp main beam |
| 6 | 5 | Electric mirrors |
| 7 | 15 | Sunroof |
| 8 | 10 | Dim-dip system |
| 9 | 15 | Stop lamps, reversing lamps, indicators |
| 10 | 10 | Rear fog lamps |
| 11 | 30 | Heated rear screen |
| 12 | 10 | Interior lamp, luggage compartment lamp, central door locking |
| 13 | 15 | Cooling fan |
| 14 | 15 | Heater blower motor |
| 15 | 15 | Windscreen wiper/wash |
| 16 | 15 | Tailgate wiper |
| 17 | 15 | Electric fuel pump |
| 18 | 10 | LH headlamp main beam |
| 19 | 30 | Horn, hazard warning lamps |
| 20 | 30 | Electric front windows |
| 21 | 15 | Turbo carburettor cooling fan |
| 22 | 10 | Radio/cassette player |
| 23 | | Spare |
| 24 | 10 | RH headlamp dip beam |
| 25 | 10 | LH headlamp dip beam |
| 26 | 5 | LH side, tail and number plate lamps |

| Relays Number | Function |
|---|---|
| 1 | Courtesy lamp delay/lights-on alarm |
| 2 | Heated rear window timer |
| 3 | Headlamp |
| 4 | Ignition |
| 5 | Auxiliary – cooling fan motor |
| 6 | Flasher relay |
| 7 | Dim-dip control |
| 8 | Front wash/wipe (non-programmed) |
| 9 | Tailgate wash/wipe (Estate) |

the heater circuits, renew the existing 15 amp fuse with one of 20 amp rating. The fuse label should be amended accordingly for future reference.

## Instrument panel switches (1989-on) – removal and refitting

68    Remove the screws from and tilt forward the instrument panel as described in Chapter 9, Section 19.
69    Pull off the multi-plug from the switch (photo).
70    Depress the clips on the side of the switch and withdraw the switch from the instrument panel (photo).

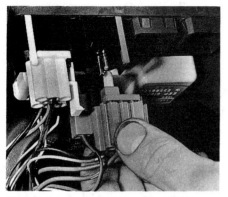

13.69 Pull the multi-plug from the switch

13.70 Depress the clips on the side of the switch and withdraw the switch

13.80 View of the rear of the digital clock with bezel withdrawn

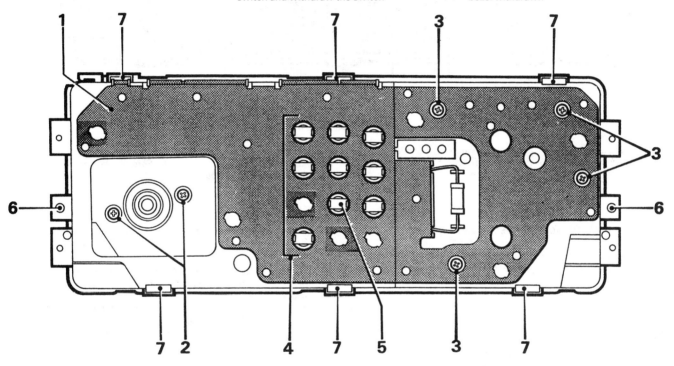

Fig 12.67 Rear view of the three-gauge instrument panel (Sec 13)

| | | |
|---|---|---|
| 1  Printed circuit | 3  Fuel and temperature gauge | 5  No-charge warning light | 7  Shroud retaining clips |
| 2  Speedometer securing screws | securing screws | 6  Shroud retaining screws | |
| | 4  Warning light cluster | | |

71  The switch illumination bulb is housed in the multi-plug connector and is a push-fit in the holder.

72  Refit in reverse order.

## Instrument panel (1989-on) – removal and refitting

73  Although of different design to earlier models, the procedure for removing the instrument panel is as described in Chapter 9, Section 19.

74  There are two types of panel in use, one with three gauges and one with four. A view of the rear of each panel appears in Figs. 12.67 and 12.68.

75  Note also that progressively from 1988, instrument panels manufactured by Nippon Seiki are being used in place of those produced by Lucas.

76  The complete instrument panels are interchangeable, but individual instruments are not.

77  The manufacturer's name is clearly marked on the rear of the panels, and to identify them once they are fitted, Nippon Seiki types have seven-digit odometers and round trip reset buttons, while Lucas have six-digit odometers and rectangular reset buttons.

## Digital clock (1989-on) – removal and refitting

78  Remove the instrument panel bezel as described earlier in this Section.

79  Disconnect the clock multi-plug.

80  Remove the retaining screws and withdraw the clock (photo).

81  Refit in reverse order.

## Dim-dip headlamp system

82  In order to comply with current regulations, all post-October 1986 models are fitted with a headlamp dim-dip system.

83  The system uses a relay-controlled resistor circuit, which is incorporated in a control unit mounted at the front of the vehicle.

84  When the parking lamps are illuminated and the ignition is switched on, the headlamps come on automatically at one-sixth of their normal power (dipped beam). This system prevents the vehicle being driven with only the parking lamps illuminated.

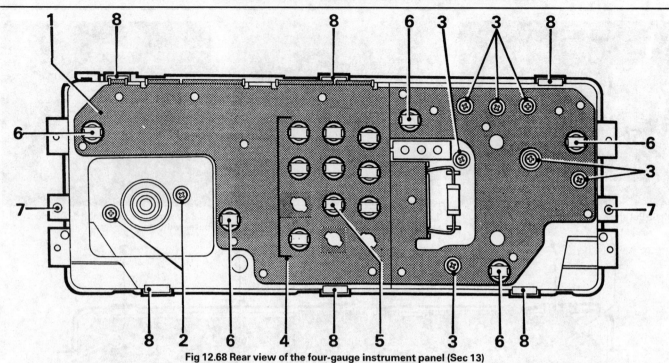

**Fig 12.68 Rear view of the four-gauge instrument panel (Sec 13)**

| | |
|---|---|
| 1 *Printed circuit* | 3 *Tachometer, fuel and* |
| 2 *Speedometer securing* |    *temperature gauge securing* |
|    *screws* |    *screws* |

| | |
|---|---|
| 4 *Warning light cluster* | 7 *Shroud retaining screws* |
| 5 *No-charge warning light* | 8 *Shroud retaining clips* |
| 6 *Panel illumination light bulb* | |

## Rear lamp cluster (Saloon models, 1989-on) – bulb renewal

85   The procedure is as described in Chapter 9, Section 9. The bulb positions are detailed in the accompanying photograph.
**Lens renewal**
86   The lens is secured to the rear panel by four nuts, accessible after

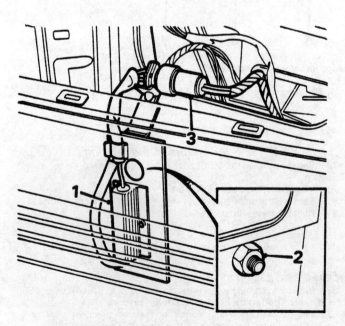

**Fig 12.69 Dim-dip headlamp system resistor (Sec 13)**

| | |
|---|---|
| 1   *Resistor* | 3   *Multi-plug* |
| 2   *Mounting nut* | |

taking down the carpeting in the luggage compartment (photo).
87   Refit in reverse order, ensuring the foam seal is in good condition.

## Rear lamp cluster (Estate models) – bulb renewal

88   Remove the rear stowage box cover, then undo the retaining screws and lift out the stowage box (photo).
89   Reach into the aperture and withdraw the relevant bulb holder by turning it anti-clockwise (photo).
90   Press and turn the bulb to remove it from the holder (photo).
91   Refitting is the reverse sequence to removal.

## Number plate lamp (Estate models) – bulb renewal

92   Undo the two lens retaining screws and withdraw the lens and bulb (photos).
93   Withdraw the bulb from the contacts.
94   Refitting is the reverse sequence to removal.

## Side repeater lamp – bulb renewal

95   Later models have side repeater lamps fitted to the front wings.
96   To renew a bulb, push the unit rearward then pull it from the wing panel (photo).
97   Pull the lens unit from the bulb-holder to gain access to the bulb, which is a push-fit in the holder (photo).
98   Refit in reverse order.

## Glovebox lamp – bulb renewal

99   Open the lid of the glovebox and gently prise the lamp from the location. Use a small screwdriver to prise the bulb end cap out of its retaining clips.

## Heater control panel illumination (1989-on) – bulb renewal

100   Remove the heater control panel as described in Section 15.
101   The bulbs are a bayonet fit in the holder (photo).
102   Refit the panel in reverse order of removal.

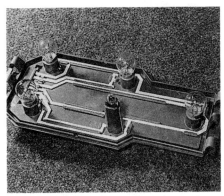

13.85 Bulb positions in rear lamp cluster

13.86 Rear lens unit securing nuts (arrowed)

13.88 Removing the rear storage box

13.89 Tailgate rear lamp cluster bulbs accessible through the stowage box aperture

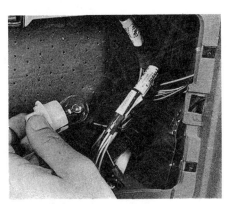

13.90 Renewing a rear lamp cluster bulb

13.92A Rear number plate lamp lens screws

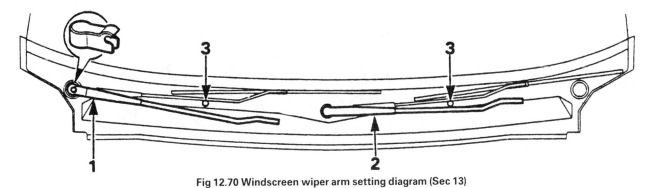

Fig 12.70 Windscreen wiper arm setting diagram (Sec 13)

1  Arm setting point – driver's side    2  Arm setting point – passenger's side    3  Parked position against stop pegs

### Windscreen wiper arms – setting position

103  Operate the wipers and switch them off by means of the wiper switch, not the ignition key. Remove the wiper arms if they require adjustment.
104  When refitting the wiper arms to their spindles, position the arms so that they contact the intake moulding on the finisher below the windscreen. After tightening the retaining nut, lift the arm and locate it against the upper face of the stop peg.
105  Operate the wipers with a wet windscreen and check that they park with the arms contacting the stop pegs.

### Tailgate wiper arm and blade (Estate models) – removal and refitting

106  To remove the wiper blade, lift the arm away from the window, release the spring retaining catch and separate the blade from the wiper arm (photo).
107  Insert the new blade into the arm, making sure that the spring retainer catch is correctly engaged.

108  To remove the wiper arm, spring back the hinged cover and withdraw the washer jet from the spindle (photo).
109  Unscrew the retaining nut, then carefully prise the arm off the spindle.
110  Before refitting the arm, switch on the ignition, then turn the wiper on and off, allowing the motor to return to the 'Park' position. Switch off the ignition.
111  Refit the arm to the spindle with the blade positioned just below the lower heater element in the window.
112  Refit and tighten the retaining nut, then refit the washer jet so that the jets are positioned on either side of the arm.
113  Close the hinged cover.

### Tailgate wiper motor (Estate models) – removal and refitting

114  Remove the tailgate wiper arm and blade as described previously.

13.92B Rear number plate lamp withdrawn

13.96 Removing the side repeater lamp from the wing

13.97 Pulling the lens unit from the bulbholder

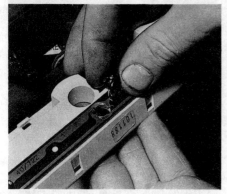

13.101 Removing a bulb from the heater control panel

13.106 Removing the wiper blade from arm

13.108 Removing tailgate washer jet

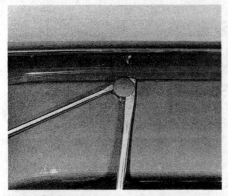

13.115 Prising out a trim moulding retaining button

13.116 Tailgate trim moulding upper retaining tag

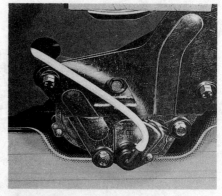

13.118 Tailgate wiper motor attachments and washer tube

115    Open the tailgate and release the interior trim moulding by prising out the retaining buttons (photo). Some of these will probably break during removal, so new buttons should be obtained before refitting.

116    Release the trim from the upper retaining tags (photo), disconnect the speaker leads (where fitted) and remove the trim.

117    Disconnect the wiring multi-plug and tailgate washer hose at the wiper motor.

118    Undo the wiper mounting bracket retaining bolts and the self-tapping screw from the support plate (photo). Withdraw the motor and mounting bracket from the tailgate.

119    With the assembly on the bench, undo the bolts and separate the wiper motor from the mounting bracket.

120    Refitting is the reverse sequence to removal.

## Continuous rear wipe (Estate models) – modification

**Note:** *The following modification should only be carried out by the home mechanic with a reasonable understanding of automobile electrics, or*

*under the guidance of, or by, an approved electrician*

121    The provision of continuous rear wipe can be obtained by fitting an on/off switch to connect together the two red/light green wires (pin numbers 1 and 8) on the rear wiper programme unit.

122    The connections are shown by the dotted line in Fig. 12.71 the switch being marked A.

123    With the switch in the open position, normal programmed wipe with delay is available.

124    With the switch closed, continuous wipe is available whenever the facia panel rear wipe switch is operated.

## Automatic rear wipe in reverse (Estate models) – modification

**Note:** *Before starting work, see note at beginning of previous paragraphs concerning continuous rear wipe*

125    Automatic rear wipe when reverse is selected can be obtained by

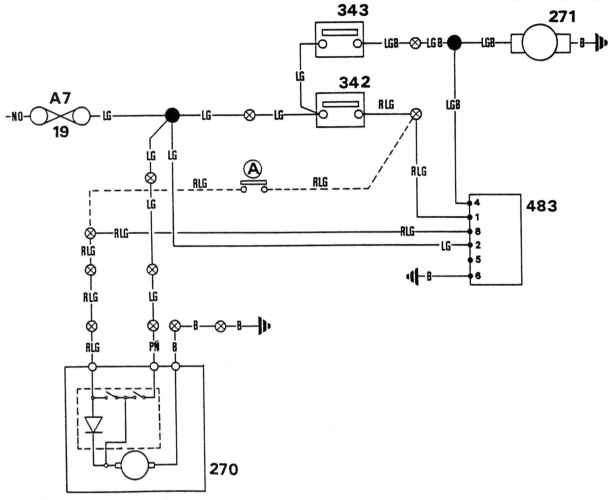

**Fig 12.71 Wiring diagram – continuous rear wipe (Sec 13)**

| | | | |
|---|---|---|---|
| 19 Fusebox | 271 Washer pump | 343 Washer switch | A New switch and wiring |
| 270 Wiper motor | 342 Wiper switch | 483 Programme unit | (dotted line) |

*For colour code see wiring diagrams at end of book*

fitting a standard 4-pin relay into the reversing light circuit, joining together the red/light green wires (see Fig. 12.72).

126   This will provide intermittent wipe when reverse is selected, irrespective of the position of the facia panel rear wipe switch.

127   By connecting the common terminal of the relay to the light green wire of the facia rear wipe switch (dotted line in Fig. 12.72), continuous rear wipe can be obtained whenever reverse is selected.

128   In both cases, the relay feed should be connected to the reverse lamp circuit.

## Horns – general

129   As from February 1986, twin horns are fitted as standard to all models.

## Electrically operated windows – general

130   Certain models are now available with electrically operated windows on both front and rear doors.

131   The procedures contained in Chapter 9 are still applicable, except that a window lift motor is now fitted to each rear door, together with a switch on the trim panel. An extra relay is located in the driver's door to control the operation of the rear windows.

132   On 1986 to 1988 models, the window lift control unit circuitry and switches are relocated in a housing on the facia. Removal and refitting of the control unit follows the same procedure as for the electronic radio/cassette player described in Chapter 9, Section 48.

133   To remove the switch from the control unit, refer to Fig. 12.73 and turn the four printed circuit retainers a quarter of a turn in the direction of least resistance.

134   Release the pair of face plate retainers by carefully depressing them with a small screwdriver. Push the harness connector pins into the housing and withdraw the face plate.

135   Disconnect the switch multi-plug, depress the switch retainers and remove the switch.

136   Refitting is the reverse sequence to removal.

137   Removal and refitting of the rear door window lift motor is essentially the same as the procedure for the front door motor except that there is only one lifting arm. The procedures for the front door are covered in Chapter 9, Section 50, and an illustration of the rear door components is shown in Fig. 12.75.

**1989-on models**

138   The control unit is located behind a panel in the roof of the glovebox.

139   The switches are located in the centre console.

## Tailgate lock solenoid (Estate models) – removal and refitting

140   To remove the tailgate lock solenoid on models equipped with

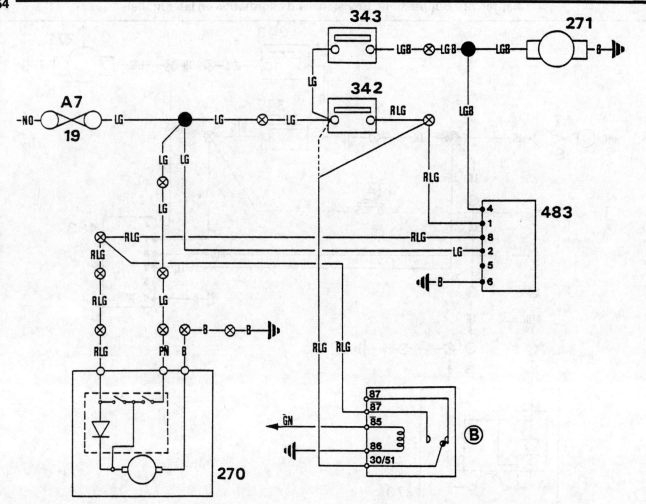

**Fig 12.72 Wiring diagram – automatic rear wipe (Sec 13) For code see Fig. 12.71 and wiring diagrams at end of book**

*B   Relay          30/51 Common terminal          85 Relay feed*

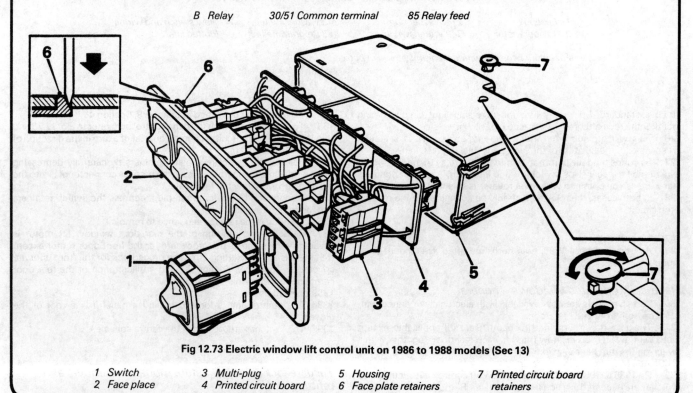

**Fig 12.73 Electric window lift control unit on 1986 to 1988 models (Sec 13)**

| | | | |
|---|---|---|---|
| *1   Switch* | *3   Multi-plug* | *5   Housing* | *7   Printed circuit board* |
| *2   Face place* | *4   Printed circuit board* | *6   Face plate retainers* | *retainers* |

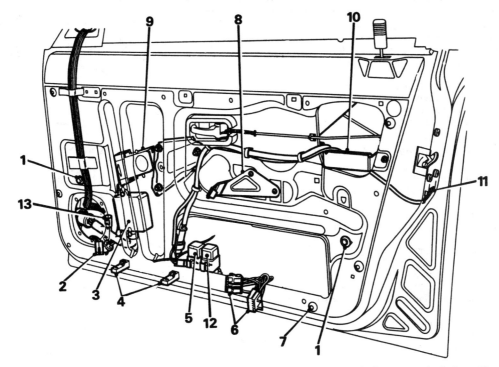

**Fig 12.74 Driver's door internal components for front and rear electrically operated windows (Sec 13)**

1 Window channel screws
2 Speaker connector
3 Window lift one-touch unit
4 Trim pad retainers
5 Window lift relay (front doors)
6 Window lift multi-plugs
7 Trim pad fixing studs
8 Trim pad bracket
9 Window lift motor
10 Central locking switch
11 Courtesy lamp door switch
12 Window lift relay (rear doors)
13 Electric door mirror connector

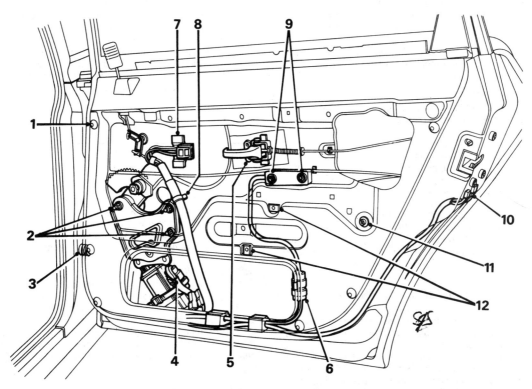

**Fig 12.75 Rear door electric window and central locking components (Sec 13)**

1 Trim panel retainer studs
2 Window lift motor retaining nuts
3 Harness sleeve
4 Harness connector
5 Release handle
6 Central locking harness connector
7 Trim retaining bracket
8 Clip
9 Central locking solenoid retaining screws
10 Door courtesy light switch
11 Window channel retaining screw
12 Trim bracket plastic nuts

central locking, open the tailgate and remove the interior trim moulding with reference to the previous sub-section.

141    Disconnect the operating rod from the private lock lever.

142    Undo the nut securing the solenoid mounting plate to the lock barrel.

143    Disconnect the wiring multi-plug and position the mounting plate in the left-hand aperture. Undo the retaining screws and withdraw the solenoid from behind the mounting plate.

144    Refitting is the reverse sequence to removal.

### Boot/tailgate central door locking solenoid (all models) – failure to operate

145    Should the boot/tailgate central door locking solenoid fail to operate, this may be due to 'bounce-back' of the mechanism and can be corrected as follows.

146    Remove the boot/tailgate central locking solenoid and motor assembly as described in Chapter 9, Section 54 (boot) or in this Section for the tailgate.

147    Ensure the slots through which the nylon plunger (which carries the lock bolt) slides are free from manufacturing burrs.

148    Using soft iron wire, wrap one turn around the groove of the motor casing as shown in Fig. 12.76, twisting the ends of the wire together using flat nosed pliers, until **slight** resistance of the plunger is felt.

149    Cut off surplus wire, and ensure that the remaining wire end cannot foul any part of the lock mechanism.

150    Refit the lock solenoid and motor assembly.

### Radio aerial – removal and refitting

151    Disconnect the battery negative terminal.

152    Carefully prise up the cover trim at the base of the aerial and undo the baseplate retaining screws. Note the screening braid location on one of the screws.

153    From inside the car, disconnect the aerial lead at the rear of the radio and withdraw the lead from the console and facia.

154    Tie a length of string to the end of the aerial lead as an aid to refitting, and withdraw the aerial and lead upwards and out of the windscreen pillar.

155    Refitting is the reverse sequence to removal. Use the string to pull the aerial lead down the pillar and through behind the facia.

### Cassette holder – general

156    The cassette holder which is located below the radio/cassette player can be removed after the facia panel has first been withdrawn as described in Chapter 11, Section 35.

157    Should a cassette tray fail to open when pushed in, it may be possible to release it using one of the following methods.

158    Insert a thin blade along the upper edge of the tray and attempt to ease the tray downwards.

159    If the tray still fails to release, eject the tray which is operating correctly, pinch its sides and withdraw it.

160    Locate the holes at the back of the cassette holder which service the jammed tray and press upwards. At the same time depress the release pad to free the catch. Withdraw the tray.

161    Failure to engage the tray guides on their rails will cause jamming of a tray.

### Service interval counter (solid-state instruments) – resetting

162    On models equipped with solid-state instruments, the service interval counter can be reset, after carrying out the specified service, using the following procedure.

163    Switch on the ignition and check that the correct date is shown on the calendar. Reset the date if necessary, then switch off the ignition.

164    Press and hold 'Select' on the message centre panel. Switch on the ignition, and the word 'English' should appear on the screen. Release 'Select', then briefly press again to clear the screen. If a visual message appears, release and press 'Select' again to clear.

165    Press and hold 'Reset' until the words 'Oil change' are shown on the display, then release 'Reset'. This cycle will take approximately six seconds.

166    Press 'Reset' again until the words 'Service Reset' appear on the display. This will also take approximately six seconds. Release 'Reset'

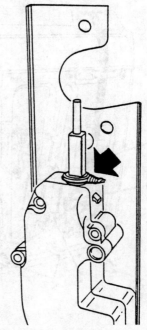

**Fig 12.76 Soft iron wire wrapped around groove of motor casing (Sec 13)**

and press 'Select' to revert to time and odometer. Switch off the ignition.

### Road speed transducer (solid-state instruments) – modification

167    EFi models with solid-state instruments are fitted with a speedometer transducer, which transmits roadspeed information to the speedometer in the form of electronic impulses.

168    Removal and refitting of the transducer is as described in Section 28 of Chapter 9, but note that two transducer types have been fitted. The earlier type has an external (female) spigot coupling, whereas the later type has an internal (male) spigot coupling. When renewing the transducer, it may be necessary to add an earth lead to the spare connector of the transducer multi-plug – refer to a Rover dealer for details.

---

### 14    Suspension and steering

---

### Front suspension – knocking

1    If a loud knock from the front suspension results when a front wheel goes sharply to the full rebound position, the noise can be minimised by fitting modified replacement cup and washer assemblies to the upper end of the suspension strut (see Fig. 10.5, Chapter 10).

2    These components are available from Austin Rover parts dealers.

### Camber angle (later models)

3    Later models have had the front suspension camber angle changed from negative to positive by moving the suspension strut lower fixing bolt hole inward by 0.04 in (1.0 mm). On earlier models with negative camber, the bolt holes were in line.

4    This modification does not alter the servicing procedures given in Chapter 10, but for the purposes of the Specifications in this Supplement, the camber able is grouped under the heading 'strut, type B', to avoid confusion.

### Front hub bearings (one-piece type) – renewal

5    One piece front hub bearings were introduced on later models. Early versions of the one-piece bearings used ball-bearings while later versions use taper-roller bearings, and are the preferred type, although both types are fully inter-changeable.

6    Note that neither type of one-piece bearing is interchangeable with the early two-piece bearings.

7    Apply the handbrake and remove the front roadwheel.

8    Have an assistant apply the footbrake hard. Unscrew the driveshaft-to-hub nut.

9    Unbolt the brake caliper and tie it up out of the way. There is no need to disconnect the hydraulic hose.

10    Remove the two screws which secure the brake disc to the hub flange. Remove the disc.

11    Extract the screws and remove both halves of the disc shield.

12    Unscrew the nut from the tie-rod balljoint taper pin, and using a suitable splitter tool disconnect the balljoint from the steering arm.

13    Unscrew and remove the two bolts which hold the base of the suspension strut to the hub carrier.

14    Unscrew the nut from the suspension track control arm balljoint, and using a lever prise the arm downwards to separate the arm from the hub carrier.

15    Remove the hub carrier assembly.

16    A press will now be required to press the drive flange from the hub. Alternatively a long bolt and nut with distance pieces can be used to draw it out, but on no account attempt to hammer it out. The bearing inner track and ballrace (outer section) will come away with the drive flange and can be removed with a puller.

17    Extract the two circlips and press the bearing from the hub. Never attempt to re-use the old bearing.

18    Press the new bearing into the hub until it contacts the circlip, then fit the second circlip.

19    Support the bearing inner track and press the drive flange into the hub.

20    Refitting the hub carrier is a reversal of removal, but observe the following points.

21    Tighten all nuts and bolts to the specified torque.

22    Fit a new driveshaft nut, tighten to the specified torque and stake the nut into the driveshaft groove (see also paragraph 24).

23    Once the brake caliper has been refitted, apply the footbrake hard two or three times.

### Front swivel hub assembly – modifications

24    On later models, the driveshaft is secured to the swivel hub using a staked retaining nut in place of the castellated nut and split pin used previously. Also, the method of tightening the castellated nut on earlier models has been changed. When removing and refitting the swivel hub refer to Section 11 of this Supplement for the revised procedures.

### Rear axle – modifications

25    From VIN 168414, the pivot bushes in the rear axle trailing arms are fitted the other way round, with the flanged shoulder towards the outside. When renewing pivot bushes, ensure that they are fitted accordingly.

### Rear anti-roll bar (MG Turbo and MG EFi models) – removal and refitting

26    Jack up the rear of the car and support it on stands.

27    Undo the nuts and washers securing each end of the anti-roll bar to the rear axle trailing arms.

28    Pull the bar forward to disengage it from the trailing arm brackets, recover the spacers and remove the bar from under the car.

29    Refitting is the reverse sequence to removal, but ensure that a spacer is fitted to each trailing arm bracket and tighten the nuts.

### Self-levelling suspension (Estate models) – description

30    Certain Estate models are fitted with self-levelling suspension which reacts to vehicle loading and automatically maintains the normal trim heights.

31    The self-levelling units are sealed dampers fitted in place on the normal rear shock absorbers. A pump in the damper operates under the action of the suspension to raise the rear of the car until normal trim height is regained. On an undulating road, this process will be carried out within one mile. When the additional load is removed, the suspension remains at the correct level.

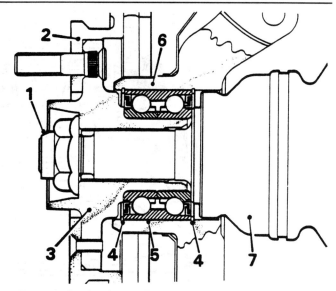

**Fig. 12.77 Sectional view of one-piece front hub bearing (Sec 14)**

| | |
|---|---|
| 1   Driveshaft retaining nut | 5   Bearing |
| 2   Brake disc | 6   Hub |
| 3   Drive flange | 7   Driveshaft |
| 4   Circlip | |

32    To test the operation of the self-levelling dampers, measure and record the rear suspension trim height on each side with the vehicle at kerbside weight. The trim height is the distance from the centre of the hub to the lower edge of the wheel arch. Compare the figures obtained with those given in the Specifications of this Supplement.

33    Evenly load the rear of the car with 440 to 550 lbs (200 to 250 kg).

34    Drive the car over undulating roads for at least 3 miles (5 km), then stop the car and measure the rear trim heights again with the driver remaining in his seat. The readings obtained should be within 0.4 in (10.0 mm) of the readings obtained previously. Both sides of the car should have risen by equal amounts and should retain their position after the car has stopped. A unit that is low, does not hold its position, or shows signs of excessive oil leakage must be renewed.

35    Removal and refitting of the self-levelling dampers is the same as for conventional rear shock absorbers, and reference should be made to Chapter 10, Sections 13 and 14. Note that on Estate models, access to the upper mounting is gained after removing the panel from the middle of the luggage compartment side trim.

### Steering wheel (all models, 1989-on) – removal and refitting

36    The procedure is as described in Chapter 10, Section 16, but the steering wheel is secured by a bolt and not a nut. Note the different torque setting figure in the Specifications.

### Steering column (all models, 1989-on) – overhaul

37    The procedure is as described in Chapter 10, Section 17 with the following differences.

38    After lifting out the inner column, the top and bottom bushes must be tapped out from the outer column using a suitable long drift.

39    Note also that there is a spacer under the top bush.

40    When fitting new bushes, apply Loctite 424 to their outer surfaces and fit them with the chamfered edge facing in to the centre of the outer column.

### Rack and pinion steering gear – modifications

41    On later versions of both the manual steering gear and the power-assisted steering gear, the centralizing hole in the rack housing has been deleted.

42    When refitting the steering gear, set the roadwheels to the straight-ahead position and centralize the steering wheel before fitting the intermediate shaft universal joint to the pinion shaft.

43    On power-assisted steering gear, lockwashers are used to retain the feed and return hose unions at the rack.

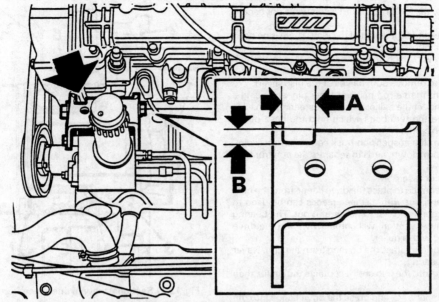

Fig 12.78 Power steering pump recess dimensions and location (Sec 14)

A = 1.00 in (25.0 mm)        B = 0.030 in (8.0 mm)

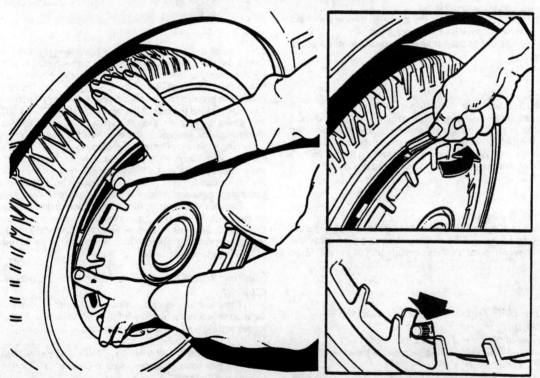

Fig 12.79 Depicting removal and refitting of later type wheel trims (Sec 14)

44   Apart from these changes and the following paragraphs, the applicable Sections of Chapter 10 are not affected.

### Power-assisted steering pump bracket – fouling cam belt

45   On some models, notably the MGi, the power steering pump bracket can come into contact with the cam belt cover when the pump is at the limit of its adjustment.

46   Modified brackets are fitted to vehicles from VIN 379553.

47   On earlier vehicles, the bracket can be modified by cutting a recess in the bracket to the dimensions shown in Fig. 12.78.

### Rack-and-pinion steering gear (power-assisted steering, 1989-on) – dismantling and reassembly

48   The procedure is as described in Chapter 10, Section 26 but note that the rack fluid seal and support bush, and the travel stop (items 19 and 20 in Fig. 10.10) are of different design.

### Power-assisted steering pump drivebelt – adjustment

49   When adjusting the tension of the belt as described in Chapter 10, Section 27, it may be found advantageous to slacken the high pressure hydraulic pipeline nut before attempting to move the pump.

50   On completion of adjustment, tighten the pipeline nut and bleed the system as described in Chapter 10, Section 30.

15.2 Remove the screws securing the upper edge of the grille to the cross-rail

15.3 Lift the grille upward to disengage the clips

15.5A Prise out the speaker grille ...

15.5B ... and undo the four screws under it

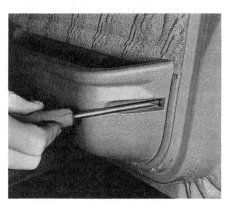

15.6A Remove the screws from the door bin ...

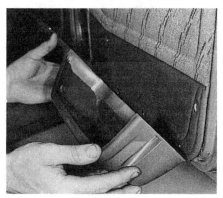

15.6B ... then lift off the door bin and panel

### Rack-and-pinion steering gear (power-assisted steering, Turbo models) – removal and refitting

51    The procedure is as described in Chapter 10, Section 24, but additionally the exhaust downpipe must be released from the manifold and the lower heat shield removed before removing the rack.

52    When undoing the rack mounting bolts, the heat shield bracket must also be removed.

### Front wheel alignment and steering angles – checking

53    When carrying out the checks and adjustments described in Chapter 10, Section 32, it is important that the suspension is 'settled' and the car is in its normal driving attitude.

54    Before carrying out any checks, ensure that the car is positioned on a level surface with the tyres correctly inflated. Set the front wheels in the straight-ahead position, rock the vehicle from side to side, then roll it backwards, then forwards at least one vehicle length. Checking can now commence. Do not move the steering wheel once the roadwheels have been set in the straight-ahead position. If the steering wheel is moved, if the car is jacked up, or if any adjustments are made, repeat this 'settling' procedure before carrying out any further checking. With whatever equipment is being used it is always advisable to take two readings, carrying out the settling procedure each time, then take an average of the two readings to determine if adjustment is required.

### Wheel trims (1989-on) – removal and refitting

55    The wheel trims fitted to 1989-on models can be difficult to remove and refit due to their more positive fit. The following procedure should be used.

56    To remove a wheel trim, insert the special tool supplied with the vehicle between the trim and the wheel rim, and using a twisting action, lever the trim outward.

57    Repeat the operation in several positions around the trim until it is released.

58    To refit the trim, position the wheel with the inflation valve at its lowest point.

59    Offer the trim to the wheel with the large cut-out over the valve stem and apply equal pressure to both sides of the wheel trim.

60    Apply pressure to the top of the trim until it is fully engaged with the wheel.

### 15  Bodywork

### Radiator grille (1989-on) – removal and refitting

1    Open the bonnet.

2    Remove the screws securing the upper edge of the grille to the cross panel (photo).

3    Lift the grille upward to disengage the clips at each lower, outer end and withdraw the grille panel (photo).

4    Refit in reverse order, ensuring that the lower clips are in full engagement.

### Front door trim panel (1989-on) – removal and refitting

5    Prise out the door speaker grille and undo the four screws underneath it (photos).

6    Remove the screws from the door bin then lift off the combined door bin and speaker trim panel (photos).

7    Prise out the plastic cover from the interior release handle escutcheon plate screw, remove the screw and lift off the escutcheon plate (photos).

8    Prise out the plastic cover from the window winder handle, remove the nut and lift off the handle and backing plate (photos). This operation is not applicable where electric windows are fitted.

9    Remove the screws from, and lift off, the armrest (photo).

10    Carefully, using a flat-bladed tool and starting from the lower edge, prise out the plastic 'screws' securing the trim panel to the door, then lift off the panel (photo).

11    Refit in reverse order.

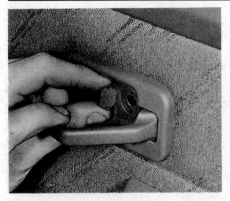

15.7A Prise out the plastic cover ...

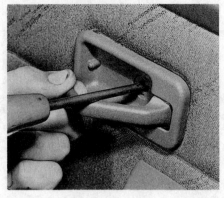

15.7B ... remove the screw ...

15.7C ... and lift off the escutcheon plate

15.8A Prise out the plastic cover ...

15.8B ... remove the nut ...

15.8C ... lift off the handle ...

15.8D ... and backing plate

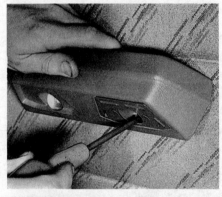

15.9 Remove the screws from the armrest

15.10 Remove the trim panel

15.14A Remove the screws securing the capping to the door ...

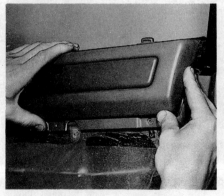

15.14B ... and lift off the capping upward

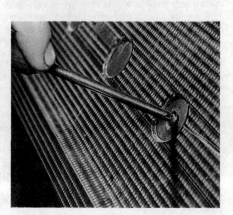

15.19 Remove the two self-tapping screws under the number plate

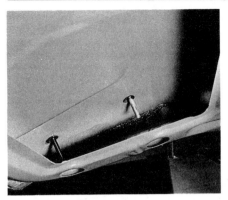

15.20 Remove the nuts from the threaded spigots

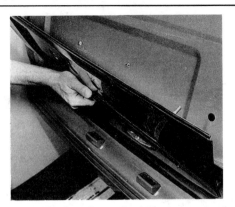

15.21 Withdraw the appliqué panel

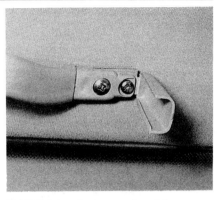

15.58 Grab handle cover and retaining screws

*Front door capping strip (1989-on) – removal and refitting*

12   Remove the door trim panel as described earlier.
13   Where fitted, prise out and disconnect the door 'tweeter' speaker.
14   Remove the screws securing the capping to the door and lift off the capping upward to disengage the seal (photos).
15   Refit in reverse order.

*Rear door trim panel (1989-on) – removal and refitting*

16   The procedure is as described for the front doors, but there is no speaker grille.

*Rear door capping strip (1989-on) – removal and refitting*

17   The procedure is as described for the front doors, but there is no 'tweeter' speaker.

*Appliqué panel (1989-on) – removal and refitting*

18   Prise off the rear number plate which is secured by double-sided tape.
19   Remove the two self-tapping screws under the number plate (photo).
20   Open the boot lid and remove the nuts from the threaded spigots at each end of the appliqué panel inside the double skin (photo).
21   Withdraw the appliqué panel (photo).
22   Refitting is a reversal of removal, but use new double-sided tape which is available from your dealer. (The use of normal double-sided tape, such as is used for carpets etc. is not recommended and would result in the number plate dropping off).

*Headlining (Estate models) – removal and refitting*

23   Disconnect the battery.
24   If fitted, fold down the rear facing seats, then remove the head restraints from the remaining seats.
25   Release the door seal from the front 'A' pillars.
26   Remove the sun visors and their brackets, the interior and load space lamps, and the front and rear grab handles.
27   Carefully release the plastic screws and buttons securing the headlining to the roof panel and the rear quarter.
28   Remove the rear edge clips then lower the lining and carefully manoeuvre it out of the open tailgate.
29   Refit in reverse.

*Headlining (Saloon models, 1989-on) – removal and refitting*

30   The procedure is similar to that described for Estate models, except that the front seats must be moved fully rearward and reclined, and the headlining lowered from the front to release the rear end from the plastic trim strip.
31   Manoeuvre the headlining from the vehicle through one of the front doors.

*Integral roof rack (Estate models) – removal and refitting*

32   Remove the headlining as described earlier.

33   Depress the spring loaded clips at the ends of the roof rack cross-rails and remove the cross-rails.
34   Remove the side-rail securing nuts and carefully ease the side-rails and gaskets from the roof. Clean all sealant from the holes in the roof and the side-rail posts.
**Caution:** *It is essential that correct and accurate refitting of the roof rack is achieved to prevent the cross-rails becoming detached in use, to avoid water leaks and prevent damage to the roof.*
35   Apply sealant (Expandite SR 51 or equivalent) to the base of each rack post, then use an adhesive (Dunlop S 758 or equivalent) to stick

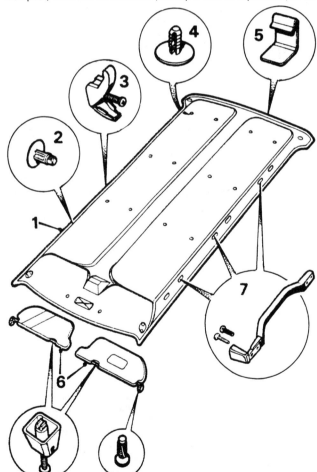

**Fig 12.80 Showing the headlining and various fixings (Sec 15)**

| | |
|---|---|
| 1  *Headlining* | 5  *Edge clip* |
| 2  *Fixing plug* | 6  *Sun visors and fixings* |
| 3  *Coat hook* | 7  *Grab handle* |
| 4  *Support button* | |

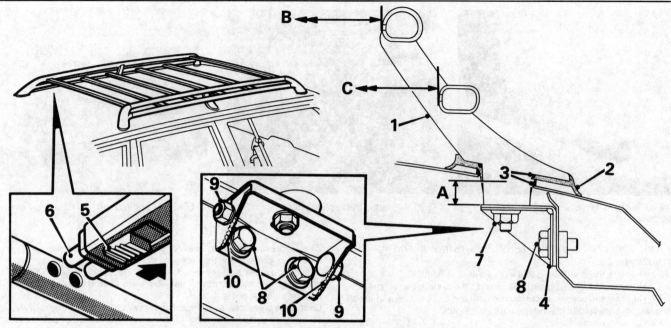

**Fig 12.81 Integral roof rack fittings and dimensions (Sec 15)**

| | | |
|---|---|---|
| 1 | Post and side-rails | |
| 2 | Gasket | |
| 3 | Sealant | |
| 4 | Mounting bracket | |
| 5 | Spring-loaded button | |
| 6 | Locating pins | |

7   Side-rail retaining nut
8   Height adjusting screw
9   Support bracket nuts
10  Tack welds (not on later models)

A   Support bracket to roof panel clearances:
Front = 0.50 ± 0.06 in (12.5 ± 1.5 mm)
Centre = 0.60 ± 0.06 in (15.2 ± 1.5 mm)
Rear = 0.44 ± 0.06 in (11.2 ± 1.5 mm)

B   Distance between upper rails = 32.8 ± 0.16 in (830 ± 4.0 mm)
C   Distance between lower rails = 35.4 ± 0.16 in (900 ± 4.0 mm)

new gaskets to the posts.
36   Apply sealant around each roof panel hole.
37   Check the distance between the roof panel and each support bracket, measuring at the centre-line of each hole. Dimensions are given in Fig. 12.81.
38   To adjust the brackets, loosen the bracket securing nuts and move the bracket up or down as necessary, tightening the nuts on completion.
39   Fit the side-rail assemblies to the roof, then fit and tighten the securing nuts only finger-tight.
40   Measure the distance between the inside faces of the rails, which must be as shown in Fig. 12.81.
41   Adjust as necessary, then tighten the securing nuts to 16 lbf ft (22 Nm).
42   If the correct dimension cannot be obtained it may be necessary to grind down the tack welds on the front edge of the front and centre brackets, loosen the nuts and reset the bracket positions. Tighten and tack weld the nuts on completion.
43   Check for water leaks before refitting the headlining, which is a reverse of removal.

*Front seats – loose backrest*
44   On some high mileage 1989-on models, the backrest-to-seat squab bolts may loosen and allow excess play between the backrest and the seat squab.
45   This problem can be overcome by fitting hexagon-headed bolts in place of the torx type bolts (available from Austin Rover dealers) and tightening them to a torque of 29 to 40 lbf ft (40 to 50 Nm).

*Door seals – inadequate retention*
46   If the door seals tend to come adrift from the top of the door frame, they can be stuck more firmly in position by using double-sided tape, available from Austin Rover dealers.

*Centre console (1989-on) – removal and refitting*
47   Disconnect the battery.
48   Remove the gear lever knob on manual transmission models.
49   Prise out the panel at the rear of the console and remove the console retaining screws under it.
50   Prise out the handbrake cover panel and remove the remaining screws securing the console to the floor pan.
51   Lift the console slightly and disconnect the plugs to the window lift switches, and the gear selector illumination bulb on automatic transmission models.
52   Lift out the console.
53   Refit in reverse order.

*Facia (1989-on) – removal and refitting*
54   The procedure is basically as described in Chapter 12, Section 35.
55   Refer to the relevant Sections of this Supplement for removal of the radio and heater controls which differ from earlier models.

*Tailgate (Estate models) – removal and refitting*
56   Open the tailgate and release the interior trim moulding by prising out the retaining buttons. Use two screwdrivers or a forked tool for this operation, but note that the buttons are easily broken and some new ones will probably be needed.
57   Release the trim from the upper retaining tags, disconnect the speaker leads (where fitted) and remove the trim.
58   Spring back the plastic covers over the grab handles, undo the retaining screws and remove the handles from the headlining (photo).
59   Peel back the rubber weatherstrip around the upper edge of the tailgate aperture, then release the clips securing the headlining (photo).
60   Unscrew the headlining retaining buttons as necessary (photo), and lower the headlining sufficiently to gain access to the tailgate hinge retaining nuts.

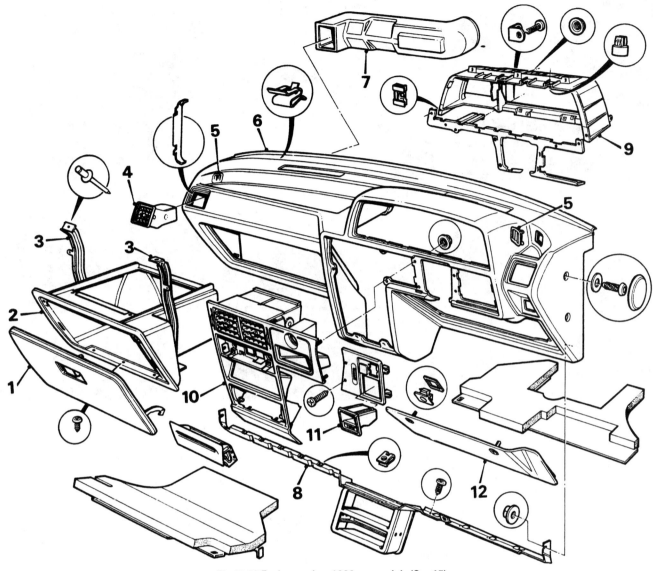

**Fig 12.82 Facia panel on 1989-on models (Sec 15)**

| | | |
|---|---|---|
| 1 Glovebox lid | 5 Side window demister escutcheon | 7 Cross duct | 10 Radio and centre vent housing |
| 2 Glovebox | 6 Facia panel | 8 Lower stiffening rail | 11 Coin tray |
| 3 Glovebox supports | | 9 Instrument carrier | 12 Fusebox cover |
| 4 Outer vent | | | |

61 Disconnect the wiring multi-plugs and cable ties at all the tailgate electrical components. Tape the multi-plugs to the wiring harness then tie drawstrings to each harness end. Similarly, disconnect the washer hose and tie a drawstring to its end.

62 Release the grommets in the tailgate and pull the wiring and washer hose out of the tailgate, until the drawstrings emerge. Untie the strings, but leave them in place.

63 Have an assistant support the tailgate, then release the support strut upper balljoints by prising off the retaining clips with a screwdriver.

64 Mark the hinge outlines on the body, then reach through the lowered headlining and undo the hinge retaining nuts. Remove the tailgate from the car.

65 Refitting is the reverse sequence to removal. Use the drawstrings to pull the wiring and washer hose through into the tailgate. Adjust the position of the hinges to achieve correct alignment and ease of closure.

### Tailgate support struts (Estate models) – removal and refitting

66 Support the tailgate in the open position and prise out the strut upper and lower balljoint retaining clips using a screwdriver (photos).

67 Withdraw the support strut from the car.

68 Refitting is the reverse sequence to removal.

### Tailgate lift assisters (Estate models)

69 Should the tailgate open only as far as the safety catch when the remote release control is operated, a pair of lift assisters may be fitted in the following way.

70 Drill holes as shown in the diagram (Fig. 12.83), on each side of the luggage area. The diameter of the holes is critical.

71 De-burr the edges of the holes, and paint. Apply underbody protective wax to the assemblies and the edges of the holes.

72 Fit the assisters, screwing them to their minimum length before closing the tailgate for the first time. Unscrew each assister half a turn and check the lift of the tailgate. Unscrew the assisters in half-turn increments until the tailgate operation is satisfactory.

### Tailgate lock (Estate models) – removal and refitting

73 Remove the tailgate interior trim moulding as described in paragraphs 56 and 57 of this Section.

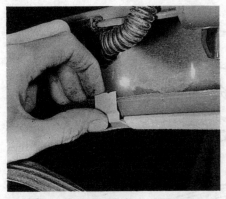
15.59 Removing a headlining retaining clip

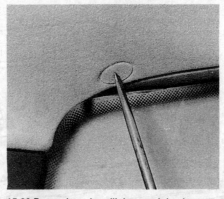
15.60 Removing a headlining retaining button

15.66A Prising off a tailgate strut upper clip

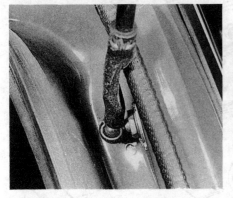
15.66B Tailgate strut lower balljoint

15.75 Tailgate lock and retaining bolts

15.78 Private lock, lock lever and retaining nuts

74    Release the retaining clip and disconnect the operating rod from the lock lever.
75    Undo the two retaining bolts and remove the lock from the tailgate (photo).
76    Refitting is the reverse sequence to removal. If necessary, adjust the striker plate position after refitting the lock, so that the tailgate shuts and locks without slamming.

### Tailgate private lock (Estate models) – removal and refitting
77    Remove the tailgate interior trim moulding as described in paragraphs 56 and 57.
78    Release the retaining clip and disconnect the operating rod from the private lock lever (photo).

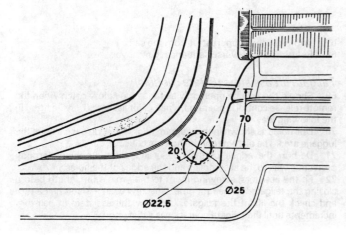

Fig 12.83 Tailgate lift assister fitting diagram (Sec 15)

*(all dimensions in mm)*

79    Undo the retaining nuts, remove the lock lever and withdraw the lock barrel from the number plate lamp housing.
80    Refitting is the reverse sequence to removal.

### Tailgate release cable and lever (Estate models) – removal and refitting
81    Fold back the rear seat cushion and the luggage compartment carpet.
82    Remove the front and rear sill carpet retainers on the driver's side.
83    Slacken all the driver's seat retaining bolts and remove the bolts from the runner nearest the door.
84    Remove the oddments tray.
85    Undo the release lever retaining screws, lift up the trim moulding and disconnect the release cable. Remove the release lever from the car.
86    Fold back the side of the front and rear carpets, disconnect the cable from the retaining clips and pull the cable into the luggage compartment.
87    Undo the screws and withdraw the tailgate aperture lower trim moulding (photo).
88    Undo the striker plate retaining bolts (photo). Lift up the striker plate and disconnect the cable from the release lever. Withdraw the cable from the car.
89    Refitting is the reverse sequence to removal, but adjust the striker plate position so that the tailgate shuts and locks without slamming.

### Rear bumper (Estate models) – removal and refitting
90    Undo the rear wheel arch liner retaining screws and remove the wheel arch liners.
91    Undo the nut securing each bumper support bracket to the rear valance.
92    From inside the car, fold the rear panel of the luggage compartment floor and remove the floor covering.
93    Undo the two bolts securing each bumper rear support bracket to the body. Lift away the rear bumper assembly.
94    Refitting is the reverse sequence to removal.

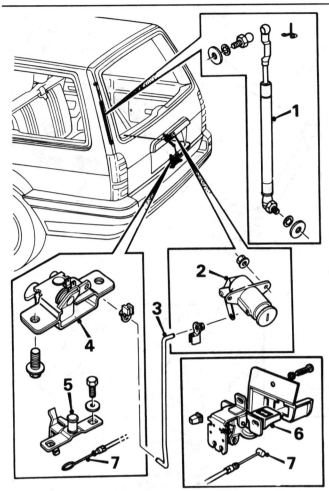

**Fig 12.84 Tailgate lock and support strut components (Sec 15)**

| | |
|---|---|
| 1  *Support strut* | 5  *Striker plate* |
| 2  *Private lock* | 6  *Release lever* |
| 3  *Operating rod* | 7  *Release cable* |
| 4  *Tailgate lock* | |

### Rear seat (Estate models) – removal and refitting

95   With the rear doors open, fold the cushion or cushions forward, undo the retaining screws and withdraw the seat cushion(s).
96   Remove the outer hinge bracket covers, release the seat squab retaining latches and lower the squab. Undo the outer bracket retaining screws.
97   If a one-piece squab is fitted, remove the squab. If a split squab is fitted, undo the retaining screw and extract the clip securing the centre pivots. Withdraw the relevant squab.
98   Refitting is the reverse sequence to removal.

### Luggage area side trim moulding (Estate models) – removal and refitting

99   If fitted, fold the rear facing seat, fold the rear seat cushion, lower the squab and remove the squab release handle.
100   Detach the rear door weatherstrip seal from along the side trim moulding.
101   Open the tailgate, and remove the guide plate and the three screws securing the tailgate aperture lower trim moulding on the side being worked on.
102   Undo the upper and lower rear seat belt mounting bolts, and detach the guide escutcheon from the side moulding.
103   Undo the upper and lower rear facing seat belt mounting bolts, detach the guide escutcheon from the side moulding and release the seat belt support strut at its bottom end.

104   Release the fastener securing the moulding to the rear of the wheel arch. Using a forked tool or suitable alternative, carefully release the four clips securing each end of the panel.
105   Pull the sill edge retaining clips from the body brackets and lift the moulding to clear the four bottom edge retainers from the body slots.
106   Disengage the support strut, pass the seat belt webbing through the side trim moulding and remove the moulding.
107   Refitting is the reverse sequence to removal.

### Rear seat belts (Saloon models) – removal and refitting
**Outer belt**
108   Remove the rear seat cushion.
109   Undo the seat belt lower mounting and rear squab extension retaining bolt.
110   Push the squab extension upwards and remove it from the car.
111   Extract the seat belt upper guide bracket plastic cover, then undo the bolt and remove the guide and spacers.
112   Release the door weatherstrip seal from the rear body pillar and remove the finisher trim.
113   Release the parcel shelf trim extension plastic retaining clips and withdraw the extension.
114   Detach the shelf trim finisher from the seat belt, and remove the shelf trim extension by sliding it along the belt.
115   From within the boot, fold back the front of the boot carpet, undo the seat belt locking buckle anchorage bolt, and remove the locking buckle.
116   Refitting is the reverse sequence to removal.
**Centre belt**
117   Remove the rear seat cushion, then from inside the boot, fold back the front of the boot carpet.
118   Undo the seat centre belt anchorage bolts and remove the belts from the car.
119   Refitting is the reverse sequence to removal.

### Rear seat belts (Estate models) – removal and refitting
120   Remove the luggage area side trim moulding as described earlier in this Section.
121   Remove the seat belt upper guide bracket plastic cover, then undo the retaining bolt and withdraw the belt guide and spacers.
122   Release the belt guide escutcheon from the side moulding, open the escutcheon and slide out the webbing. Pass the belt through the side moulding.
123   Undo the reel retaining bolt and remove the belt assembly.
124   Undo the seat belt locking buckle retaining bolt and remove the locking buckle. Note the position of the rear facing seat belt where these are fitted.
125   Refitting is the reverse sequence to removal.
126   To remove the centre belt, fold the rear facing seat if fitted, fold the rear seat and remove the backrest(s) as necessary.
127   Undo the locking buckle and anchor bracket retaining bolts, and remove the complete centre seat belt.
128   Refitting is the reverse sequence to removal.

### Rear facing seat belts (Estate models) – removal and refitting
129   Fold the rear seat and the rear facing seat, then remove the rear seat backrest.
130   Remove the upper guide bracket plastic cover, then undo the retaining bolt.
131   Remove the plastic cover, then undo the lower anchorage retaining bolt.
132   Release the belt guide escutcheon from the side moulding, open the escutcheon and slide it off the belt.
133   Remove the relevant grab handle and the rearmost headlining retaining buttons.
134   Release the tailgate weatherstrip seal from the upper edge of the aperture, then extract the headlining retaining clips.
135   Undo the bolt at the top of the support strut, then support the headlining by refitting the retaining clips.
136   Remove the luggage area side trim moulding as described earlier in this Section.
137   Undo the seat belt support strut lower retaining bolt and the belt reel retaining bolt from inside the side panel.
138   Pass the seat belt through the side moulding and remove the belt assembly.

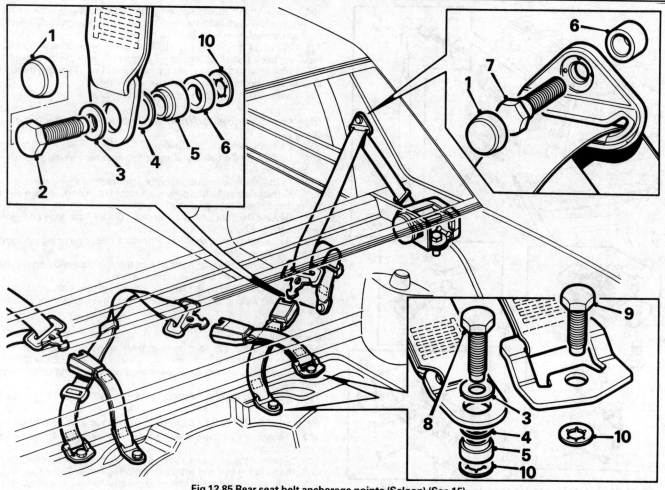

**Fig 12.85 Rear seat belt anchorage points (Saloon) (Sec 15)**

1  *Plastic cover*
2  *Lower mounting bolt*
3  *Plain washer*

4  *Anti-rattle washer*
5  *Shouldered distance piece*
6  *Distance piece*

7  *Upper mounting bolt*
8  *Locking buckle bolt*

9  *Centre belt bolt*
10 *Transit washer*

139    Undo the locking buckle anchorage bolt and remove the locking buckle.
140    Refitting is the reverse sequence to removal.

*Remote control door mirror glass – renewal*

141    Refer to Chapter 11, Section 25 for removal.
142    If the mirror glass is to be renewed, this can be done without the need to remove the mirror from the door.
143    Wear a protective glove and push the outboard end of the mirror towards the front of the car, until the mirror casing begins to move away from the glass. Insert the fingers between the glass and the casing and prise the glass gently. When the gap is wide enough, insert a thin screwdriver to release the socket from the pivot bolt.
144    Disconnect the tension spring and remove the glass.
145    To fit the glass, first apply silicone grease to the socket, then fit the glass to the retaining arm and connect the spring.
146    Gently apply even pressure to push the glass into position, engaging the socket with the ball.

*Electrically operated rear view mirror*

147    Later top of the range models have electrically operated and demisted exterior rear view mirrors. The heating elements are automatically energised as soon as the ignition is switched on.
148    Removal is similar to the procedure described in earlier paragraphs for the manually operated type, except that the wiring harness must be disconnected as the mirror is withdrawn.

149    If the glass is to be removed, operate the control switch until the top of the glass is tilted as far as possible into the casing.
150    Insert a forked tool into the gap at the centre of the lower edge of the glass and release the glass retainers.
151    Again operate the control switch until the bottom of the glass is tilted as far as possible into the casing. Release the glass upper retainers.
152    Disconnect the wiring connectors and withdraw the glass.
153    Before refitting the glass, centralise the control switch (ignition on) and connect the wiring. Check that the insulated connector is uppermost.
154    Align the pegs on the retaining ring and press the glass firmly to engage the securing clips.

*Fuel filler flap remote control lever and cable – removal and refitting*

155    On Saloon models, remove the rear seat cushion and peel back the carpet from the right-hand side of the luggage compartment.
156    On Estate models, fold the rear seat and peel back the luggage area floor covering. Where fitted, erect the rear facing seats and remove the right-hand side trim from the luggage area.
157    Remove the sill carpet retainers from the driver's side.
158    Slacken the driver's seat bolts and then remove the bolts from the outboard seat runner.
159    Remove the oddments tray.
160    Extract the fuel flap release lever fixing screws, raise the moulding and disconnect the release lever from the cable.

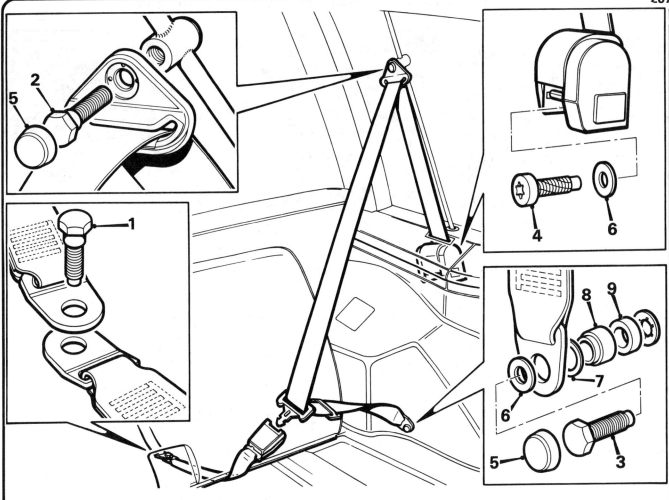

**Fig 12.86 Rear facing seat belt anchorage points (Sec 15)**

| | | | |
|---|---|---|---|
| 1 Locking buckle bolt | 4 Belt reel bolt | 6 Washer | 8 Shouldered spacer |
| 2 Upper fixing bolt | 5 Bolt plastic cover | 7 Anti-rattle washer | 9 Spacer |
| 3 Lower fixing bolt | | | |

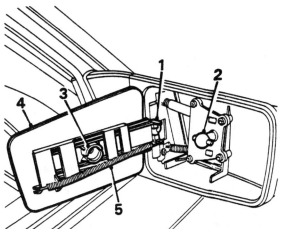

**Fig 12.87 Door mirror glass details (Sec 15)**

| | |
|---|---|
| 1 Arm | 4 Glass |
| 2 Control ball | 5 Spring |
| 3 Socket | |

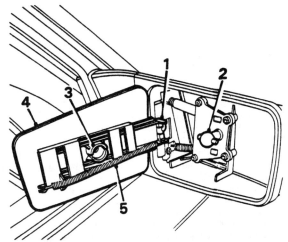

**Fig 12.88 Electrically operated door mirror (Sec 15)**

| | |
|---|---|
| 1 Internal finisher | 4 Edge finisher |
| 2 Retaining plate | 5 Mirror flange seal |
| 3 Wiring harness | |

15.87 Removing the tailgate aperture lower trim moulding

15.88 Tailgate lock striker plate and retaining bolts

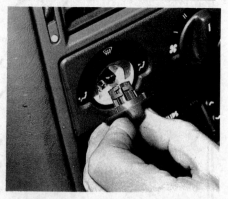

15.236 Pull off the control knobs

161   Peel back the carpets to free the cable by releasing the retaining clips. Withdraw the cable into the boot (Saloon) or luggage area (Estate).
162   Open the fuel filler flap, release the retaining clip and then pull the cable from the rear of the wheel arch into the boot or luggage area.
163   Refitting is a reversal of removal.

### Asymmetrical split rear seat
164   On later Estates, the rear seat is split 60/40 in order that, by folding the sections independently, load carrying area is provided while still retaining rear passenger accommodation. Removal operations are similar to those described in paragraphs 95 to 98 of this Section.

### Rear facing child seats
165   Optionally available on Estate models, the seats should be regarded as being for occasional use only. They must only be used with the forward facing rear seat erected and locked. Seat belts must always be used. The seats are not suitable for adults.

### Spare wheel and tools (Estate models)
166   The spare wheel and tools are located below the luggage compartment floor. Access is obtained by raising the floor or seat backrest if rear facing seats are fitted.
167   The tools are located in a plastic moulding within the spare wheel.

### Grab handles – removal and refitting
168   To remove a grab handle from the headlining above the door opening, flip back the end covers with a small screwdriver to expose the fixing screws. Remove the screws and the handle.
169   Refitting is a reversal of removal.

### Tilt/slide sunroof components (1987-on) – removal and refitting
**Glass panel and seal**
170   Open the panel to the half way position, then release the screws which secure the side shield and rear drain channel to the glass panel.
171   Slide the cover, drain channel and visor into the recess above the headlining.
172   Close the glass panel then tilt the rear edge.
173   Extract the screws which hold the glass panel to the side arms.
174   Lift off the glass panel and remove the seal.
175   When refitting, first push the seal onto the glass panel with the rubber surface facing inwards. Cut the seal to form a butt joint.
176   Set the operating mechanism in the closed position and locate the glass panel on the side arms. Fit the screws, align the panel with the roof and then tighten the screws.
177   Open and close the panel and then check that in the closed position, the front edge is flush with, or up to 0.04 in (1.0 mm) below, the level of the roof. The rear edge should be within the same tolerance above the level of the roof.
178   Open the panel to the halfway position locate the side shield and rear drain channel and fit the fixing screws.
**Sunroof visor**
179   Remove the glass panel as previously described.
180   Remove the pilot plates and side arms.

181   Remove the slide and collect the flat washer from under the front of each side-rail.
182   Disengage the rear guides from their channels and allow them to hang in the roof aperture.
183   Press the wind deflector downwards, then slide the side cover and rear drain channel forward and out through the roof.
184   Slide the visor forwards and out through the roof aperture, taking care not to damage the guides.
185   Refitting is a reversal of removal; remember to fit the flat washer under each side rail.
**Sunroof drivegear (manually operated type)**
186   Remove the glass panel as previously described, then set the operating mechanism in the closed position.
187   Check that the handle is in the free play position, and then remove the handle bezel and escutcheon.
188   Remove the finisher from around the sunroof aperture, and both windscreen sun visors and clips.
189   Remove the grab handle from the passenger side headlining, and the stud from the driver's side.
190   Release and pull down the front edge of the headlining and extract the screws which hold the operating handle.
191   Extract the fixing screw and withdraw the drivegear.
192   To refit, position the side arm assembly so that the hole in the lifting lever and the triangular cut-out in the side arm are aligned with the slot in the pilot plate, as shown in Fig. 12.90.
193   Carry out the alignment on the opposite side.
194   Smear the drivegear with silicone grease and engage the cables. Fit and tighten the mounting screw.
195   Fit the operating handle, secure the headlining and fit the grab handle and visors.
196   Fit the finisher, handle escutcheon and bezel, and the glass panel.
**Sunroof drivegear (electrically operated type)**
197   Remove the glass panel as previously described.
198   Check that the sunroof mechanism is switched off in the closed position. If there is any doubt about closure, use the emergency hand crank which is stored in the glovebox.
199   Open the cover, release the switch panel, disconnect the switch leads and remove the switch panel.
200   Release and pull away the headlining.
201   Extract the screw and release the relay from the motor bracket (Fig. 12.91).
202   Remove the mounting screws and withdraw the motor/gearbox drive assembly, noting that the gear timing hole is adjacent to the terminals when the mechanism is in the roof-closed mode.
203   Commence refitting by aligning the guides to the pilot plates using 5.0 mm diameter pins to retain the alignment.
204   Lubricate the motor drive gears.
205   Support the motor, connect the multi-plug and switch on the ignition. Actuate the rod switch to bring the gear timing hole to the position shown in Fig. 12.91. Switch off the ignition.
206   Align the gearbox slides and engage the drive cables, then fit the motor/gearbox drive assembly. Withdraw the temporary pins.
207   Refit the headlining, switch panel and glass panel.
**Rear guide and cable assemblies**
208   Remove the glass panel as previously described.
209   Set the operating mechanism in the closed position.

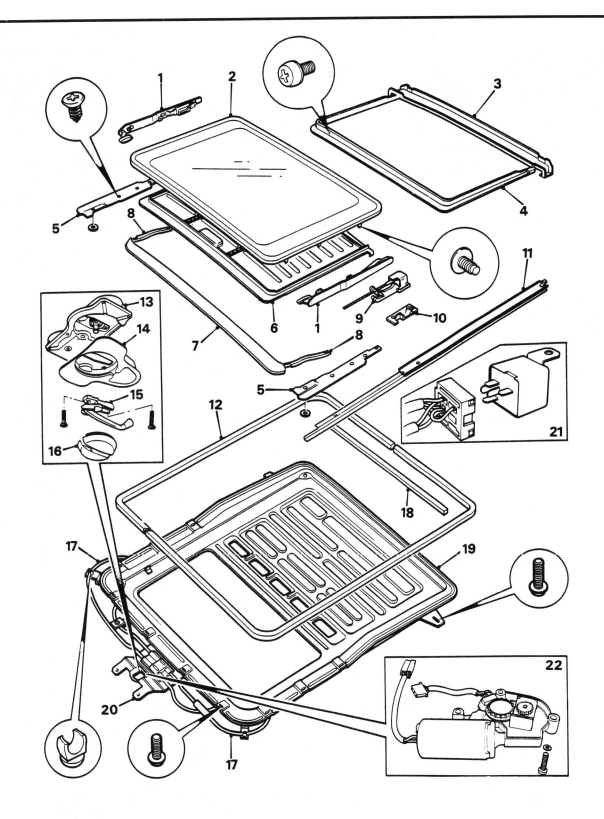

**Fig 12.89 Tilt/slide sunroof components (Sec 15)**

1 Side arm assembly
2 Glass panel
3 Rear drain channel
4 Side shield
5 Slide rail
6 Sliding visor
7 Wind deflector
8 Deflector lifting arm
9 Rear guide and cable assembly
10 Pilot plate
11 Guide rail
12 Tray seal
13 Drivegear (manual)
14 Escutcheon
15 Operating crank
16 Bezel
17 Cable guide and run-out tubes
18 Rear tray seal
19 Sunroof tray
20 Front bracket and mounting plate
21 Relay
22 Electric motor

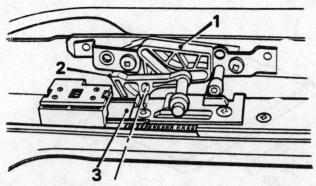

**Fig 12.90 Side arm/lifting lever alignment (Sec 15)**

1  *Side arm*          3  *Pilot plate*
2  *Lifting arm*

210    Remove the drivegear (manual) or motor/gearbox drive (electrically operated).
211    Extract the screws and remove the pilot plates and front side rails.
212    Remove both side arm assemblies and withdraw the cable and rear guides.
213    Before refitting, inject grease into the cable guide tubes and insert the cables.
214    Locate the rear guides, fit the side arm assemblies, front slide rails and pilot plates.
215    Align the side arms as described in paragraphs 192 and 193.
216    Fit the drivegear or motor/gearbox drive according to type. Refit the glass panel.

**Sunroof – complete**
217    Remove the glass panel as previously described.
*Manually operated type*
218    Remove the handle bezel, escutcheon and the handle itself.
*Electrically operated type*
219    Open the cover and release the switch panel. Disconnect the switch and relay and take off the switch panel.
*All models*
220    Disconnect the drain tubes from the sunroof tray.
221    Locate a block of wood above the front bracket and drill out the two pop-rivets.

222    Fully recline the two front seats and then extract all the screws which hold the sunroof tray to the roof.
223    Lower the tray and withdraw it through the front passenger door. Discard the tray seal.
224    Refitting is a reversal of removal but before starting, fit a strip of 20.0 x 20.0 mm compriband seal (available from your dealer) around the edge of the sunroof tray.

*Plastic components*
225    With the use of more and more plastic body components by the vehicle manufacturers (eg bumpers, spoilers, and in some cases major body panels), rectification of more serious damage to such items has become a matter of either entrusting repair work to a specialist in this field, or renewing complete components. Repair of such damage by the DIY owner is not really feasible owing to the cost of the equipment and materials required for effecting such repairs. The basic technique involves making a groove along the line of the crack in the plastic using a rotary burr in a power drill. The damaged part is then welded back together by using a hot air gun to heat up and fuse a plastic filler rod into the groove. Any excess plastic is then removed and the area rubbed down to a smooth finish. It is important that a filler rod of the correct plastic is used, as body components can be made of a variety of different types (eg polycarbonate, ABS, polypropylene).
226    Damage of a less serious nature (abrasions, minor cracks etc) can be repaired by the DIY owner using a two-part epoxy filler repair material, like Holts Body + Plus or Holts No Mix which can be used directly from the tube. Once mixed in equal proportions (or applied direct from the tube in the case of Holts No Mix), this is used in similar fashion to the bodywork filler used on metal panels. The filler is usually cured in twenty to thirty minutes, ready for sanding and painting.
227    If the owner is renewing a complete component himself, or if he has repaired it with epoxy filler, he will be left with the problem of finding a suitable paint for finishing which is compatible with the type of plastic used. At one time the use of a universal paint was not possible owing to the complex range of plastics encountered in body component applications. Standard paints, generally speaking, will not bond to plastic or rubber satisfactorily, but Holts Professional Spraymatch paints to match any plastic or rubber finish can be obtained from dealers. However, it is now possible to obtain a plastic body parts finishing kit which consists of a pre-primer treatment, a primer and coloured top coat. Full instructions are normally supplied with a kit, but basically the method of use is to first apply the pre-primer to the component concerned and allow it to dry for up to 30 minutes. Then the primer is applied and left to dry for about an hour before finally applying the special coloured top coat. The result is a correctly

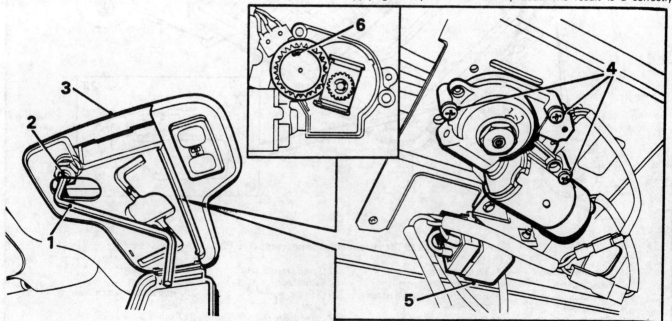

**Fig 12.91 Sunroof drivegear – electrically operated type (Sec 15)**

1  *Emergency crank*          3  *Switch panel*          5  *Relay*
2  *Hexagonal drive socket*   4  *Motor mounting screws*  6  *Gearwheel timing hole*

15.237 Prise the cover panel forward

15.238A Remove the screws securing the back panel ...

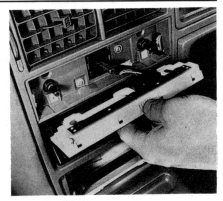

15.238B ... and drop the back panel forward

coloured component where the paint will flex with the plastic or rubber, a property that standard paint does not normally possess.

### Non-vacuum type heater – general

228    Later models are equipped with a modified type of heater unit, whereby the air intake and temperature flaps are operated by cables instead of the vacuum actuators used previously.

229    Apart from the cable locations shown in Fig. 12.92 and the cable adjustment procedure contained in the following paragraphs, the heater is the same in construction and layout as the earlier type. All operations contained in Chapter 11 are still applicable, but note the cable locations and attachments when carrying out the work.

### Non-vacuum type heater – adjustments

230    Remove the fusebox cover and, where fitted, the access panel above the pedals.

231    Move the heater control levers fully upward.

232    With the levers held upward, slacken the clamp bolt and move the air temperature flap fully downwards, then tighten the clamp bolt.

233    Similarly slacken the fresh air and air intake flap lever clamp bolts, and move both levers fully rearwards. Tighten the clamp bolts, refit the fusebox cover, and the access panel where fitted.

### Heater (1989-on) – general description

234    The heater in 1989-on models is basically the same as the non-vacuum type fitted to earlier models, but has rotary controls as opposed to levers.

### Heater control panel (1989-on) – removal and refitting

235    Disconnect the battery.
236    Pull off the control knobs (photo).
237    Prise the cover panel forward (photo).

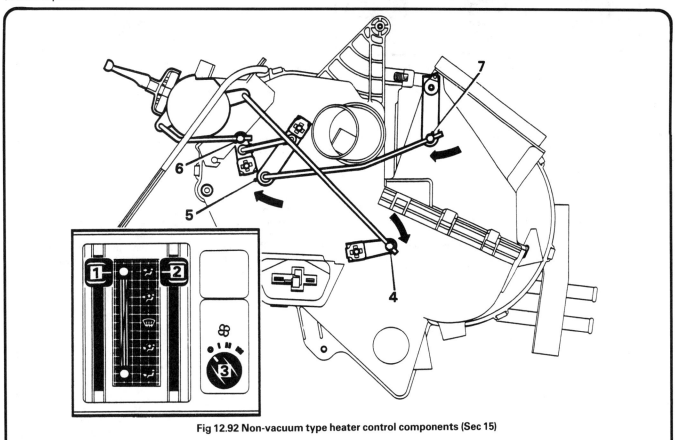

**Fig 12.92 Non-vacuum type heater control components (Sec 15)**

| | | |
|---|---|---|
| 1   Air temperature control | 3   Blower switch | 5   Fresh air flap lever | 7   Air intake flap lever |
| 2   Air distribution control | 4   Air temperature flap lever | 6   Air flap lever demisters | |

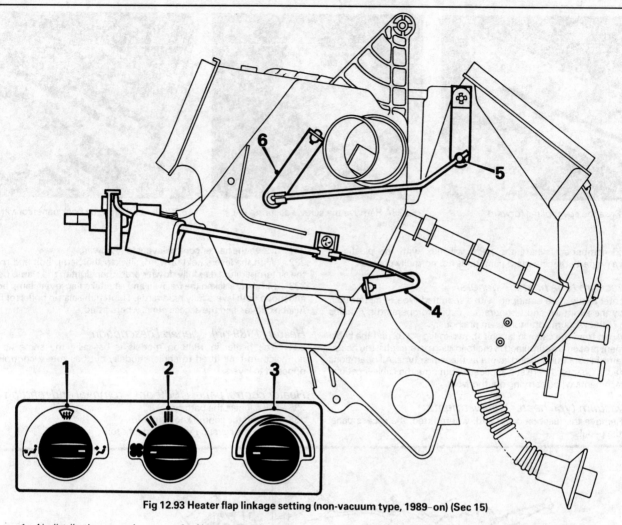

**Fig 12.93 Heater flap linkage setting (non-vacuum type, 1989–on) (Sec 15)**

1  *Air distribution control*        3  *Air temperature control*        5  *Air inlet flap lever and*        6  *Matrix air outlet flap lever*
2  *Blower switch*                   4  *Air distribution flap lever*          *pinchbolt*

238    Remove the screws securing the back panel to the facia panel and drop the back panel forward (photos).
239    Disconnect the lead and withdraw the panel.
240    Refit in reverse order.

### Heater control cables (1989-on) – removal and refitting

241    Disconnect the battery.
242    Remove the centre console and facia as described in Chapter 11, Sections 34 and 35 respectively.
243    Remove the heater control cover as described in paragraphs 232 to 236.
244    Remove the outer cable clips from the control panel.

245    Release the cable operating racks from the control panel, then disconnect the cables from the racks.
246    Remove the cable clips from the heater casing, disconnect the cables from the flap levers and withdraw the cables.
247    Disconnect the blower motor switch multi-plug and withdraw the control panel.
248    Refitting is a reversal of removal, fully closing the heater flaps before locating the control cables in the racks, which must also be in the fully closed position before fitting the cables to them.
249    Adjust the heater flap by loosening the lever pinchbolt and pushing the air intake flap lever fully rearward. Push the matrix air outlet flap lever rearward and tighten the pinchbolt.

**Key to main wiring diagram – carburettor models (up to 1986)**

| No | Description | Grid reference |
|---|---|---|
| 1 | Alternator | A1 |
| 3 | Battery | A1 |
| 4 | Starter motor solenoid | A1 |
| 5 | Starter motor | A1 |
| 6 | Lighting switch – main | B1 |
| 7 | Headlamp dip switch | B1 |
| 8 | Headlamp dip beam | B3 |
| 9 | Headlamp main beam | B3 |
| 10 | Main beam warning lamp | A4 |
| 11 | RH sidelamp | B3 |
| 12 | LH sidelamp | B3 |
| 14 | Panel illumination lamps | A4 |
| 15 | Number plate illumination lamps | B4 |
| 16 | Stop-lamps | B4 |
| 17 | RH tail lamp | B4 |
| 18 | Stop-lamp switch | B4 |
| 19 | Fusebox | A2, B2 |
| 20 | Interior lamps | B3 |
| 21 | Interior lamp door switches | C3 |
| 22 | LH tail lamp | B4 |
| 23 | Horns | B1 |
| 24 | Horn push | B1 |
| 26 | Direction indicator switch | B1 |
| 27 | Direction indicator warning lamp(s) | A4 |
| 28 | RH front direction indicator lamp | B3 |
| 29 | LH front direction indicator lamp | B3 |
| 30 | RH rear direction indicator lamp | B4 |
| 31 | LH rear direction indicator lamp | B4 |
| 32 | Heater/fresh air motor switch | B3 |
| 33 | Heater/fresh air motor | B3 |
| 34 | Fuel level gauge | A4 |
| 35 | Fuel level gauge tank unit | B1 |
| 37 | Windscreen wiper motor | B3 |
| 38 | Ignition/starter switch | B1 |
| 39 | Ignition coil | A1 |
| 40 | Distributor | B1 |
| 42 | Oil pressure switch | A4 |
| 43 | Oil pressure warning lamp | A4 |
| 44 | Ignition/no-charge warning lamp | A3 |
| 45 | Headlamp flash switch | B1 |
| 46 | Coolant temperature gauge | A4 |
| 47 | Coolant temperature thermistor | C1 |
| 49 | Reverse lamp switch | C4 |
| 50 | Reverse lamps | B4 |
| 56 | Clock | A2 |
| 57 | Cigar lighter | B2 |
| 65 | Luggage area lamp switch | B3 |
| 66 | Luggage area lamp | B3 |
| 75 | Automatic gearbox starter inhibitor switch | A1 |
| 76 | Automatic gearbox selector indicator lamp | B4 |
| 77 | Windscreen washer motor | B1 |
| 82 | Switch illumination lamp(s) | A2 |
| 95 | Tachometer | A4 |
| 115 | Heated rear screen switch | C2 |
| 116 | Heated rear screen | C2 |
| 118 | Windscreen washer/wiper switch | B3 |
| 150 | Heated rear screen warning lamp | A2 |
| 152 | Hazard warning lamp | B1 |
| 153 | Hazard warning switch | B1 |
| 165 | Handbrake warning lamp switch | B4 |
| 166 | Handbrake warning lamp | A4 |
| 174 | Starter solenoid relay | A1 |
| 178 | Radiator cooling fan thermostat | C2 |
| 179 | Radiator cooling fan motor | C2 |
| 182 | Brake fluid level switch | B4 |
| 208 | Cigar lighter illumination lamp | B3 |
| 210 | Panel illumination lamp rheostat/resistor | C3 |
| 211 | Heater control illumination | B2 |
| 231 | Headlamp relay | B1 |
| 232 | Sidelamp warning lamp | A4 |
| 240 | Heated rear screen relay | A2 |
| 242 | Two way interior lamp switch | B3 |
| 246 | Glovebox illumination lamp | B3 |
| 247 | Glovebox illumination switch | B3 |
| 256 | Brake warning blocking diode | A4 |
| 265 | Ambient air temperature sensor | C1 |
| 267 | Headlamp washer motor | B1 |
| 286 | Fog rearguard lamp switch | A2 |
| 287 | Fog rearguard warning lamp | A2 |
| 288 | Fog rearguard lamp(s) | B4 |
| 294 | Fuel cut-off solenoid | C2 |
| 297 | Brake failure warning lamp | A4 |
| 308 | Direction indicator/hazard flasher unit | C2 |
| 325 | Brake pad wear warning lamp | A4 |
| 326 | Brake pad wear sensor | A4 |
| 347 | ECU – fuel management | C2 |
| 355 | Accelerator pedal switch | C1 |
| 359 | Stepper motor | C2 |
| 367 | Trailer indicator warning light | A3 |
| 368 | Spare warning lamp | A4 |
| 381 | Knock sensor | B1 |
| 382 | Crankshaft sensor | C1 |
| 386 | One-touch control – window lift | C2 |
| 389 | Column switch illumination | B3 |
| 393 | ECU – Programmed ignition | C1 |
| 398 | Manifold heater relay | B1 |
| 399 | Manifold heater | B1 |
| 400 | Temperature switch | B1 |
| 401 | Interior lamp delay unit | C3 |
| 402 | Windscreen wipe/wash programmed control | B3 |
| 403 | Auxiliary ignition relay | A1 |
| 410 | Centre console illumination | B2 |
| 413 | Fusible link | A1 |
| 415 | Fuel level LED | A4 |
| 416 | Coolant temperature LED | A4 |
| 426 | Heater vacuum solenoid | C3 |
| 427 | Heater vacuum solenoid switch | C3 |

**Supplementary diagram connections**

| | | |
|---|---|---|
| A | Headlamp wash | B1 |
| B | Radio cassette player | C2 |
| C | Electric windows | C3 |
| D | Central locking | B4 |
| E | Rear washer/wiper | A2 |
| F | Electric mirror | C3 |

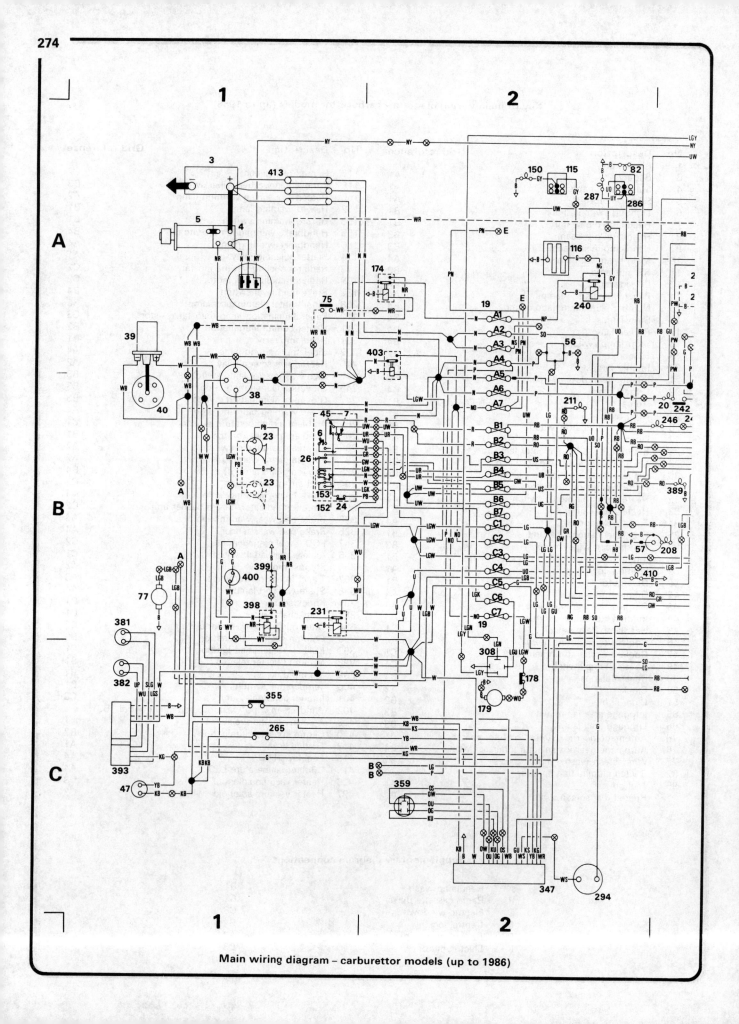

**Main wiring diagram – carburettor models (up to 1986)**

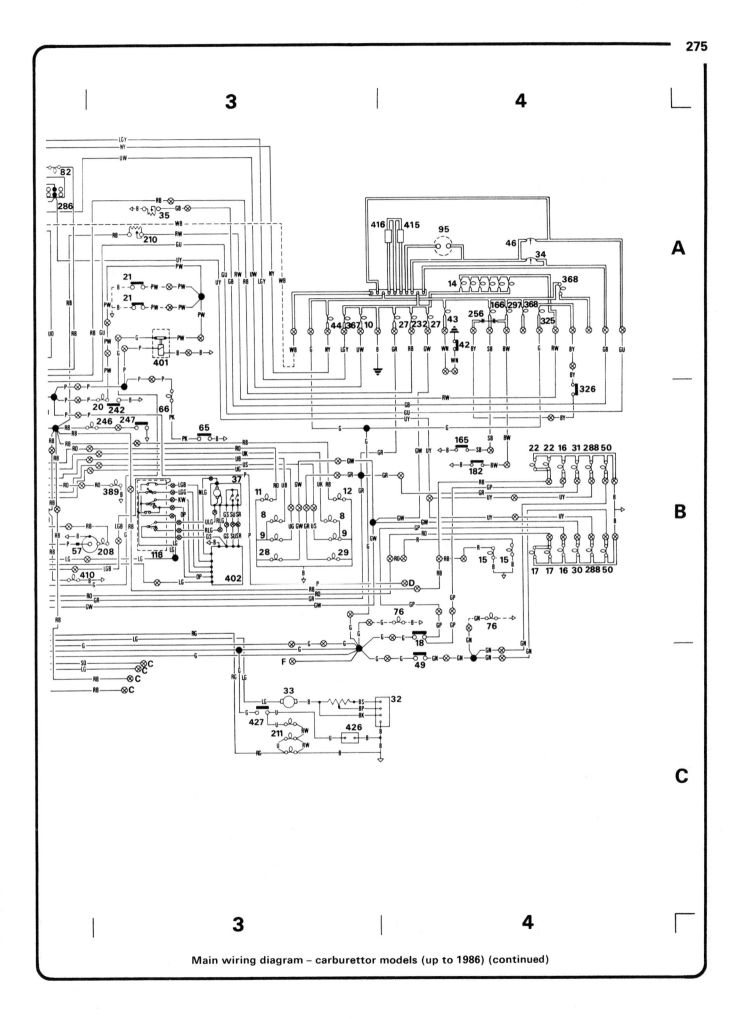

Main wiring diagram – carburettor models (up to 1986) (continued)

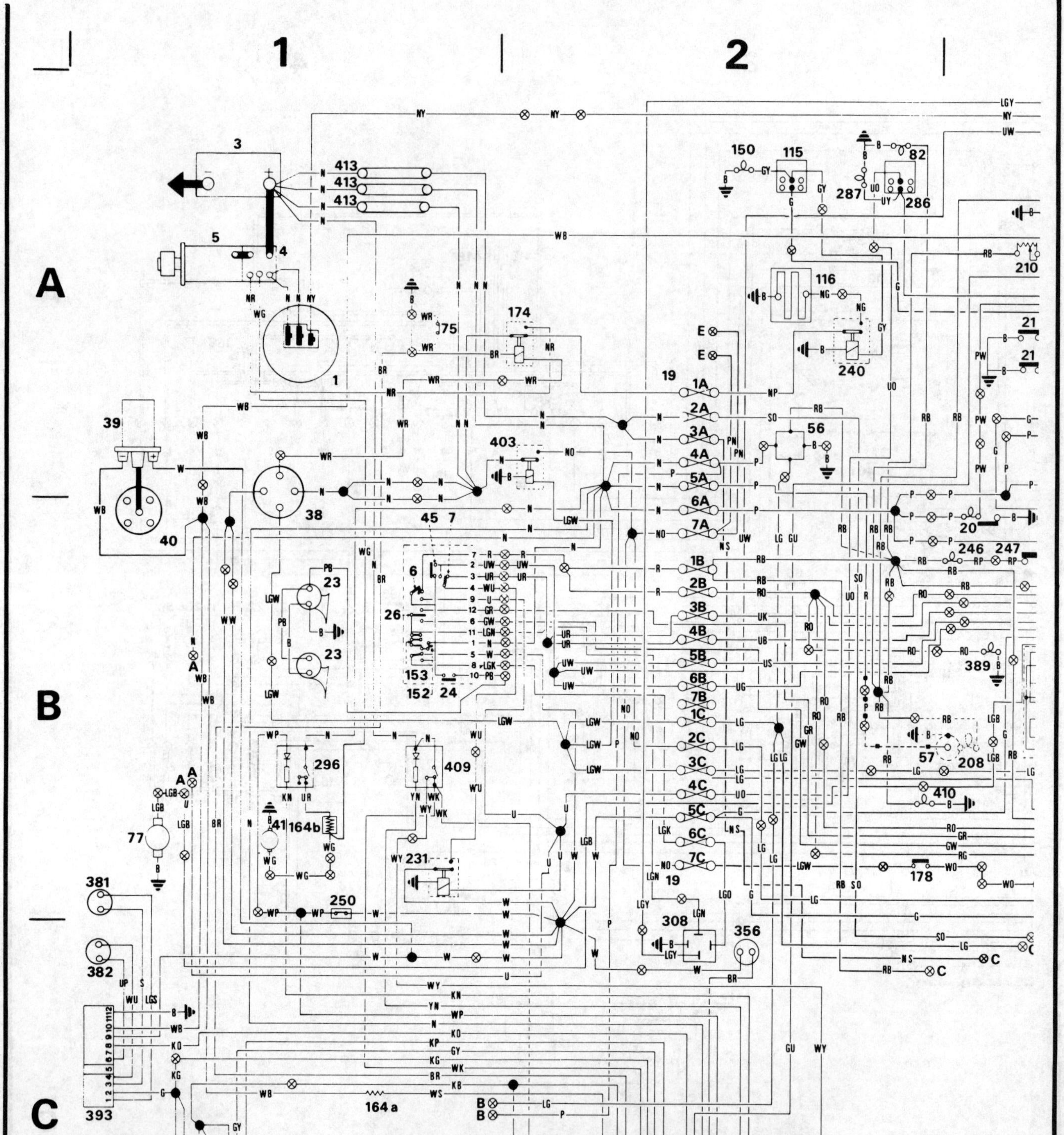

1

2

A

B

C

Main wiring diagram – EFi models with conventional instruments (up to 1986)

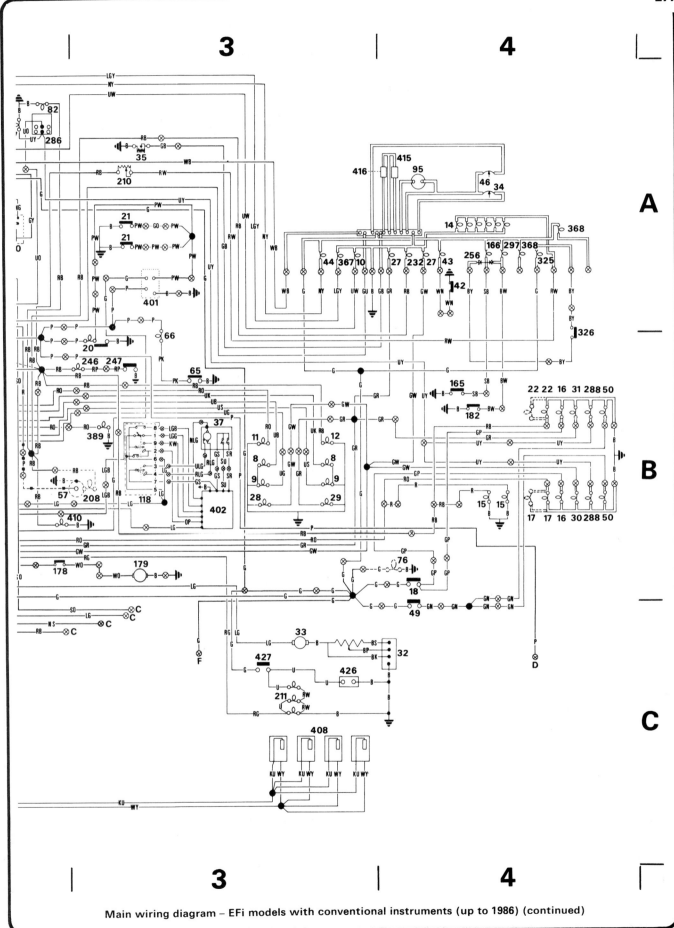

**Main wiring diagram – EFi models with conventional instruments (up to 1986) (continued)**

## Key to main wiring diagram – EFi models with conventional instruments (up to 1986)

| No | Description | Grid reference | No | Description | Grid reference |
|----|-------------|----------------|----|-------------|----------------|
| 1 | Alternator | A1 | 116 | Heated rear screen | A2 |
| 3 | Battery | A1 | 118 | Windscreen washer/wiper switch | B3 |
| 4 | Starter motor solenoid | A1 | 150 | Heated rear screen warning lamp | A2 |
| 5 | Starter motor | A1 | 152 | Hazard warning lamp | B1 |
| 6 | Lighting switch – main | B1 | 153 | Hazard warning switch | B1 |
| 7 | Headlamp dip switch | B1 | 164a | In-line resistor engine speed | C1 |
| 8 | Headlamp dip beam | B3 | 164b | In-line resistor fuel pump | B1 |
| 9 | Headlamp main beam | B3 | 165 | Handbrake warning lamp switch | B4 |
| 10 | Main beam warning lamp | A3 | 166 | Handbrake warning lamp | A4 |
| 11 | RH sidelamp | B3 | 174 | Starter solenoid relay | A2 |
| 12 | LH sidelamp | B3 | 178 | Radiator cooling fan thermostat | B2 |
| 14 | Panel illumination lamps | A4 | 179 | Radiator cooling fan motor | B3 |
| 15 | Number plate illumination lamps | B4 | 182 | Brake fluid level switch | B4 |
| 16 | Stop-lamps | B4 | 208 | Cigar lighter lamp | B3 |
| 17 | RH tail lamp | B4 | 210 | Panel lamp rheostat/resistor | A3 |
| 18 | Stop-lamp switch | B4 | 211 | Heater control lamp | C3 |
| 19 | Fusebox | A2, B2 | 231 | Headlamp relay | B1 |
| 20 | Front interior lamp | B3 | 232 | Sidelamp warning lamp | A4 |
| 21 | Interior lamp door switches | A3 | 240 | Heated rear screen relay | A2 |
| 22 | LH tail lamp | B4 | 246 | Glovebox lamp | B3 |
| 23 | Horns | B1 | 247 | Glovebox lamp switch | B3 |
| 24 | Horn push | B1 | 250 | Inertia switch | B1 |
| 26 | Direction indicator switch | B1 | 256 | Brake blocking diode | A4 |
| 27 | Direction indicator warning light | A4 | 286 | Fog rearguard lamp switch | A2 |
| 28 | RH front indicator lamp | B3 | 287 | Fog rearguard warning light | A2 |
| 29 | LH front indicator lamp | B3 | 288 | Fog rearguard lamps | B4 |
| 30 | RH rear indicator lamp | B4 | 296 | Fuel pump relay | B1 |
| 31 | LH rear indicator lamp | B4 | 297 | Brake failure warning lamp | A4 |
| 32 | Heater/fresh air motor switch | C4 | 308 | Direction indicator/hazard flasher unit | C2 |
| 33 | Heater/fresh air motor | C3 | 325 | Brake pad wear warning lamp | A4 |
| 34 | Fuel level indicator | C4 | 326 | Brake pad wear sensor | A4 |
| 35 | Fuel level gauge tank unit | A3 | 356 | Speed transducer | C2 |
| 37 | Windscreen wiper motor | B3 | 367 | Trailer indicator warning light | A3 |
| 38 | Ignition/starter switch | B1 | 368 | Spare warning light | A4 |
| 39 | Igniton coil | A1 | 381 | Knock sensor | B1 |
| 40 | Distributor | B1 | 382 | Crankshaft sensor | C1 |
| 41 | Fuel pump | B1 | 389 | Column switch lamp | B3 |
| 42 | Oil pressure switch | A4 | 393 | ECU Programmed ignition | C1 |
| 43 | Oil pressure warning lamp | A4 | 401 | Interior lamp delay unit | A3 |
| 44 | Ignition/no-charge warning lamp | A3 | 402 | Windscreen wipe/wash programmed control | B3 |
| 45 | Headlamp flash switch | B1 | 403 | Auxiliary ignition relay | A2 |
| 46 | Coolant temperature indicator | A4 | 404 | Fuel temperature switch | C1 |
| 47 | Coolant temperature thermistor | C1 | 405 | Throttle potentiometer | C1 |
| 49 | Reverse lamp switch | C4 | 406 | Stepper motor air valve | C2 |
| 50 | Reverse lamps | B4 | 407 | Airflow meter | C2 |
| 56 | Clock | A2 | 408 | Fuel injectors | C3 |
| 57 | Cigar lighter | B2 | 409 | Main relay – EFi | B1 |
| 65 | Luggage area lamp switch | B3 | 410 | Centre console illumination | B2 |
| 66 | Luggage area lamp | B3 | 413 | Fusible links | A1 |
| 75 | Automatic gearbox inhibitor switch | A1 | 414 | ECU Electronic fuel injection | C2 |
| 76 | Automatic gearbox selector lamps | B4 | 415 | Low fuel LED | A4 |
| 77 | Windscreen washer motor | B1 | 416 | High engine temperature LED | A4 |
| 82 | Switch illumination lamps | A2 | 426 | Heater vacuum solenoid | C3 |
| 95 | Tachometer | A4 | 427 | Heater vacuum solenoid switch | C3 |
| 115 | Heated rear screen switch | A2 | | | |

## Supplementary diagram connections

| | | | | | |
|---|---|---|---|---|---|
| A | Headlamp wash | B1 | D | Central locking | C4 |
| B | Radio cassette player | C1 | E | Rear wiper/washer | A2 |
| C | Electric window | C3 | F | Electric mirror | C3 |

## Key to main wiring diagram – EFi models (1986 to 1988)

| No | Description | Grid reference | No | Description | Grid reference |
|----|-------------|----------------|----|-------------|----------------|
| 1 | Alternator | A1 | 153 | Hazard warning switch | B1 |
| 3 | Battery | A1 | 164 | Fuel pump ballast resistor | B1 |
| 4 | Starter motor solenoid | A1 | 165 | Handbrake warning lamp switch | B4 |
| 5 | Starter motor | A1 | 166 | Handbrake warning lamp | A4 |
| 6 | Lighting switch – main | B1 | 174 | Starter solenoid relay | A1 |
| 7 | Headlamp dip switch | B1 | 178 | Radiator cooling fan thermostat | C3 |
| 8 | Headlamp dip beam | B3 | 179 | Radiator cooling fan motor | C3 |
| 9 | Headlamp main beam | B3 | 182 | Brake fluid level switch | B4 |
| 10 | Main beam warning lamp | A3 | 208 | Cigar lighter lamp | B2, C3 |
| 11 | RH side lamp | B3 | 209 | Blocking diode | C4 |
| 12 | LH side lamp | B3 | 210 | Panel lamp rheostat/resistor | A2 |
| 14 | Panel illumination lamps | A4 | 211 | Heater control lamp | C3 |
| 15 | Number plate illumination lamps | B4 | 231 | Headlamp relay | C1 |
| 16 | Stop-lamps | B4 | 232 | Sidelamp warning lamp | A3 |
| 17 | RH tail lamp | B4 | 246 | Glovebox lamp | B2 |
| 18 | Stop-lamp switch | C3 | 247 | Glovebox lamp switch | B2 |
| 19 | Fusebox | A2, B2, C2 | 250 | Inertia switch | B1 |
| 20 | Front interior lamp | C4 | 256 | Brake blocking diode | A4 |
| 21 | Interior lamp door switches | C4 | 267 | Headlamp wash motor | B1 |
| 22 | LH tail lamp | B4 | 286 | Fog rearguard lamp switch | A2 |
| 23 | Horns | B1 | 287 | Fog rearguard warning light | A2 |
| 24 | Horn push | B1 | 288 | Fog rearguard lamps | B4 |
| 26 | Direction indicator switch | B1 | 296 | Fuel pump relay | B1 |
| 27 | Direction indicator warning light | A3 | 297 | Brake failure warning lamp | A4 |
| 28 | RH front indicator lamp | B3 | 300 | Ignition switch relay | C1 |
| 29 | LH front indicator lamp | B3 | 308 | Direction indicator/hazard flasher unit | C2 |
| 30 | RH rear indicator lamp | B4 | 313 | Headlamp wash relay | B1 |
| 31 | LH rear indicator lamp | B4 | 325 | Brake pad wear warning lamp | A4 |
| 32 | Heater/fresh air motor switch | C3 | 326 | Brake pad wear sensor | A4 |
| 33 | Heater/fresh air motor | C3 | 327 | In-line resistance – 6.8k ohm | C1 |
| 34 | Fuel level indicator | A4 | 339 | Automatic gearbox quadrant illumination | C4 |
| 35 | Fuel level gauge tank unit | A2 | 356 | Speed transducer | C2 |
| 37 | Windscreen wiper motor | B3 | 367 | Trailer indicator warning light | A3 |
| 38 | Ignition/starter switch | A1 | 368 | Spare warning light | A4 |
| 39 | Ignition coil | A1 | 381 | Knock sensor | C1 |
| 40 | Distributor | A1 | 382 | Crankshaft sensor | C1 |
| 41 | Fuel pump | B1 | 389 | Column switch lamp | B2 |
| 42 | Oil pressure switch | A4 | 393 | ECU Programmed ignition | C1 |
| 43 | Oil pressure warning lamp | A4 | 401 | Interior lamp delay unit | C4 |
| 44 | Ignition/no charge warning lamp | A3 | 402 | Windscreen wipe/wash programmed control | B3 |
| 45 | Headlamp flash switch | B1 | 403 | Auxiliary ignition relay | A1 |
| 46 | Coolant temperature indicator | A4 | 404 | Fuel temperature switch | C1 |
| 47 | Coolant temperature thermistor | C1 | 405 | Throttle potentiometer | C1 |
| 49 | Reverse lamp switch | C3 | 406 | Stepper motor air valve | C2 |
| 50 | Reverse lamps | B4 | 407 | Airflow meter | C2 |
| 56 | Clock | A2 | 408 | Fuel injectors | C3 |
| 57 | Cigar lighter | B2, C3 | 409 | Main relay – EFi | B1 |
| 65 | Luggage area lamp switch | C4 | 410 | Centre console illumination | B2 |
| 66 | Luggage area lamp | C4 | 413 | Fusible links | A1 |
| 77 | Windscreen washer motor | B2 | 414 | ECU Electronic fuel injection | C2 |
| 82 | Switch illumination lamps | A2 | 415 | Low fuel LED | A3 |
| 95 | Tachometer | A3 | 416 | High engine temperature LED | A3 |
| 110 | Side repeater flasher | B3 | 488 | Heated rear screen relay and timer | A2 |
| 115 | Heated rear screen switch | A2 | | | |
| 116 | Heated rear screen | A2 | | **Supplementary diagram connections** | |
| 118 | Windscreen washer/wiper switch | B2 | A | Radio cassette player | A2 |
| 131 | Combined automatic gearbox inhibitor and reverse light switch | A1, C3 | B | Electric windows | C2 |
| | | | C | Central locking | B4 |
| 150 | Heated rear screen warning lamp | A2 | D | Rear wiper/washer | A2 |
| 152 | Hazard warning lamp | B1 | E | Electric mirror | C2 |

**Main wiring diagram – EFi models (1986 to 1988)**

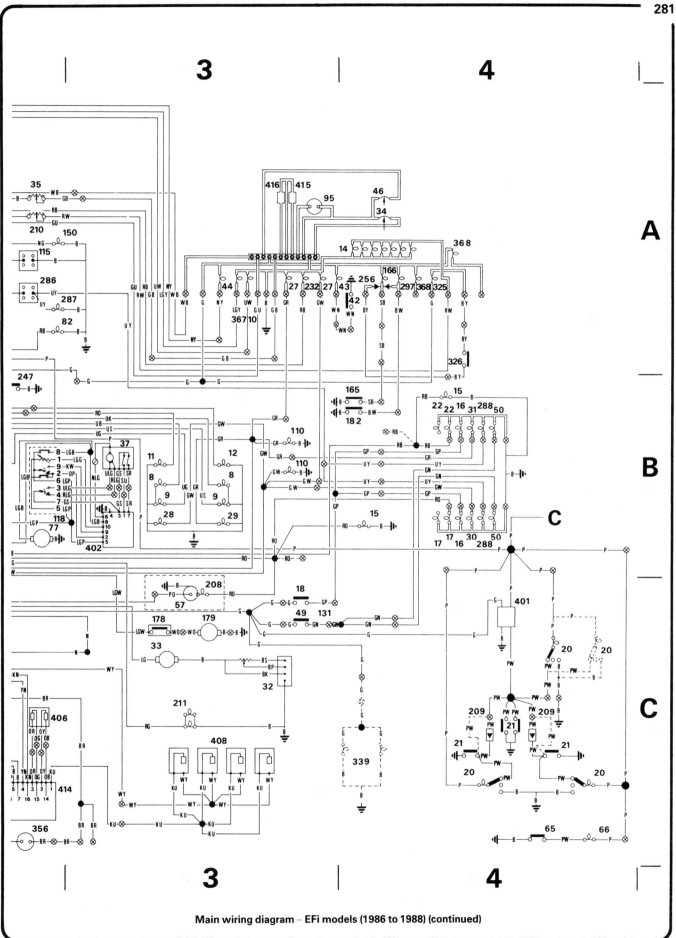

**Main wiring diagram – EFi models (1986 to 1988) (continued)**

**Main wiring diagram – MG Turbo models (1986 to 1988)**

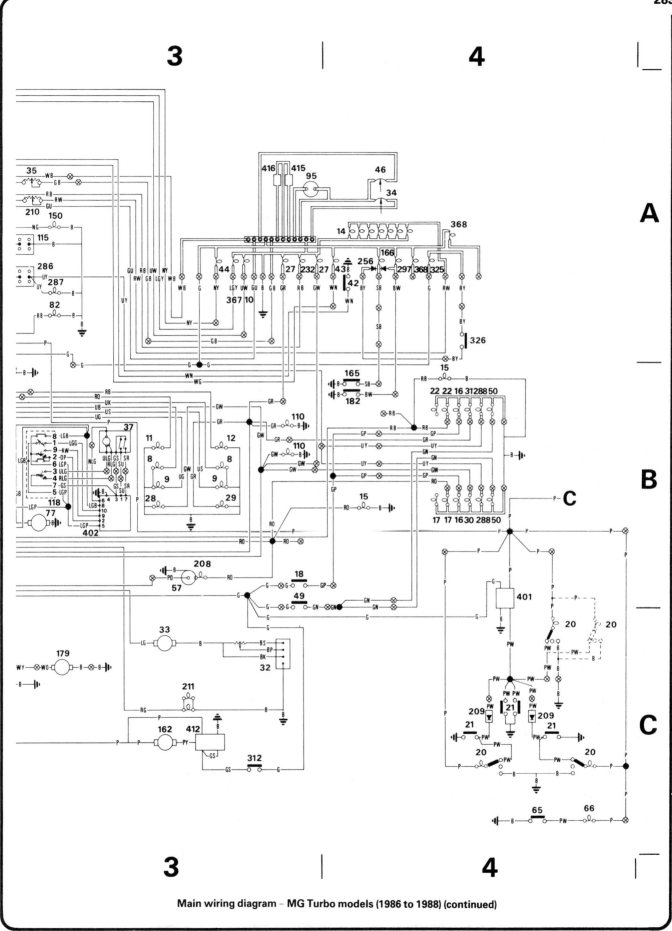

**Main wiring diagram – MG Turbo models (1986 to 1988) (continued)**

## Key to main wiring diagram – MG Turbo models (1986 to 1988)

| No. | Description | Grid reference | No. | Description | Grid reference |
|---|---|---|---|---|---|
| 1 | Alternator | A1 | 153 | Hazard warning switch | B1 |
| 3 | Battery | A1 | 162 | Carburettor cooling fan motor | C3 |
| 4 | Starter motor solenoid | A1 | 165 | Handbrake warning lamp switch | B4 |
| 5 | Starter motor | A1 | 166 | Handbrake warning lamp | A4 |
| 6 | Lighting switch – main | B1 | 174 | Starter solenoid relay | A1 |
| 7 | Headlamp dip switch | B1 | 177 | Radiator cooling fan relay | C2 |
| 8 | Headlamp dip beam | B3 | 178 | Radiator cooling fan thermostat | C2 |
| 9 | Headlamp main beam | B3 | 179 | Radiator cooling fan motor | C3 |
| 10 | Main beam warning lamp | A3 | 182 | Brake fluid level switch | B4 |
| 11 | RH side lamp | B3 | 208 | Cigar lighter illumination lamp | B2, B3 |
| 12 | LH side lamp | B3 | 209 | Blocking diode | C4 |
| 14 | Panel illumination lamps | A4 | 210 | Panel illumination lamp rheostat | A3 |
| 15 | Number plate illumination lamps | B4 | 211 | Heater control illumination | C3 |
| 16 | Stop-lamps | B4 | 231 | Headlamp relay | B1 |
| 17 | RH tail lamp | B4 | 232 | Side lamp warning lamp | A3 |
| 18 | Stop-lamp switch | B3 | 246 | Glovebox illumination lamp | B2 |
| 19 | Fusebox | A2, B1, B2 | 247 | Glovebox illumination switch | B2 |
| 20 | Interior lamps | C4 | 250 | Inertia switch | B1 |
| 21 | Interior lamp door switches | C4 | 256 | Brake warning blocking diode | A4 |
| 22 | LH tail lamp | B4 | 265a | Intake air temperature sensor | C1 |
| 23 | Horns | B1 | 265b | Ambient air temperature sensor | C1 |
| 24 | Horn push | B1 | 267 | Headlamp wash motor | B1 |
| 26 | Direction indicator switch | B1 | 286 | Fog rearguard lamp switch | A3 |
| 27 | Direction indicator warning lamp(s) | A3 | 287 | Fog rearguard warning lamp | A3 |
| 28 | RH front direction indicator lamp | B3 | 288 | Fog rearguard lamp(s) | B4 |
| 29 | LH front direction indicator lamp | B3 | 296 | Fuel pump relay | B1 |
| 30 | RH rear direction indicator lamp | B4 | 297 | Brake failure warning lamp | A4 |
| 31 | LH rear direction indicator lamp | B4 | 300 | Ignition switch relay | B1 |
| 32 | Heater/fresh air motor switch | C3 | 308 | Direction indicator/hazard flasher unit | C2 |
| 33 | Heater/fresh air motor | C3 | 312 | Carburettor cooling fan thermostatic switch | C3 |
| 34 | Fuel level gauge | A4 | 313 | Headlamp wash relay | B1 |
| 35 | Fuel level gauge tank unit | A3 | 325 | Brake pad wear warning lamp | A4 |
| 37 | Windscreen wiper motor | B3 | 326 | Brake pad wear sensor | A5 |
| 38 | Ignition/starter switch | A1 | 347 | Fuel ECU | C2 |
| 39 | Ignition coil | A1 | 355 | Accelerator pedal switch | C1 |
| 40 | Distributor | A1 | 359 | Stepper motor | C2 |
| 41 | Fuel pump | B1 | 363 | Carburettor vent valve | C1 |
| 42 | Oil pressure switch | A4 | 367 | Trailer indicator warning light | A3 |
| 43 | Oil pressure warning lamp | A4 | 368 | Spare warning lamp | A4 |
| 44 | Ignition/no charge warning lamp | A3 | 381 | Knock sensor | B1 |
| 45 | Headlamp flash switch | B1 | 382 | Crankshaft sensor | C1 |
| 46 | Coolant temperature gauge | A4 | 389 | Column switch illumination | B2 |
| 47 | Coolant temperature thermistor | C1 | 393 | Programmed ignition ECU | C1 |
| 49 | Reverse lamp switch | B3 | 401 | Interior lamp delay unit | B4 |
| 50 | Reverse lamps | B4 | 402 | Windscreen wipe/wash programmed control | B3 |
| 56 | Clock | A2 | 403 | Auxiliary ignition relay | A1 |
| 57 | Cigar lighter | B2, B3 | 410 | Centre console illumination | B2 |
| 65 | Luggage area lamp switch | C4 | 412 | Carburettor cooling fan delay unit | C3 |
| 66 | Luggage area lamp | C4 | 413 | Fusible links | A1 |
| 77 | Windscreen washer motor | B3 | 415 | Low fuel LED | A3 |
| 82 | Switch illumination lamp(s) | A3 | 416 | High engine temperature LED | A3 |
| 95 | Tachometer | A3 | 488 | Heated rear screen timer and relay | A2 |
| 110 | Indicator repeater lamps | B3 | | | |
| 115 | Heated rear screen switch | A3 | | **Supplementary diagram connections** | |
| 116 | Heated rear screen | A2 | A | Radio cassette player | C1 |
| 118 | Combined windscreen wiper/washer switch | B3 | B | Electric windows | C2 |
| 150 | Heated rear screen warning lamp | A3 | C | Central locking | B4 |
| 152 | Hazard warning lamp | B1 | D | Electric mirrors | B2 |

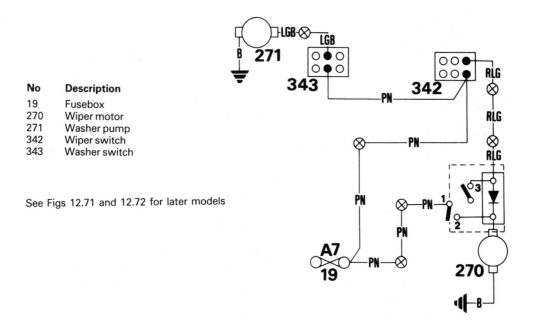

| No | Description |
|----|-------------|
| 19 | Fusebox |
| 270 | Wiper motor |
| 271 | Washer pump |
| 342 | Wiper switch |
| 343 | Washer switch |

See Figs 12.71 and 12.72 for later models

**Supplementary wiring diagram – rear screen wash/wipe (up to 1986)**

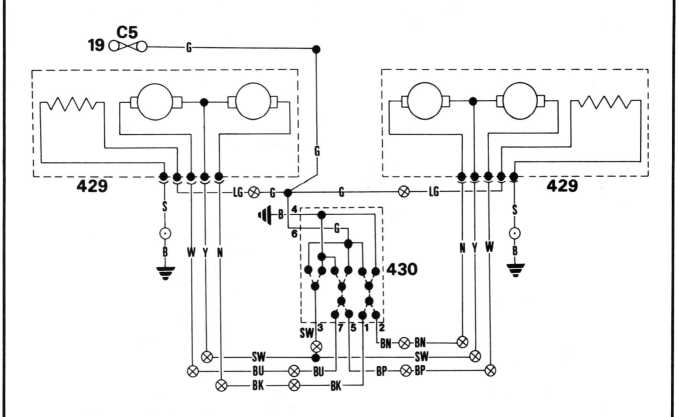

**Supplementary wiring diagram – electric door mirrors (up to 1986)**

| No | Description |
|----|-------------|
| 19 | Fusebox |
| 429 | Mirror |
| 430 | Switch |

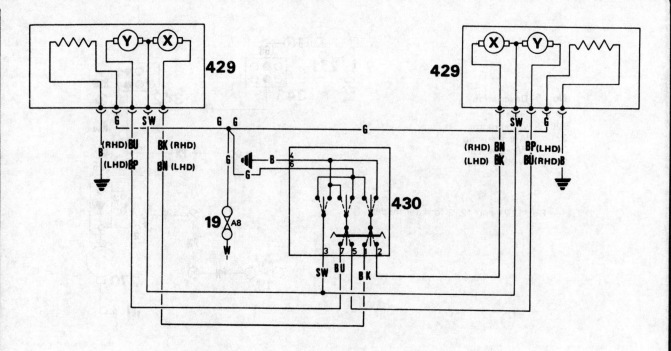

**Supplementary wiring diagram – electric door mirrors (1986 on)**

| 19 | Fusebox | 430 | Switch | Y | Horizontal movement motor |
| 429 | Mirror | X | Vertical movement motor | | |

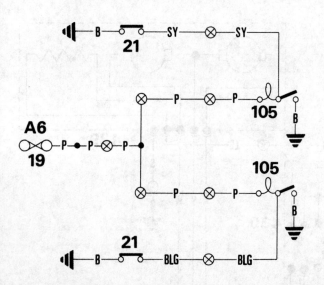

**Supplementary wiring diagram – rear courtesy lamps**

19 Fusebox
21 Door switches
105 Courtesy lamps

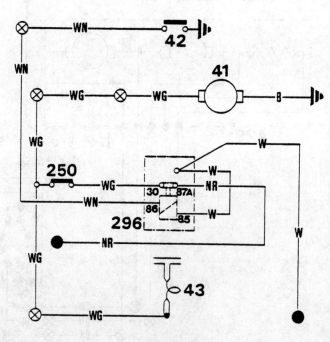

**Supplementary wiring diagram – electric fuel pump**

41 Fuel pump
42 Oil pressure switch
43 Oil pressure warning lamp
250 Fuel cut-off switch
296 Fuel pump relay

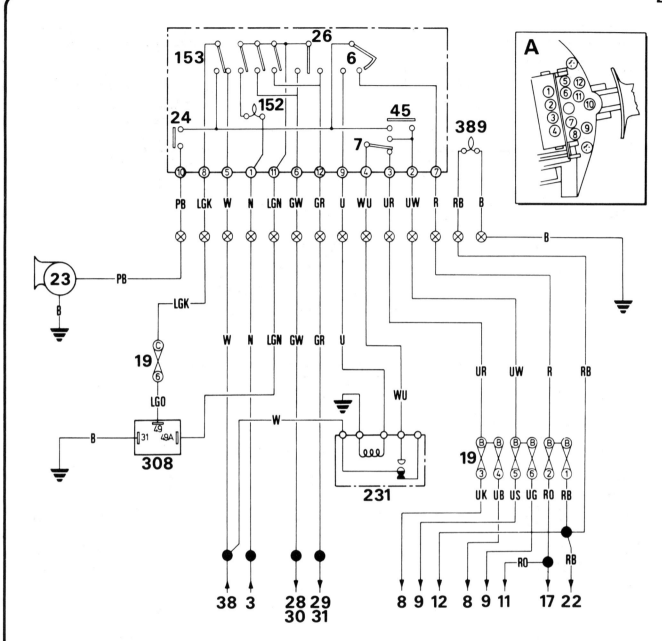

**Supplementary wiring diagram – direction indicator and lighting switch circuit**

| | | | |
|---|---|---|---|
| 3 | Battery | 28 | RH front direction indicator lamp |
| 6 | Lighting switch – main | 29 | LH front direction indicator lamp |
| 7 | Headlamp dip switch | 30 | RH rear direction indicator lamp |
| 8 | Headlamp dip beam | 31 | LH rear direction indicator lamp |
| 9 | Headlamp main beam | 38 | Ignition switch |
| 11 | RH sidelamp | 45 | Headlamp flasher switch |
| 12 | LH sidelamp | 152 | Hazard warning light |
| 17 | RH tail lamp | 153 | Hazard warning switch |
| 19 | Fusebox | 231 | Headlamps relay |
| 22 | LH tail lamp | 308 | Flasher unit |
| 23 | Horn | 389 | Column switch lamp – fibre optic |
| 24 | Horn push switch | A | Switch connectors |
| 26 | Direction indicator switch | | |

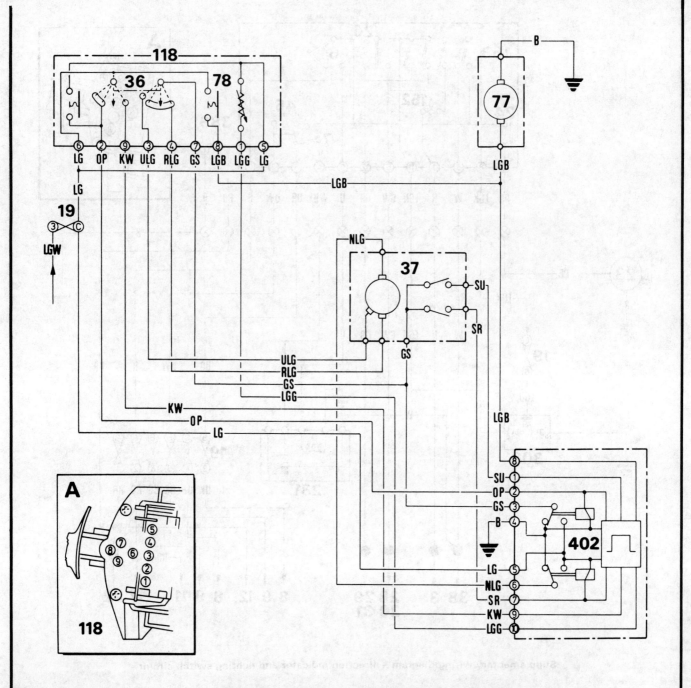

**Supplementary wiring diagram – wash/wipe with variable intermittent delay circuit**

19 Fusebox
36 Windscreen wiper switch and flick wipe
37 Windscreen wiper motor
77 Screen washer pump
78 Screen washer switch with intermittent wiper control

118 Combined windscreen wipe/wash and intermittent wipe switch
402 Windscreen wipe/wash programmed control
A Switch connectors – wipe/wash and intermittent wipe only

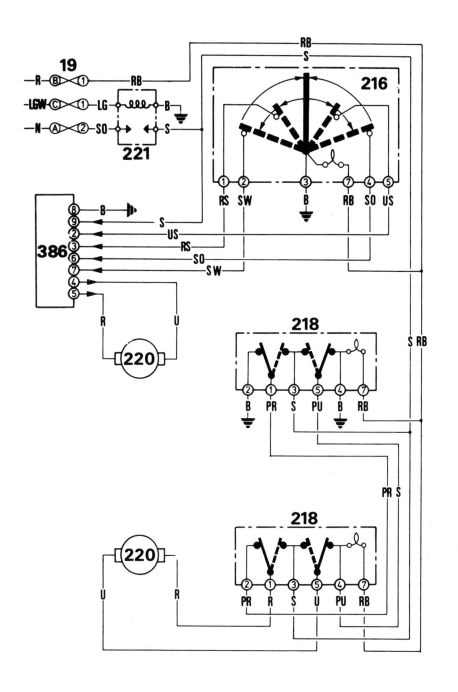

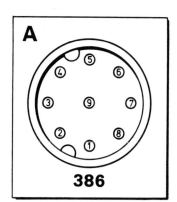

**Supplementary wiring diagram – front door electric windows (up to 1986)**

19   Fusebox
216  Driver's switch
218  Passenger's switch
220  Window lift motor

221  Relay
386  One-touch lift unit
A    Multi-plug – one-touch unit

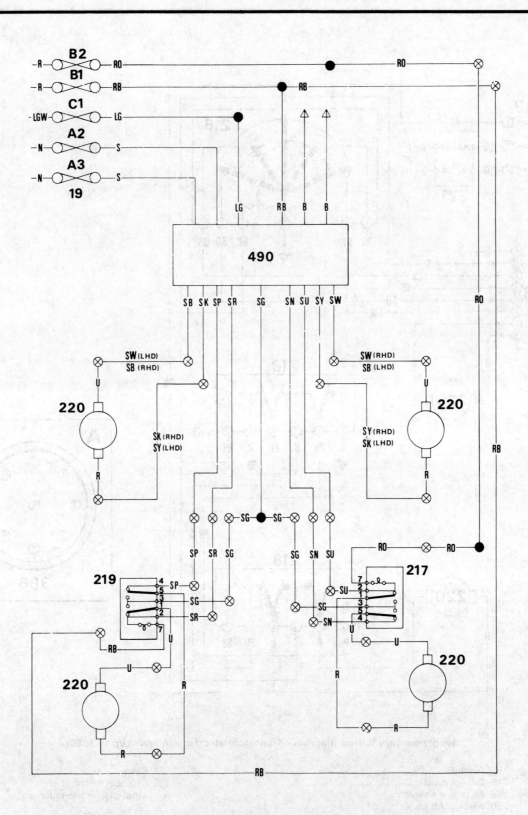

**Supplementary wiring diagram – front and rear electric windows (1986 on)**

| | | |
|---|---|---|
| 19 Fusebox | 219 LH rear switches | 490 Control unit |
| 217 RH rear switches | 220 Lift motors | |

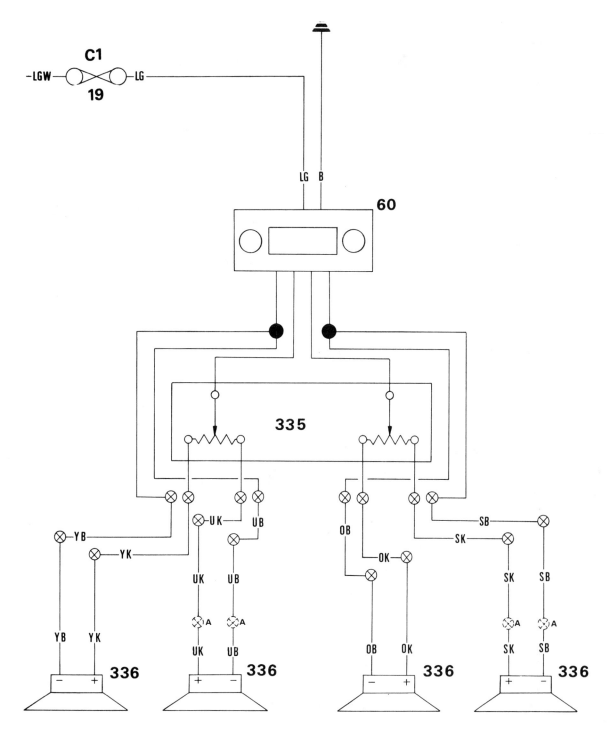

**Supplementary wiring diagram – four-speaker manually-tuned radio/cassette player (1986 on)**

| | | |
|---|---|---|
| 19  Fusebox | 335  Balance control | A    Connectors (Estate models only) |
| 60  Radio/cassette player | 336  Speakers | |

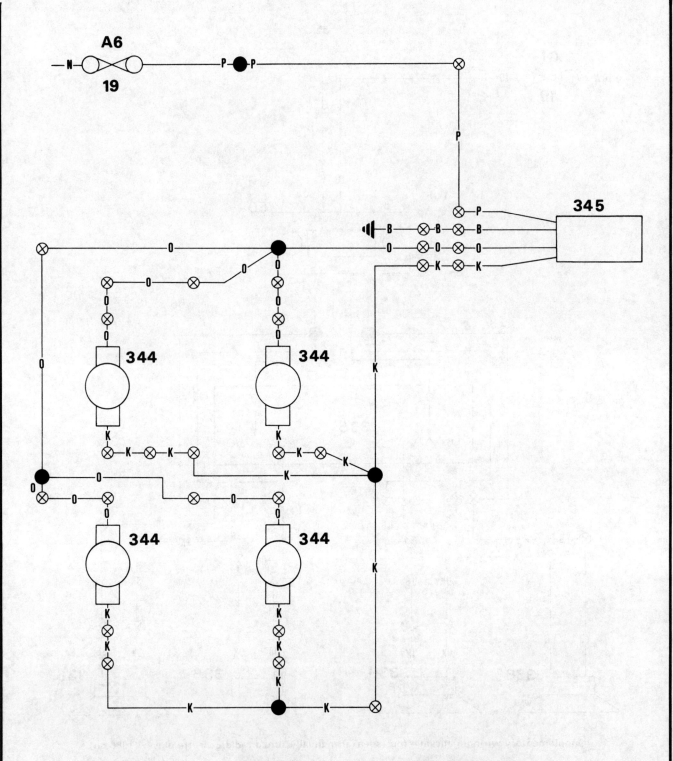

**Supplementary wiring diagram – central door locking system (1986 on)**

19  Fusebox                344  Door lock motor                345  Driver's door lock control unit

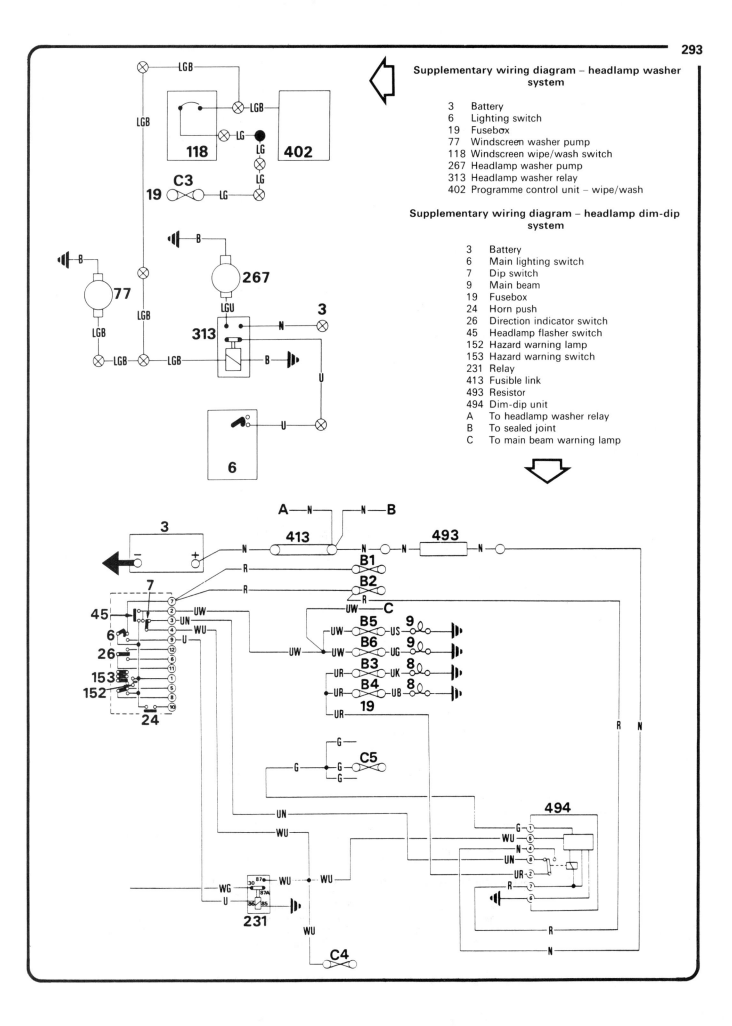

**Supplementary wiring diagram – headlamp washer system**

3 Battery
6 Lighting switch
19 Fusebox
77 Windscreen washer pump
118 Windscreen wipe/wash switch
267 Headlamp washer pump
313 Headlamp washer relay
402 Programme control unit – wipe/wash

**Supplementary wiring diagram – headlamp dim-dip system**

3 Battery
6 Main lighting switch
7 Dip switch
9 Main beam
19 Fusebox
24 Horn push
26 Direction indicator switch
45 Headlamp flasher switch
152 Hazard warning lamp
153 Hazard warning switch
231 Relay
413 Fusible link
493 Resistor
494 Dim-dip unit
A To headlamp washer relay
B To sealed joint
C To main beam warning lamp

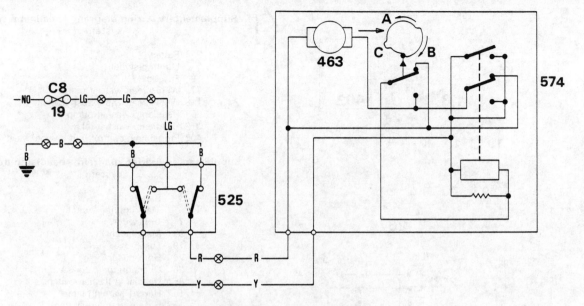

**Supplementary wiring diagram – electrically-operated sunroof**

| | | | | | |
|---|---|---|---|---|---|
| 19 | Fusebox | 525 | Sunroof switch | A | Open |
| 463 | Motor | 574 | Changeover relay | B | Close |
| | | | | C | Tilt |

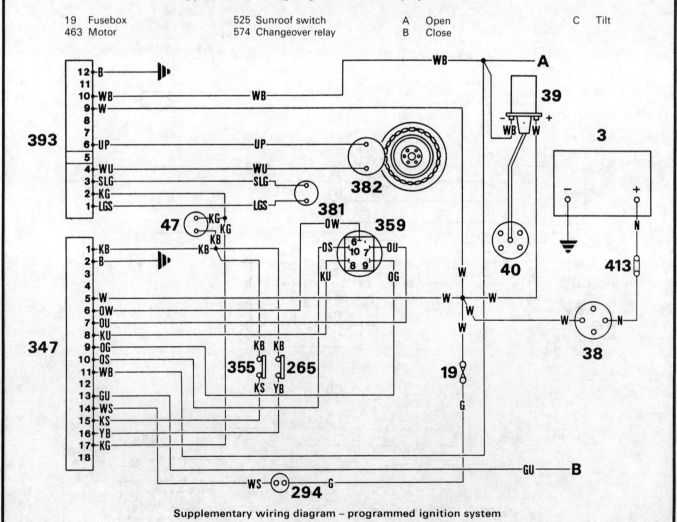

**Supplementary wiring diagram – programmed ignition system**

| | | | | | | | |
|---|---|---|---|---|---|---|---|
| 3 | Battery | 47 | Coolant temperature thermistor | 355 | Throttle switch | 393 | Programmed ignition ECU |
| 19 | Fusebox | | | 359 | Stepper motor | 413 | Fusible link |
| 38 | Ignition switch | 265 | Ambient temperature sensor | 381 | Knock sensor | A | To tachometer |
| 39 | Ignition coil | | | 382 | Crankshaft sensor and reluctor disc | B | To coolant temperature gauge |
| 40 | Distributor cap | 294 | Fuel cut-off solenoid | | | | |
| | | 347 | Fuel ECU | | | | |

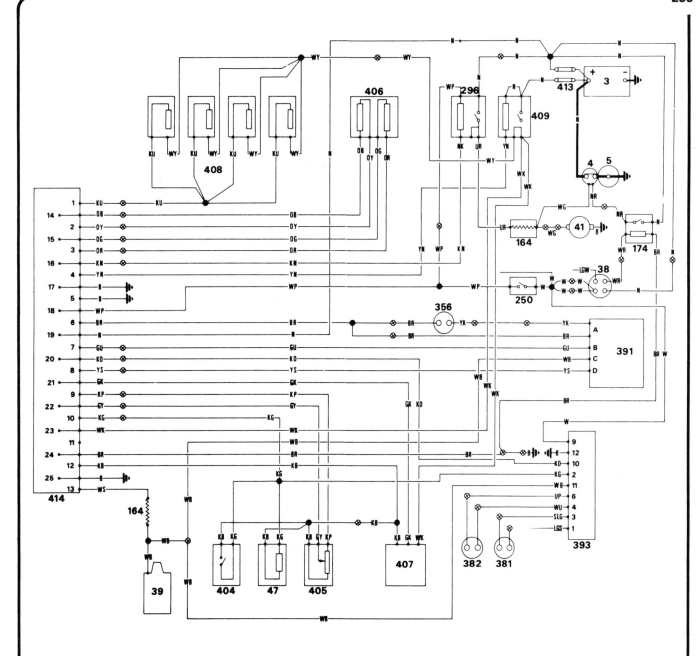

**Supplementary wiring diagram – electronic fuel injection system**

| | | | |
|---|---|---|---|
| 3 | Battery | 391 | LCD solid state instruments |
| 4 | Starter motor solenoid | A | Speedometer |
| 5 | Starter motor | B | Fuel gauge |
| 38 | Ignition/starter switch | C | Tachometer |
| 39 | Ignition coil | D | Trip computer |
| 41 | Fuel pump | 393 | Ignition ECU |
| 47 | Coolant temperature thermistor | 404 | Fuel temperature switch |
| 164 | Ballast solenoid relay | 405 | Throttle potentiometer |
| 174 | Starter solenoid relay | 406 | Air valve stepper motor |
| 250 | Inertia switch | 407 | Airflow meter |
| 296 | Fuel pump relay | 408 | Fuel injectors |
| 356 | Speed transducer | 409 | Main relay |
| 381 | Knock sensor | 413 | Fusible links |
| 382 | Crankshaft sensor | 414 | Fuel ECU |

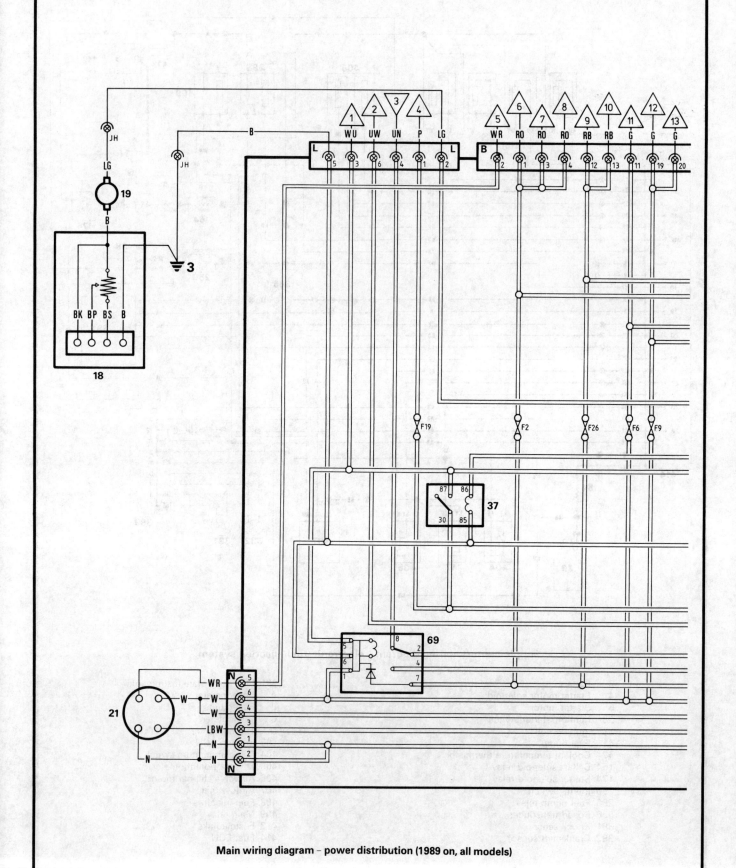

Main wiring diagram – power distribution (1989 on, all models)

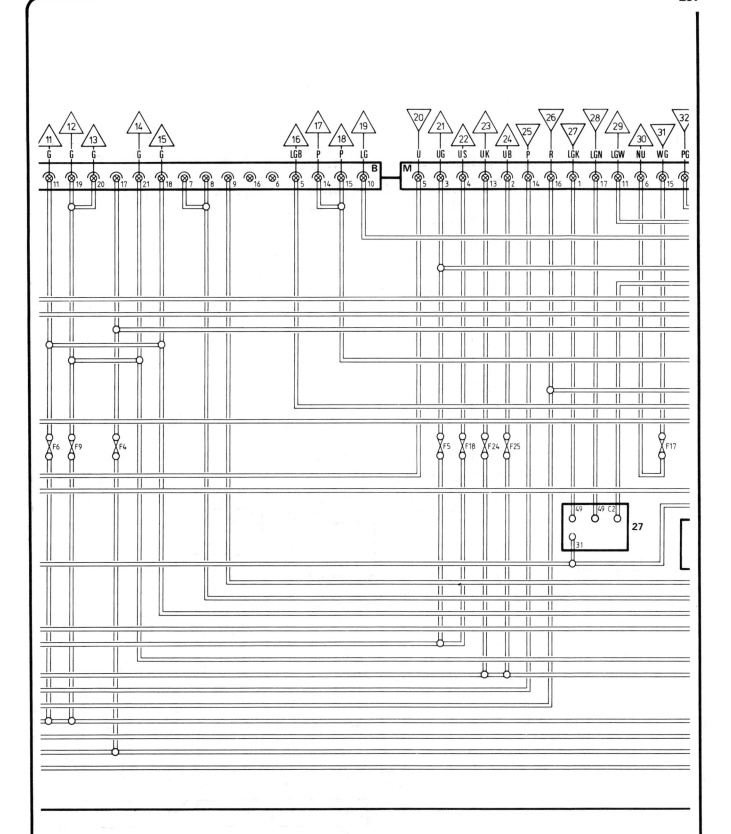

**Main wiring diagram – power distribution (1989 on, all models) (continued)**

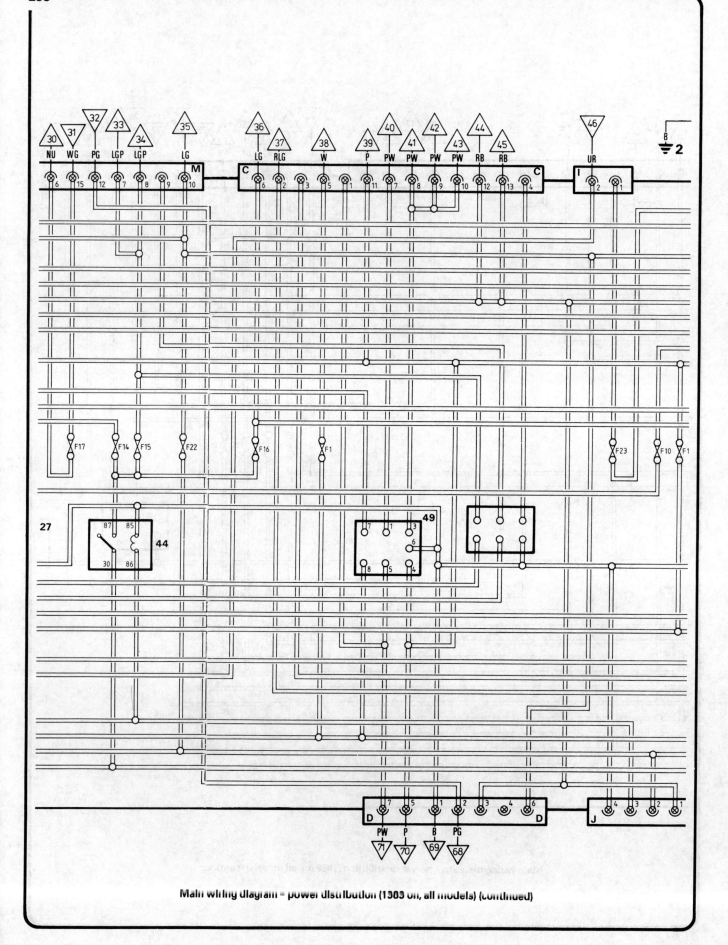

Main wiring diagram - power distribution (1983 on, all models) (continued)

**Main wiring diagram – power distribution (1989 on, all models) (continued)**

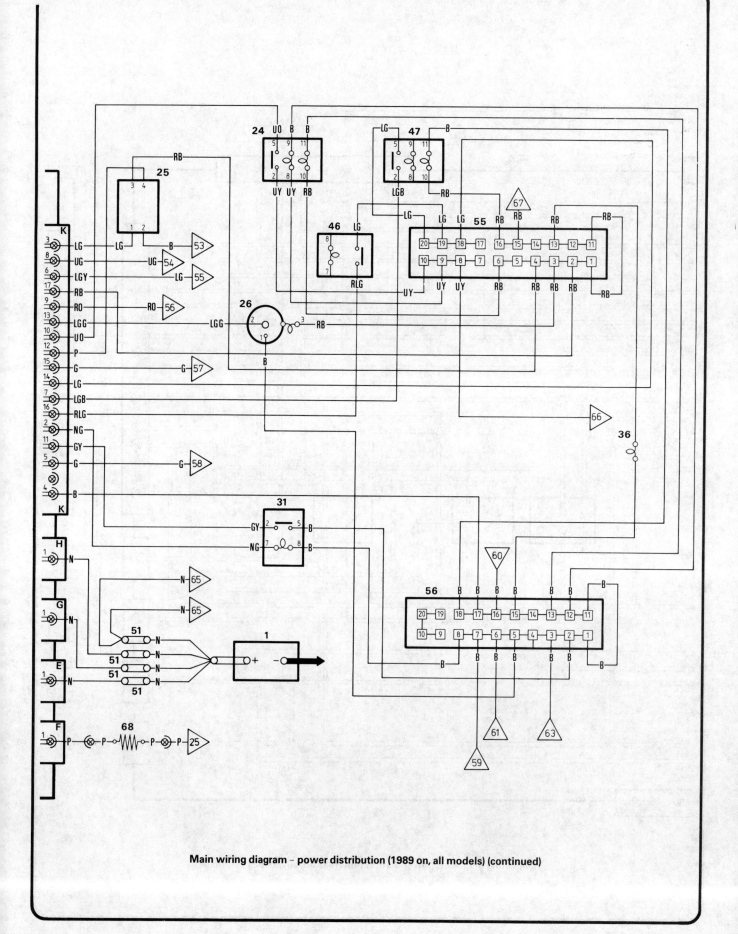

**Main wiring diagram – power distribution (1989 on, all models) (continued)**

## Key to wiring diagram – power distribution (1989 on, all models)

| No | Description |
|----|-------------|
| 1 | Battery |
| 11 | Fusebox |
| V18 | Heater or fresh air motor switch(s) or rheostat |
| 19 | Heater motor |
| 21 | Ignition/start switch or diesel master switch |
| 24 | Foglamp switch |
| 25 | Clock |
| 26 | Cigar lighter |
| 27 | Flasher relay |
| 31 | Heated rear screen switch |
| 32 | Heated rear screen element |
| 35 | Radiator cooling fan motor |
| 36 | Heater control illumination |

| No | Description |
|----|-------------|
| 37 | Head lamp relay |
| 38 | Heated rear screen timer |
| 44 | Ignition switch relay |
| 45 | Auxiliary circuits relay |
| 46 | Rear screen wiper switch (Estate only) |
| 47 | Rear screen wash switch (Estate only) |
| 49 | Interior lamp delay unit |
| 51 | Fusible links |
| 55 | Facia lamp header |
| 56 | Facia earth header |
| 68 | Dim-dip resistor |
| 69 | Dim-dip unit |

## Connections to other circuits

| No | Description |
|----|-------------|
| 1 | Exterior lamps/wipers |
| 2 | Exterior lamps/wipers |
| 3 | Exterior lamps/wipers – RHD only |
| 4 | Exterior lamps/wipers |
| 5 | Engine – 2.0 litre carburettor |
| 6 | Exterior lamps/wipers |
| 7 | Radio/cassette |
| 8 | Exterior lamps/wipers |
| 9 | Exterior lamps/wipers |
| 10 | Exterior lamps/wipers |
| 11 | Mirrors |
| 12, 13 and 14 | Exterior lamps/wipers |
| 15 | Mirrors |
| 16 | Exterior lamps/wipers |
| 17 | Central door locking |
| 18 | Exterior lamps/wipers |
| 19 | All window lift circuits |
| 20, 21, 22, 23 and 24 | Exterior lamps/wipers |
| 25 | Power distribution – Dim-dip resistor |
| 26, 27, 28 and 29 | Exterior lamps/wipers |
| 30 and 31 | Engine 2.0 Carburettor, EFi and Turbo |
| 32 | All window lift circuits |
| 33 and 34 | Exterior lamps/wipers |

| No | Description |
|----|-------------|
| 35 | Radio/cassette |
| 36 and 37 | Exterior lamps/wipers – Estate only |
| 38 | Engine 2.0 Carburettor, EFi and Turbo |
| 39 | Radio/cassette |
| 40, 41, 42 and 43 | Interior lamps |
| 44 | All window lift circuits |
| 45 | Exterior lamps/wipers |
| 46 | Exterior lamps/wipers – LHD only |
| 49 and 50 | All window lift circuits |
| 51 | Two-door window lift – Turbo only |
| 52 | Four-door window lift/two-door window lift – Turbo only |
| 53 | Exterior lamps/wipers |
| 54, 55, 56 and 57 | Instruments |
| 58 | Mirrors |
| 59 | Instruments |
| 60 | Mirrors |
| 61 | Four-door window lift |
| 63 | Exterior lamps/wipers |
| 65 | Engine 2.0 Carburettor, EFi and Turbo |
| 66 | Exterior lamps/wipers |
| 67 | Instruments |
| 68 | All window lift circuits |
| 69, 70 and 71 | Interior lamp circuit |

**Wiring diagram – exterior lamps/wipers (Saloon, 1989 on)**

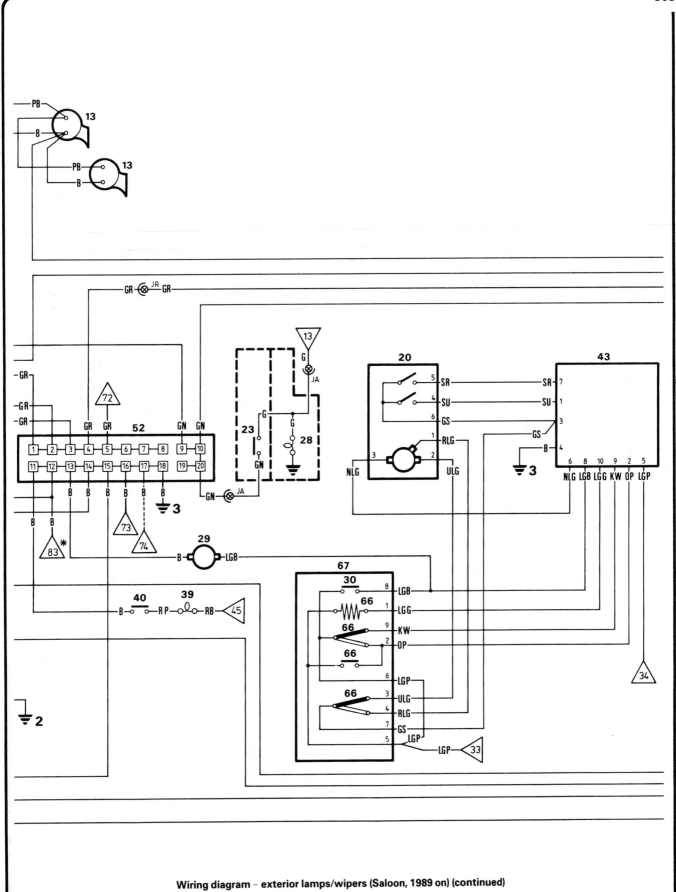

**Wiring diagram – exterior lamps/wipers (Saloon, 1989 on) (continued)**

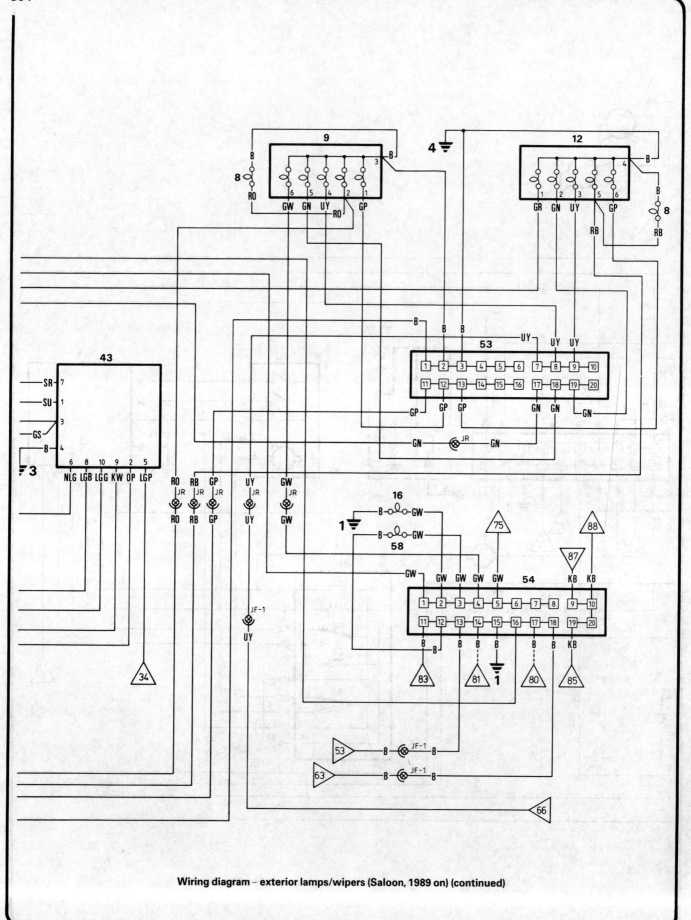

**Wiring diagram – exterior lamps/wipers (Saloon, 1989 on) (continued)**

## Key to wiring diagram – exterior lamps/wiper (Saloon, 1989-on)

| No | Description | No | Description |
|----|-------------|----|-------------|
| 2 | Main lighting switch | 30 | Windscreen washer switch |
| 3 | Headlamp dip switch | 33 | Hazard warning lamp |
| 4 | Dip beam | 34 | Hazard warning switch |
| 5 | Main beam | 39 | Glovebox illumination lamp |
| 6 | Sidelamp – RH | 40 | Glovebox illumination switch |
| 7 | Sidelamp – LH | 43 | Windscreen wiper delay unit |
| 8 | Number plate illumination lamp(s) | 48 | Radiator fan switch |
| 9 | Tail lamp – RH | 52 | Passenger header |
| 10 | Stop-lamp switch | 53 | Rear lamp header |
| 12 | Tail lamp – LH | 54 | Fusebox header |
| 13 | Horn(s) | 57 | LH repeater lamp |
| 14 | Horn push | 58 | RH repeater lamp |
| 15 | Direction indicator switch | 66 | Wash/wipe switch |
| 16 | RH front indicator lamp | 67 | Column switch |
| 17 | LH front indicator lamp | 70 | Boot lamp |
| 20 | Windscreen wiper motor | 71 | Boot lamp switch |
| 22 | Headlamp flash switch | 75 | Reversing lamp switch (all models except 1.6 automatic) |
| 23 | Reverse lamp switch in centre console (1.6 automatic only) | 186 | LH headlamp |
| 28 | Automatic gearbox selector indicator lamp | 187 | RH headlamp |
| 29 | Windscreen washer pump | | |

### Connections to other circuits

| | | | |
|----|-------------|----|-------------|
| 1 to 71 | Power distribution circuit | 80 | Central locking |
| 72 | Instruments | 81 | Mirrors |
| 73 | Radio/cassette | 83, 85, | |
| 74 | Mirrors | 87 and | |
| 75 | Instruments | 88 | Engine |

## Key to Wiring diagram – exterior lamps/wipers (Estate models, 1989 on)

| No | Description | No | Description |
|----|-------------|----|-------------|
| 2 | Main lighting switch | 33 | Hazard warning lamp |
| 3 | Headlamp switch | 34 | Hazard warning switch |
| 4 | Dip beam | 39 | Glovebox illumination lamp |
| 5 | Main beam | 40 | Glovebox illumination switch |
| 6 | Sidelamp – RH | 41 | Rear screen wiper motor |
| 7 | Sidelamp – LH | 43 | Windscreen wiper delay unit |
| 8 | Number plate illumination lamp(s) | 48 | Radiator cooling fan switch |
| 9 | Tail lamp – RH | 52 | Passenger header |
| 10 | Stop-lamp switch | 53 | Rear lamp cluster |
| 12 | Tail lamp – LH | 54 | Fusebox header |
| 13 | Horns | 57 | LH flasher |
| 14 | Horn push | 58 | RH flasher |
| 15 | Direction indicator switch | 63 | Loadspace lamp |
| 16 | RH front direction indicator lamp | 64 | Latch switch |
| 17 | LH front direction indicator lamp | 65 | Rear screen washer motor |
| 20 | Windscreen wiper motor | 66 | Wash/wipe switch |
| 22 | Headlamp flash switch | 67 | Column switch |
| 23 | Reverse lamp switch | 75 | Reversing lamp switch |
| 28 | Automatic transmission selector indicator lamp | 186 | LH headlamp |
| 29 | Windscreen washer pump | 187 | RH headlamp |
| 30 | Windscreen washer switch | | |

### Connections to other circuits – as for Saloon models

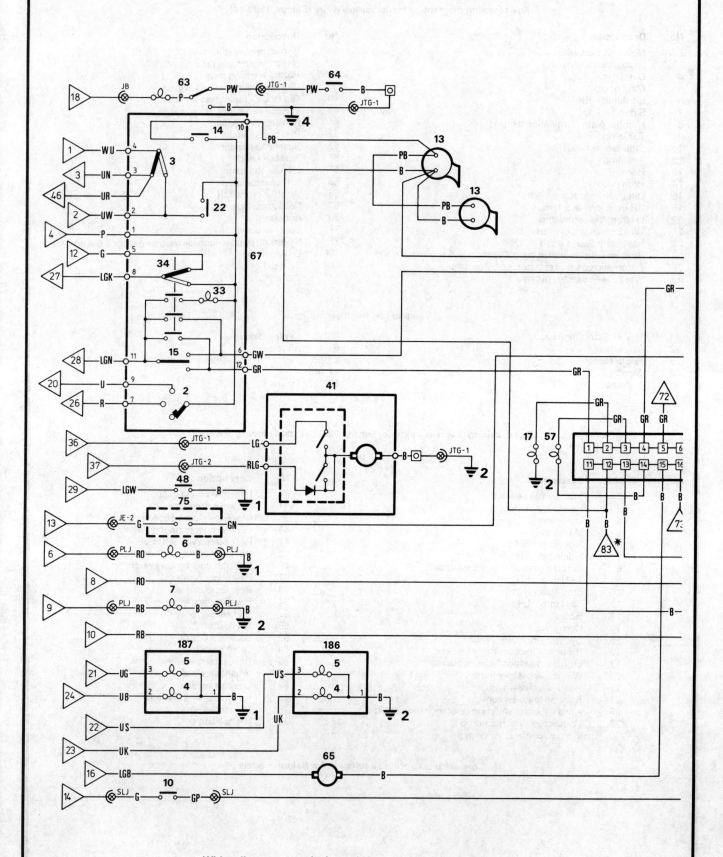

**Wiring diagram – exterior lamps/wipers (Estate models, 1989 on)**

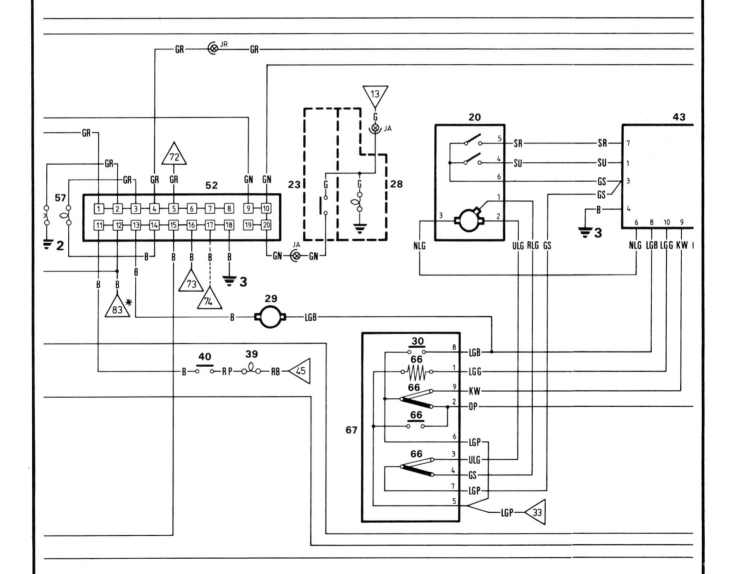

**Wiring diagram – exterior lamps/wipers (Estate models, 1989 on) (continued)**

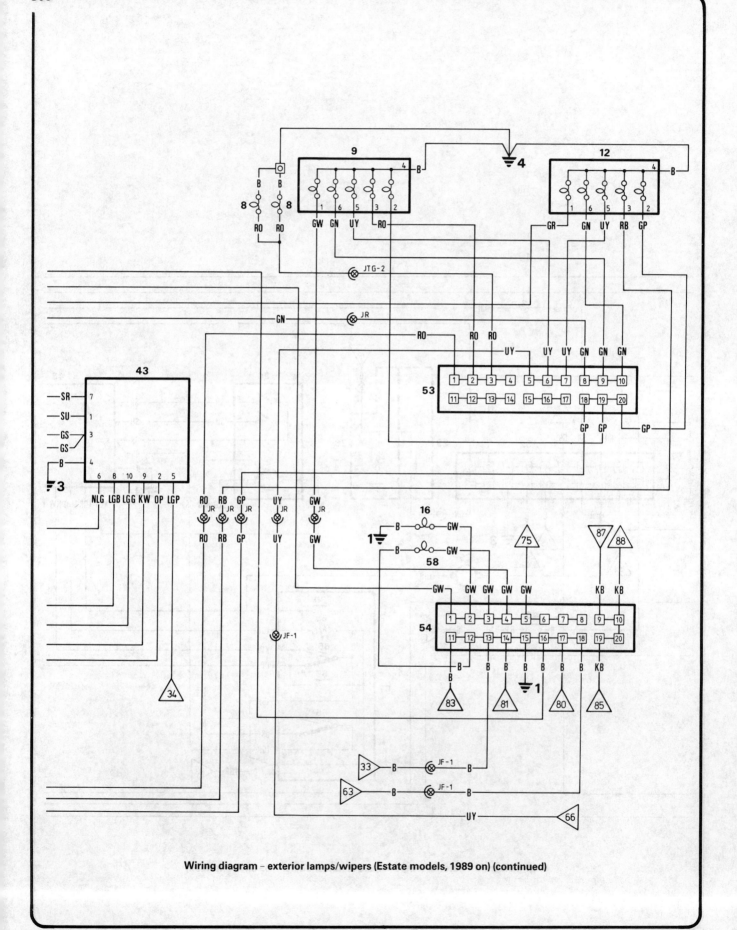

**Wiring diagram – exterior lamps/wipers (Estate models, 1989 on) (continued)**

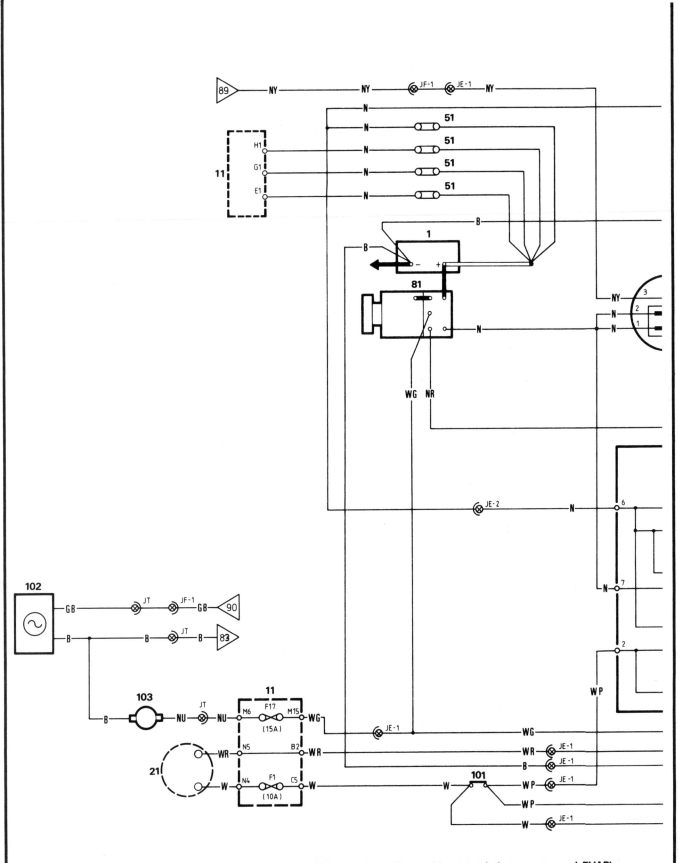

Supplementary wiring diagram – engine circuit (EFi models, 1989-on without catalytic converter and EVAP)

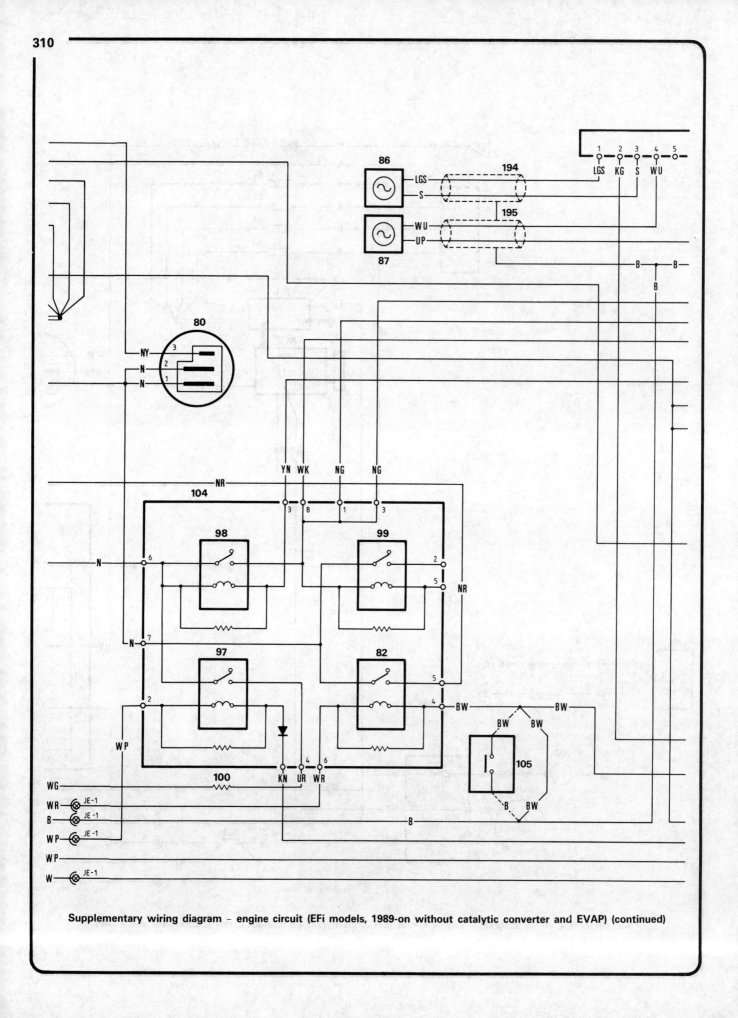

**Supplementary wiring diagram – engine circuit (EFi models, 1989-on without catalytic converter and EVAP) (continued)**

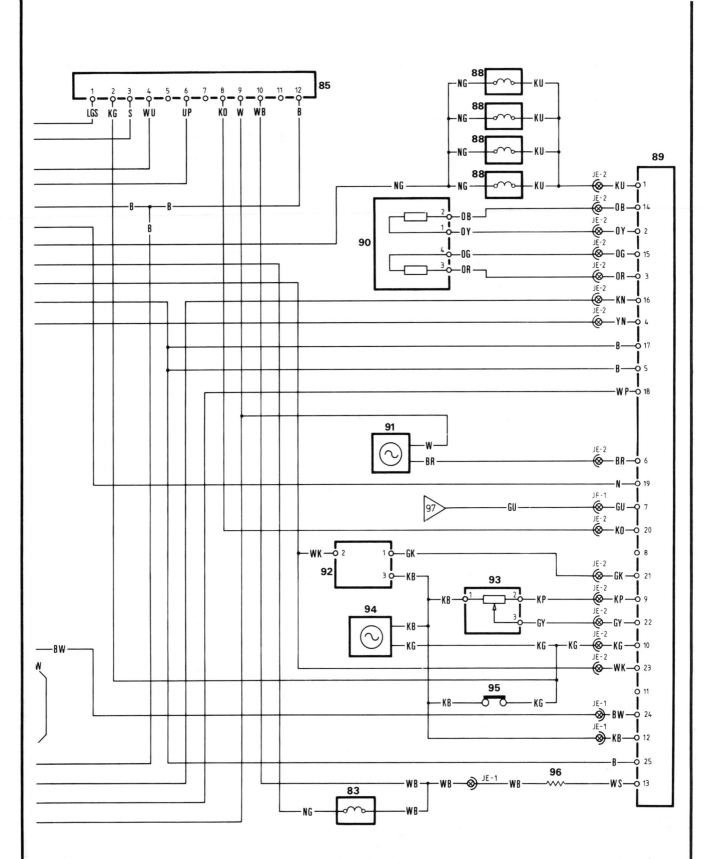

**Supplementary wiring diagram – engine circuit (EFi models, 1989-on without catalytic converter and EVAP) (continued)**

**Key to supplementary wiring diagram – engine circuit (EFi models, 1989 on without catalytic converter and EVAP)**

| No | Description | No | Description |
|----|-------------|-----|-------------|
| 1 | Battery | 93 | Throttle potentiometer |
| 11 | Fusebox | 94 | Coolant temperature sensor |
| 21 | Ignition switch | 95 | Hot start switch |
| 51 | Fusible links | 96 | In-line 1 ohm resistor |
| 80 | Alternator | 97 | Fuel pump relay |
| 81 | Starter motor/solenoid | 98 | Main relay |
| 82 | Starter relay | 99 | Manifold heater relay |
| 83 | Ignition coil | 100 | Ballast resistor |
| 85 | Programmed ignition unit | 101 | Inertia switch |
| 86 | Knock sensor | 102 | Fuel gauge sender unit |
| 87 | Crankshaft sensor | 103 | Fuel pump |
| 88 | Injector | 104 | Multi-function unit |
| 89 | EFI electronic control unit | 105 | Inhibitor switch |
| 90 | Air valve | 194 | Knock sensor screen |
| 91 | Speed sensor | 195 | Crank sensor screen |
| 92 | Airflow meter | | |

**Connections to other circuits**

| No | Description |
|----|-------------|
| 83 | Exterior lamps/wipers – RHD only |
| 89, 90 and 97 | Instruments |

**Key to supplementary wiring diagram – engine circuit (Carburettor models, 1989 on)**

| No | Description | No | Description |
|----|-------------|-----|-------------|
| 1 | Battery | 102 | Fuel sender unit |
| 11 | Fusebox | 103 | Fuel pump |
| 21 | Ignition switch | 104 | Multi-function unit |
| 51 | Fusible links | 105 | Inhibitor switch |
| 54 | Fusebox header | 106 | ERIC ECU |
| 80 | Alternator | 107 | Coolant thermistor |
| 81 | Starter motor solenoid | 108 | Ambient thermistor |
| 82 | Starter relay | 109 | Throttle pedal switch |
| 83 | Ignition coil | 110 | Diagnostic socket |
| 86 | Knock sensor | 111 | Stepper motor |
| 87 | Crankshaft sensor | 112 | Fuel cut-off solenoid |
| 97 | Fuel pump relay | 113 | Manifold heater |
| 98 | Main relay | 194 | Knock sensor screen |
| 99 | Manifold heater relay | 195 | Crank sensor screen |
| 101 | Inertia switch | | |

**Connections to other circuits**

| No | Description |
|----|-------------|
| 83 | Exterior lamps/wipers |
| 89, 90, 91 and 92 | Instruments |

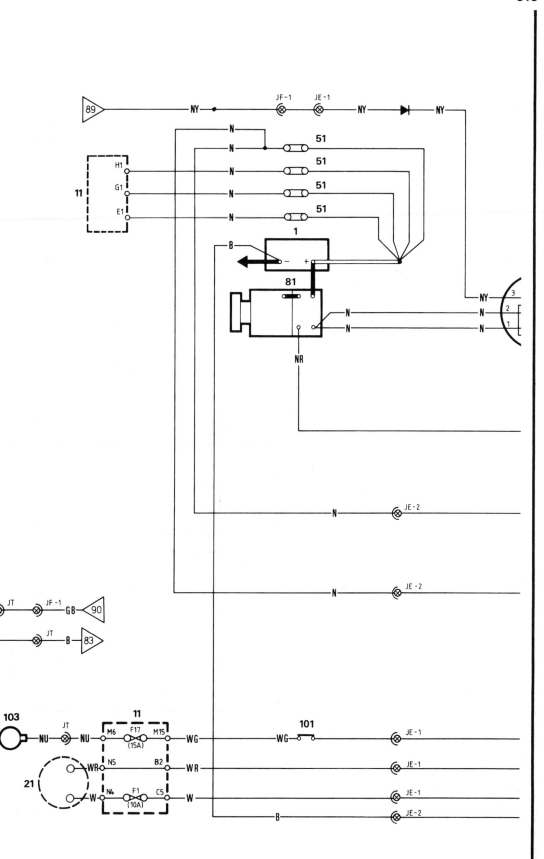

**Supplementary wiring diagram – engine circuit (Carburettor models, 1989 on)**

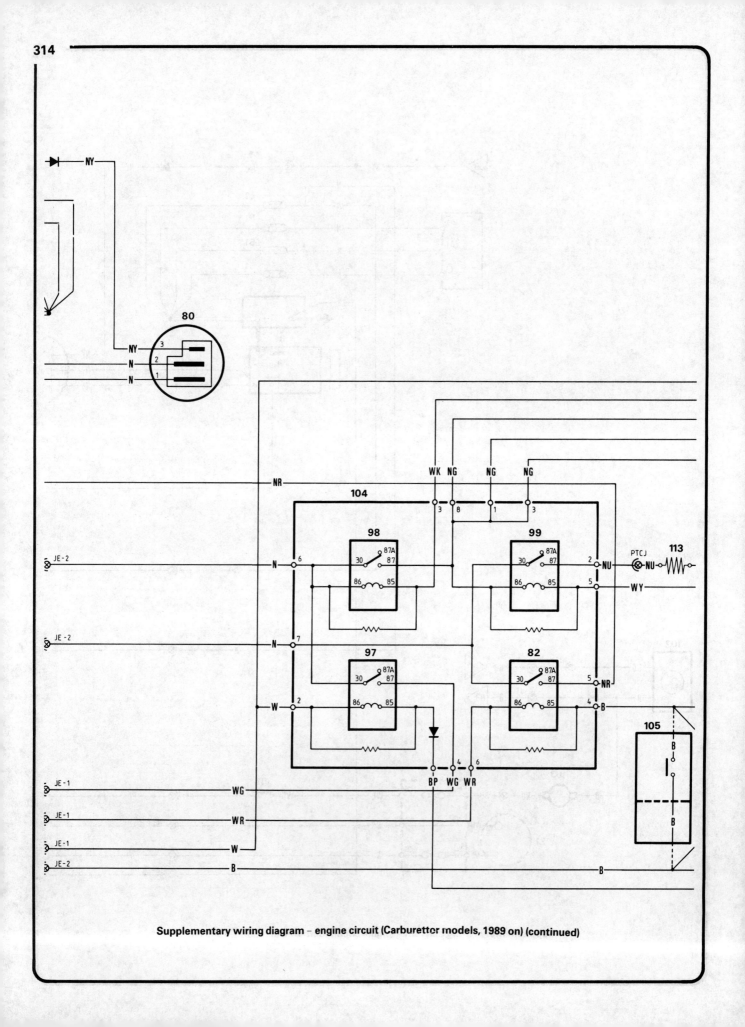

**Supplementary wiring diagram – engine circuit (Carburettor models, 1989 on) (continued)**

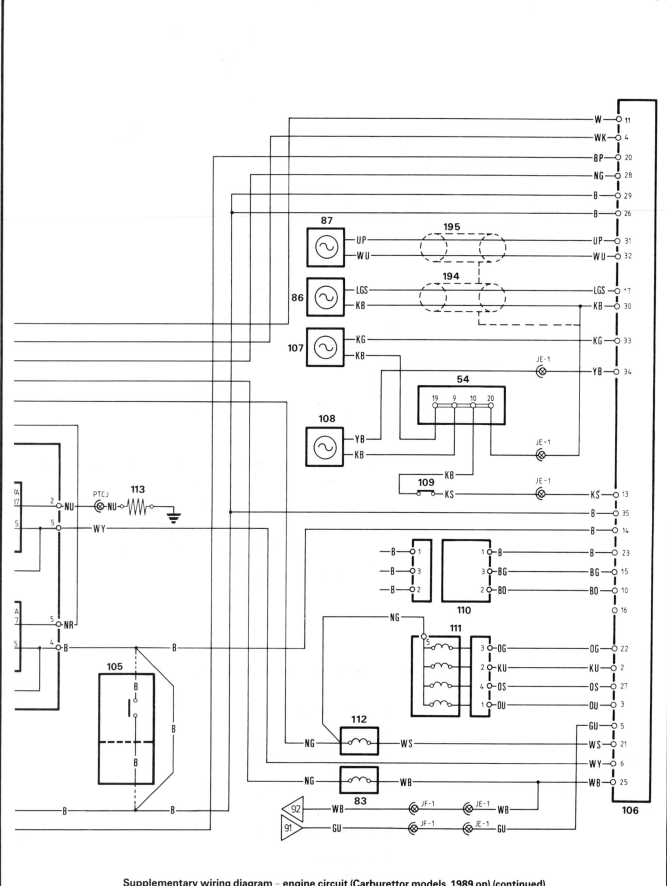

**Supplementary wiring diagram – engine circuit (Carburettor models, 1989 on) (continued)**

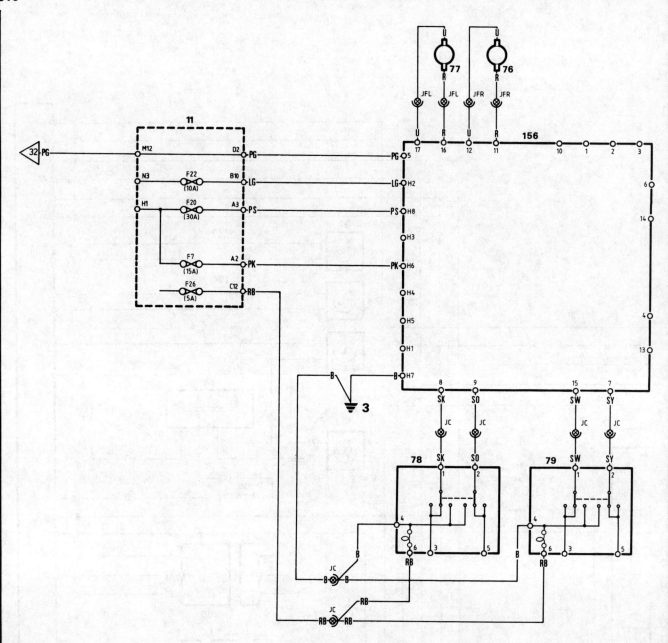

**Supplementary wiring diagram – electric front windows (non-Turbo, 1989 on)**

| No | Description |
|----|-------------|
| 11 | Fusebox |
| 76 | RH front window lift motor |
| 77 | LH front window lift motor |
| 78 | LH front window switch |
| 79 | RH front window switch |
| 156 | Two-door window lift control unit |

**Connections to other circuits**

| No | Description |
|----|-------------|
| 32 | Interior lamps/sunroof |

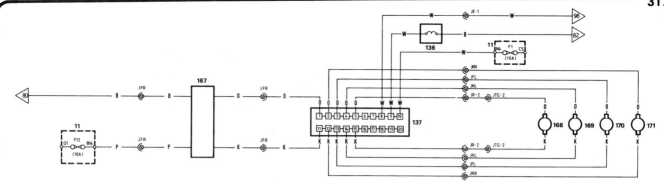

**Supplementary wiring diagram – central door locking (all models, 1989 on)**

| No | Description |
|----|-------------|
| 11 | Fusebox |
| 64 | Tailgate lock switch (Estate) |
| 136 | Carburettor vent valve (Turbo only) |
| 137 | Central door lock header |
| 167 | Driver's door control unit |
| 168 | Tailgate lock motor (Estate) |
| 169 | LH rear motor |
| 170 | Passenger door motor |
| 171 | RH rear motor |

**Connections to other circuits**

| No | Description |
|----|-------------|
| 80 | Exterior lamps/wipers |
| 82 | Engine (Turbo only) |
| 98 | Engine (Turbo only) |

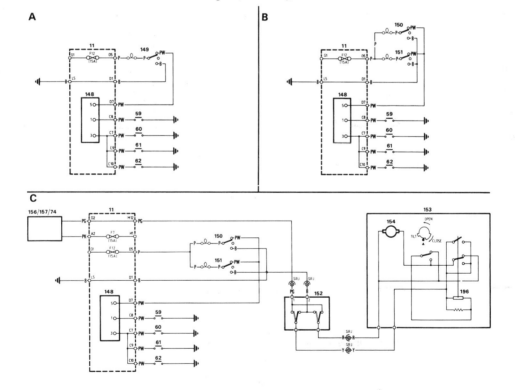

**Supplementary wiring diagram – interior lamps/sunroof (all models, 1989 on)**

| A | Without sunroof | | 149 | Front interior lamp |
|---|-----------------|---|-----|---------------------|
| B | With mechanical sunroof | | 150 | RH interior lamp |
| C | With electric sunroof | | 151 | LH interior lamp |
| 11 | Fusebox | | 152 | Sunroof switch |
| 59 | RH door switch | | 153 | Sunroof unit |
| 60 | LH door switch | | 154 | Sunroof motor |
| 61 | RH door switch | | 156 | Two door window lift control |
| 62 | LH rear door switch | | 157 | Four door window lift control |
| 74 | Carburettor cooling fan and window lift timer (Turbo only) | | 196 | Change-over relay |
| 148 | Interior lamp unit | | | |

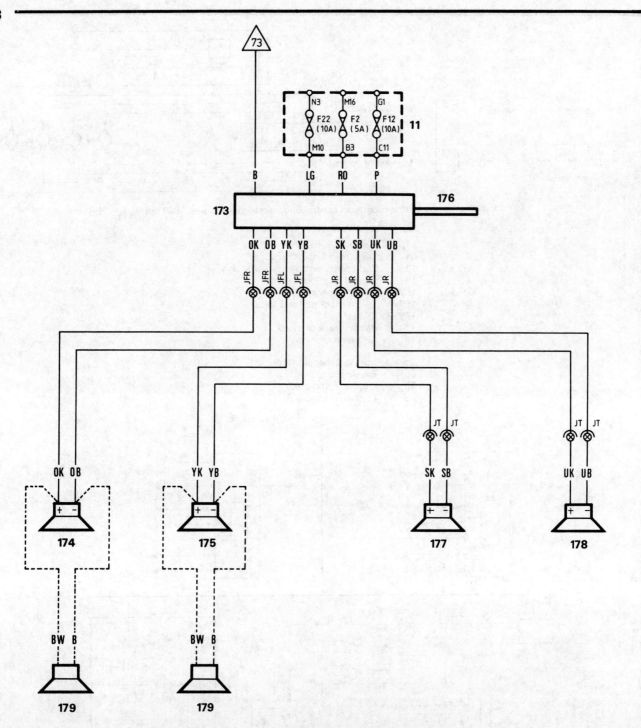

**Supplementary wiring diagram – radio/cassette player (all models, 1989 on)**

| No | Description |
|----|-------------|
| 11 | Fusebox |
| 173 | Radio/cassettre |
| 174 | Front right speaker |
| 175 | Front left speaker |
| 176 | Aerial |
| 177 | Rear right speaker |
| 178 | Rear left speaker |
| 179 | Tweeter |

**Connections to other circuits**
| | |
|----|-------------|
| 73 | Power distribution circuit |

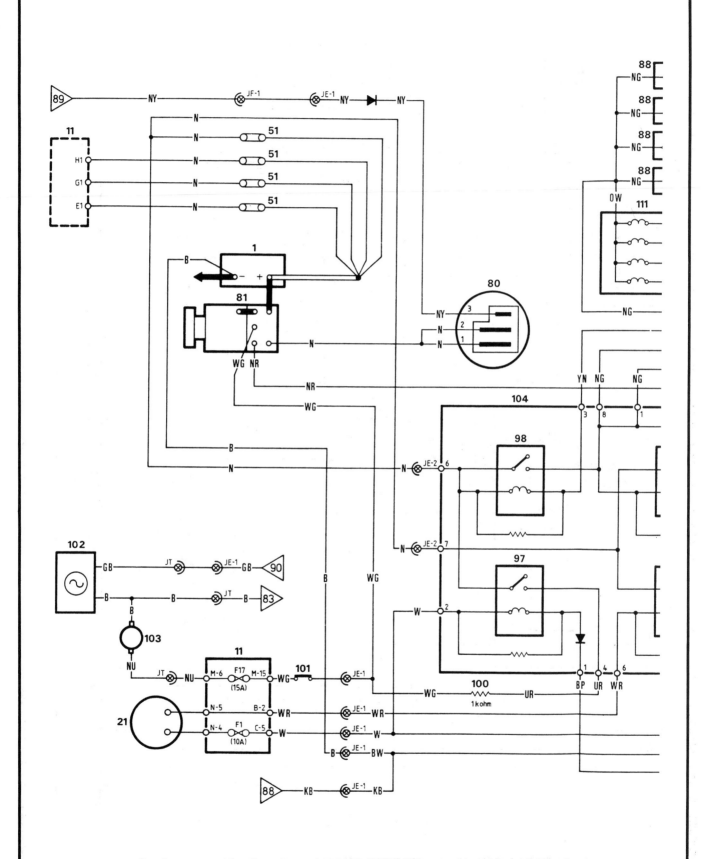

Supplementary wiring diagram – engine circuit, MEMS MPi, manual transmission (1989 on)

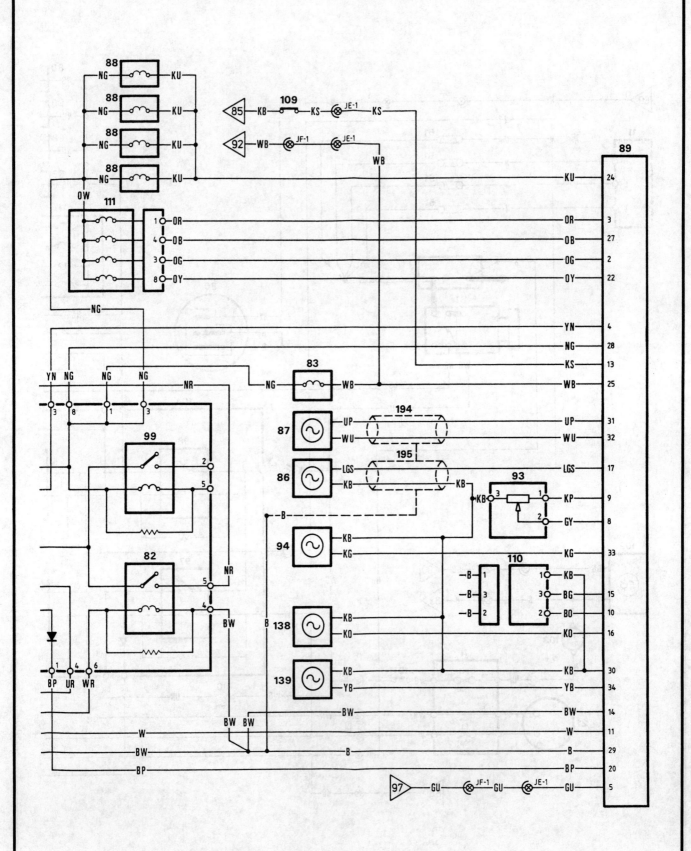

Supplementary wiring diagram – engine circuit, MEMS MPi, manual transmission (1989 on) (continued)

**Key to supplementary wiring diagram – engine circuit MEMS MPi, manual transmission (1989 on)**

| No | Description |
|---|---|
| 1 | Battery |
| 11 | Fusebox |
| 21 | Ignition switch |
| 51 | Fusible links |
| 80 | Alternator |
| 81 | Starter motor/solenoid |
| 82 | Starter relay |
| 83 | Ignition coil |
| 86 | Knock sensor |
| 87 | Crankshaft sensor |
| 88 | Injector |
| 89 | MEMS electronic control unit |
| 93 | Throttle potentiometer |
| 94 | Coolant temperature sensor |
| 97 | Fuel pump relay |
| 98 | Main relay |
| 99 | Manifold heater relay |
| 100 | Ballast resistor |
| 101 | Inertia switch |
| 102 | Fuel gauge sender unit |
| 103 | Fuel pump |
| 104 | Multi-function unit |
| 109 | Throttle pedal switch |
| 110 | Diagnostic socket |
| 111 | Stepper motor |
| 138 | Inlet air temperature sensor |
| 139 | Fuel rail temperature sensor |
| 194 | Knock sensor screen |
| 195 | Crankshaft sensor screen |

**Connections to other circuits**

| No | Description |
|---|---|
| 83 | Exterior lamps/wipers – Estates/Saloons |
| 85 | Exterior lamps/wipers – Estates/Saloons |
| 88 | Exterior lamps/wipers – Estates/Saloons |
| 89 | Instruments |
| 90 | Instruments |
| 92 | Instruments |
| 97 | Instruments |

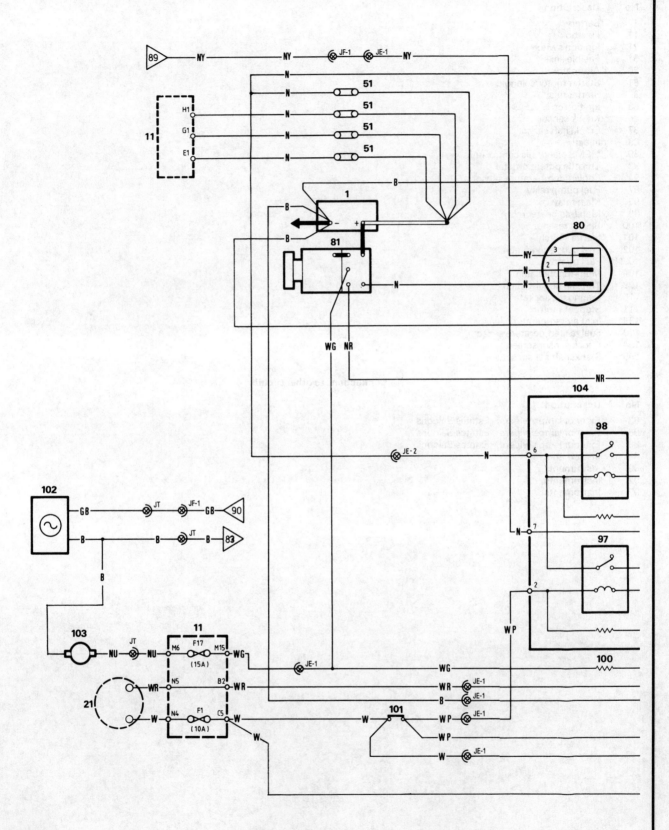

**Supplementary wiring diagram – engine circuit, 1991-on models with catalytic converter and fuel evaporative loss control (EVAP) system**

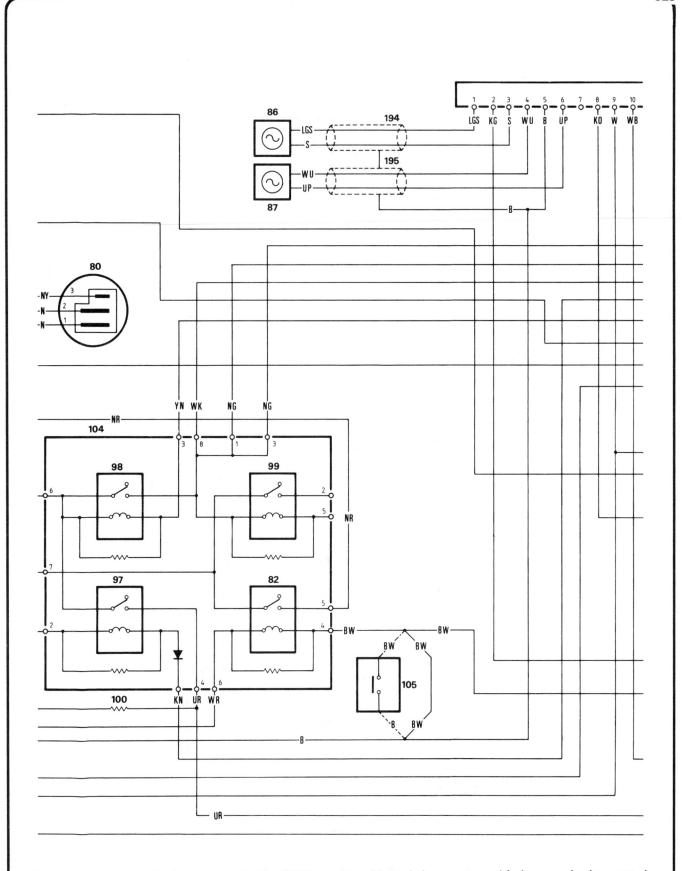

**Supplementary wiring diagram – engine circuit, 1991-on models with catalytic converter and fuel evaporative loss control (EVAP) system (continued)**

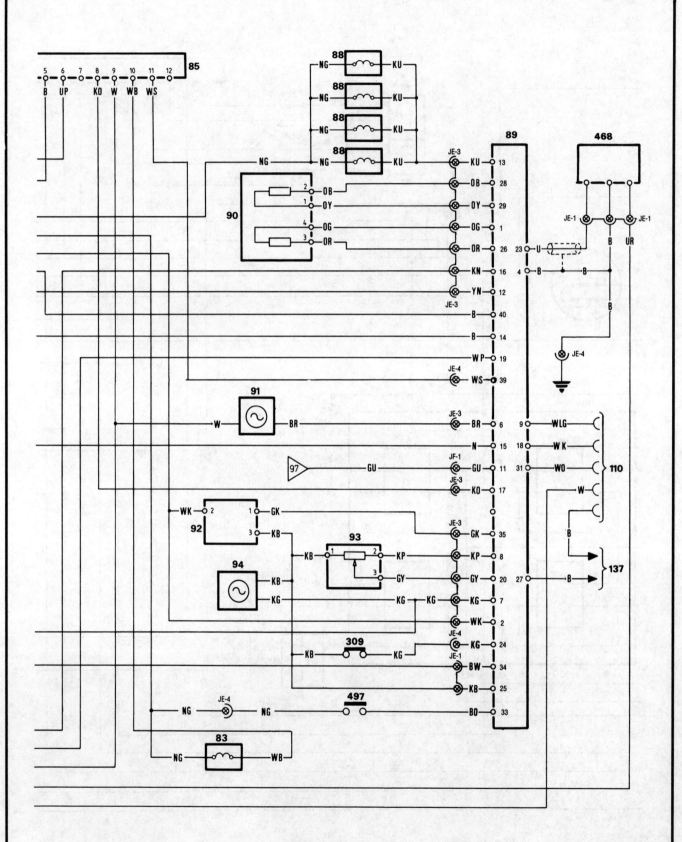

**Supplementary wiring diagram – engine circuit, 1991-on models with catalytic converter and fuel evaporative loss control (EVAP) system (continued)**

**Key to supplementary wiring diagram – engine circuit, 1991-on models with catalytic converter and fuel evaporative loss control (EVAP) system**

| No | Description |
|----|-------------|
| 1 | Battery |
| 11 | Fuse box |
| 21 | Ignition switch |
| 51 | Fusible links |
| 80 | Alternator |
| 81 | Starter motor/solenoid |
| 82 | Starter relay |
| 83 | Ignition coil |
| 85 | Programmed ignition unit |
| 86 | Knock sensor |
| 87 | Crankshaft sensor |
| 88 | Injector |
| 89 | EFi electronic control unit |
| 90 | Air valve |
| 91 | Speed sensor |
| 92 | Airflow meter |
| 93 | Throttle potentiometer |
| 94 | Coolant temperature sensor |
| 97 | Fuel pump relay |
| 98 | Main relay |
| 99 | Manifold heater relay |
| 100 | Ballast resistor |
| 101 | Inertia switch |
| 102 | Fuel gauge sender unit |
| 103 | Fuel pump |
| 104 | Multi-function unit |
| 105 | Inhibitor switch |
| 110 | Diagnostic socket |
| 137 | CDL header |
| 194 | Knock sensor screen |
| 195 | Crankshaft sensor screen |
| 309 | Fuel temperature switch |
| 468 | Oxygen sensor |
| 497 | Purge valve |

**Connections to other circuits**

| No | Circuit |
|----|---------|
| 83 | Exterior lamps/wipers |
| 89 | Instruments |
| 90 | Instruments |
| 97 | Instruments |

## Symbols used in early wiring diagrams (up to 1988)

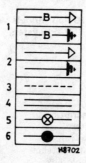

M8702

| 1 | Component earthed by a lead | 3 | If fitted | 5 | Line connector |
|---|---|---|---|---|---|
| 2 | Component earthed through its mounting | 4 | Printed circuit | 6 | Sealed joint |

## Symbols used in later wiring diagrams (1989 on)

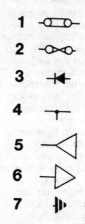

| 1 | Fusible link | 3 | Diode | 5 | Input connection |
|---|---|---|---|---|---|
| 2 | Fuse | 4 | Sealed joint | 6 | Output connection |
| | | | | 7 | Earthing point |

## Colour code used in wiring diagrams

**Note:** *when two colour code letters are shown together the first denotes the main wire colour and the second denotes the tracer colour*

| B | Black | LG | Light green | P | Purple | U | Blue |
|---|---|---|---|---|---|---|---|
| G | Green | N | Brown | R | Red | W | White |
| K | Pink | O | Orange | S | Slate | Y | Yellow |

## Earthing point locations

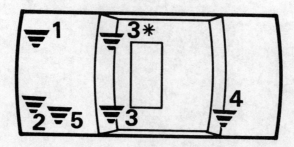

| No | Description |
|---|---|
| 1 | Front RH inner wing |
| 2 | Front LH inner wing |
| 3 | Top LH bulkhead behind facia panel (RHD only) |
| 3 | Top RH bulkhead behind facia panel (LHD only) |
| 4 | Rear LH wing behind trim panel |
| 5 | Battery negative terminal |

# Conversion factors

**Length (distance)**

| | | | | | |
|---|---|---|---|---|---|
| Inches (in) | X | 25.4 | = Millimetres (mm) | X 0.0394 | = Inches (in) |
| Feet (ft) | X | 0.305 | = Metres (m) | X 3.281 | = Feet (ft) |
| Miles | X | 1.609 | = Kilometres (km) | X 0.621 | = Miles |

**Volume (capacity)**

| | | | | | |
|---|---|---|---|---|---|
| Cubic inches (cu in; in³) | X | 16.387 | = Cubic centimetres (cc; cm³) | X 0.061 | = Cubic inches (cu in; in³) |
| Imperial pints (Imp pt) | X | 0.568 | = Litres (l) | X 1.76 | = Imperial pints (Imp pt) |
| Imperial quarts (Imp qt) | X | 1.137 | = Litres (l) | X 0.88 | = Imperial quarts (Imp qt) |
| Imperial quarts (Imp qt) | X | 1.201 | = US quarts (US qt) | X 0.833 | = Imperial quarts (Imp qt) |
| US quarts (US qt) | X | 0.946 | = Litres (l) | X 1.057 | = US quarts (US qt) |
| Imperial gallons (Imp gal) | X | 4.546 | = Litres (l) | X 0.22 | = Imperial gallons (Imp gal) |
| Imperial gallons (Imp gal) | X | 1.201 | = US gallons (US gal) | X 0.833 | = Imperial gallons (Imp gal) |
| US gallons (US gal) | X | 3.785 | = Litres (l) | X 0.264 | = US gallons (US gal) |

**Mass (weight)**

| | | | | | |
|---|---|---|---|---|---|
| Ounces (oz) | X | 28.35 | = Grams (g) | X 0.035 | = Ounces (oz) |
| Pounds (lb) | X | 0.454 | = Kilograms (kg) | X 2.205 | = Pounds (lb) |

**Force**

| | | | | | |
|---|---|---|---|---|---|
| Ounces-force (ozf; oz) | X | 0.278 | = Newtons (N) | X 3.6 | = Ounces-force (ozf; oz) |
| Pounds-force (lbf; lb) | X | 4.448 | = Newtons (N) | X 0.225 | = Pounds-force (lbf; lb) |
| Newtons (N) | X | 0.1 | = Kilograms-force (kgf; kg) | X 9.81 | = Newtons (N) |

**Pressure**

| | | | | | |
|---|---|---|---|---|---|
| Pounds-force per square inch (psi; lbf/in²; lb/in²) | X | 0.070 | = Kilograms-force per square centimetre (kgf/cm²; kg/cm²) | X 14.223 | = Pounds-force per square inch (psi; lbf/in²; lb/in²) |
| Pounds-force per square inch (psi; lbf/in²; lb/in²) | X | 0.068 | = Atmospheres (atm) | X 14.696 | = Pounds-force per square inch (psi; lbf/in²; lb/in²) |
| Pounds-force per square inch (psi; lbf/in²; lb/in²) | X | 0.069 | = Bars | X 14.5 | = Pounds-force per square inch (psi; lbf/in²; lb/in²) |
| Pounds-force per square inch (psi; lbf/in²; lb/in²) | X | 6.895 | = Kilopascals (kPa) | X 0.145 | = Pounds-force per square inch (psi; lbf/in²; lb/in²) |
| Kilopascals (kPa) | X | 0.01 | = Kilograms-force per square centimetre (kgf/cm²; kg/cm²) | X 98.1 | = Kilopascals (kPa) |
| Millibar (mbar) | X | 100 | = Pascals (Pa) | X 0.01 | = Millibar (mbar) |
| Millibar (mbar) | X | 0.0145 | = Pounds-force per square inch (psi; lbf/in²; lb/in²) | X 68.947 | = Millibar (mbar) |
| Millibar (mbar) | X | 0.75 | = Millimetres of mercury (mmHg) | X 1.333 | = Millibar (mbar) |
| Millibar (mbar) | X | 0.401 | = Inches of water (inH₂O) | X 2.491 | = Millibar (mbar) |
| Millimetres of mercury (mmHg) | X | 0.535 | = Inches of water (inH₂O) | X 1.868 | = Millimetres of mercury (mmHg) |
| Inches of water (inH₂O) | X | 0.036 | = Pounds-force per square inch (psi; lbf/in²; lb/in²) | X 27.68 | = Inches of water (inH₂O) |

**Torque (moment of force)**

| | | | | | |
|---|---|---|---|---|---|
| Pounds-force inches (lbf in; lb in) | X | 1.152 | = Kilograms-force centimetre (kgf cm; kg cm) | X 0.868 | = Pounds-force inches (lbf in; lb in) |
| Pounds-force inches (lbf in; lb in) | X | 0.113 | = Newton metres (Nm) | X 8.85 | = Pounds-force inches (lbf in; lb in) |
| Pounds-force inches (lbf in; lb in) | X | 0.083 | = Pounds-force feet (lbf ft; lb ft) | X 12 | = Pounds-force inches (lbf in; lb in) |
| Pounds-force feet (lbf ft; lb ft) | X | 0.138 | = Kilograms-force metres (kgf m; kg m) | X 7.233 | = Pounds-force feet (lbf ft; lb ft) |
| Pounds-force feet (lbf ft; lb ft) | X | 1.356 | = Newton metres (Nm) | X 0.738 | = Pounds-force feet (lbf ft; lb ft) |
| Newton metres (Nm) | X | 0.102 | = Kilograms-force metres (kgf m; kg m) | X 9.804 | = Newton metres (Nm) |

**Power**

| | | | | | |
|---|---|---|---|---|---|
| Horsepower (hp) | X | 745.7 | = Watts (W) | X 0.0013 | = Horsepower (hp) |

**Velocity (speed)**

| | | | | | |
|---|---|---|---|---|---|
| Miles per hour (miles/hr; mph) | X | 1.609 | = Kilometres per hour (km/hr; kph) | X 0.621 | = Miles per hour (miles/hr; mph) |

**Fuel consumption***

| | | | | | |
|---|---|---|---|---|---|
| Miles per gallon, Imperial (mpg) | X | 0.354 | = Kilometres per litre (km/l) | X 2.825 | = Miles per gallon, Imperial (mpg) |
| Miles per gallon, US (mpg) | X | 0.425 | = Kilometres per litre (km/l) | X 2.352 | = Miles per gallon, US (mpg) |

**Temperature**

Degrees Fahrenheit = (°C x 1.8) + 32

Degrees Celsius (Degrees Centigrade; °C) = (°F - 32) x 0.56

*It is common practice to convert from miles per gallon (mpg) to litres/100 kilometres (l/100km), where mpg (Imperial) x l/100 km = 282 and mpg (US) x l/100 km = 235

# Index